Food product development

Related titles:

Auditing in the food industry (ISBN: 978-1-85573-450-0)

Increasing consumer expectations, government legislation and levels of competition have led to a growing number of standards and benchmarks against which food processors may be measured and audited. This unique book provides food processors with a guide to the main types of audit, their importance, how to prepare for them and how to use them to gain competitive advantage. It will be an essential resource for all quality assurance and production managers in the food industry.

Food process modelling (ISBN: 978-1-85573-565-1)

A major trend within the food industry over the past decade has been the concern to measure, predict and control food processes more accurately in search of greater consistency, quality and safety in the final product. This book explores the current trends in modelling, their strengths and weaknesses and applications across the supply chain. It will be a valuable guide for production and technical managers within the food industry.

Instrumentation and sensors for the food industry (ISBN: 978-1-85573-560-6)

The first edition of this book quickly established itself as a standard work in its field, providing an authoritative and practical guide to the range of instrumentation and sensors available to the food industry professional. This new edition has been comprehensively revised to include new developments and techniques. Instrumentation for food quality assurance; principles of colour measurement; chemosensors, biosensors and immunosensors and instrumental techniques in quality control are all discussed in detail.

Details of these books and a complete list of Woodhead's food science, technology and nutrition titles can be obtained by:

- visiting our website at www.woodheadpublishing.com
- contacting Customer Services (email: sales@woodheadpublishing.com; fax: +44 (0) 1223 893694; tel.:+44 (0) 1223 891358 ext. 130; address: Woodhead Publishing Limited, Abington Hall, Granta Park, Great Abington, Cambridge CB21 6AH, UK)

If you would like to receive information on forthcoming titles in this area, please send your address details to: Francis Dodds (address, tel. and fax as above; e-mail: francis. dodds@woodheadpublishing.com). Please confirm which subject areas you are interested in.

Food product development

Mary Earle, Richard Earle and Allan Anderson

CRC Press
Boca Raton Boston New York Washington, DC

WOODHEAD PUBLISHING LIMITED

Oxford   Cambridge   New Delhi

Published by Woodhead Publishing Limited, Abington Hall, Granta Park,
Great Abington, Cambridge CB21 6AH, UK
www.woodheadpublishing.com

Published in North America by CRC Press LLC, 6000 Broken Sound Parkway, NW
Suite 300, Boca Raton, FL 33487, USA

First published 2001, Woodhead Publishing Limited and CRC Press LLC
Paperback edition 2009
© 2001, Woodhead Publishing Limited
The authors have asserted their moral rights.

British Library Cataloguing in Publication Data
A catalogue record for this book is available from the British Library.

Library of Congress Cataloging in Publication Data
A catalog record for this book is available from the Library of Congress.

Woodhead Publishing ISBN 978-1-84569-722-8 (paperback)
Woodhead Publishing ISBN 978-1-85573-468-5 (hardback)
Woodhead Publishing ISBN 978-1-85573-639-9 (e-book)
CRC Press ISBN 978-1-4398-2038-4 (paperback); order number N10127
CRC Press ISBN 978-0-8493-1209-0 (hardback); order number WP1209

Contents

Preface

Managing innovation is a necessary skill for senior management of all food companies producing new raw materials, new ingredients or new consumer products. Company growth and even survival depends on the introduction of successful new products into old and new markets. The dividing line between product success and failure depends on many factors, but the most important are new product qualities, skills and resources of the company, market and marketing proficiency, and an organised product development process. There is a need to understand consumers' behaviour and attitudes and to be able to design a product to meet the users' needs. But it is also necessary to have the technological knowledge and the skills, and the organisational ability to bring a product to a successful commercial conclusion in the marketplace. This book studies some of these key issues in product development and outlines the methods of managing them.

The book started on a day in 1956 when Mary Earle joined the product development team at Unilever Limited, Colworth House, Sharnbrook, Bedfordshire. Jack Savage, the leader of this team, was a pioneer of product development in the food industry. It was his understanding of product development as a coordination of technology and marketing, always aimed at the final consumer, that laid the basis for Mary's work in product development during the next 40 years. She tried to put these ideas into practice in the food industry in Britain and the meat industry in New Zealand, and quickly realised that there was a real need for education in product development for all people entering the food industry particularly technologists, engineers and marketers. In 1965, at Massey University, New Zealand, she introduced courses in product development and food marketing in the Bachelor of Technology degree in food technology. These courses combined theory and projects, so that the students not only could learn basic techniques in product development but also the philosophy and the practical application in industry. Gradually product development (PD) was developed as an academic discipline and Bachelor and Masterate degree programmes in product development were introduced for all industries at Massey University. Allan Anderson was involved in much of this development over the years. Dick Earle's expertise is in process development, particularly in the New Zealand meat industry, and he developed the combination of product design and process development in a new venture producing pharmaceutical industrial products.

At the same time, particularly in the last 20 years, there has been a great deal of research in other industries on product development, and today it is recognised as a particular industrial research discipline. The book has been built on this research and also the research at Massey University and with the food industries in New Zealand, Thailand, Malaysia, Singapore, Canada and Australia. We need to thank the hundreds of undergraduate students for their work in their product development projects with industrial companies, the postgraduate students in their research on the activities and techniques in product development, and in particular the people in the companies who collaborated in these projects. Examples from some of these projects are used in the chapters. The staff at Massey University involved in product development in the Food Technology Department, Food Technology Research Centre and the Department of Consumer Technology, built up the multidisciplinary, consumer directed, systematic PD Process used throughout the book.

Part I starts by looking at the different categories of new food products. It then identifies past reasons for product success and product failure, and asks the reader to identify some of the specific reasons for product success and product failure in their company. This is to stimulate thoughts on PD in particular companies and to lay the basis for studying the core elements in PD and relating these to the problems of PD management. It ends by identifying specific aspects of product development in the food industry.

Part II in four chapters studies the core elements of product development:

1. Developing an innovation strategy
2. PD Process(es)
3. Knowledge base for product development
4. Consumer in product development.

Product development at both the programme and the project levels needs to be based on the business strategy. It is the responsibility of top management and they need to set the strategies for the product development programme for the present and future years, and also the aims for the individual projects. Top management needs to ensure that there are systematic PD Processes for the different levels of innovation and types of products. Having set the strategy and the PD Process, they need to ensure that there is the necessary product, processing, distribution and marketing knowledge in the company, and also the ability to create new knowledge in design, development and commercialisation.

Finally there needs to be an understanding and consideration of the final consumer – their needs, wants, behaviour and attitudes – as well as the other customers in the food system between the producer and the consumer. Part II studies these core areas, so that the reader develops a basic understanding of product development.

Part III studies PD in general and then in the food system. Managing product development in the food industry varies with the types of markets (industrial, food service and consumer) and with the place of the company in the food system (primary producer, food processor, food manufacturer, retailer). Top management needs to recognise these variations and also identify their level of risk and the company's resources in skills and knowledge. From this, they can build an organisation for product development and also identify the decisions that they have to make in the programme building and during the individual projects. They can then specify for the middle management the critical points in the PD Process and the knowledge that must be available for their decisions at the critical points. From this, the middle management, called product development manager, product manager, R&D manager, can identify the aims, activities and outcomes for the individual stages in projects, and also the coordinated plan for the product development programme. The project leader can identify the techniques to be used in the project, the resources needed and the time schedule. For successful product development management, these three layers of management need to be coordinated and aiming for the same outcomes from the product launch. Four case studies illustrate management at different stages of the food system, including fresh products, industrial products and manufactured consumer products. The book ends with a chapter on evaluating and improving product development. Product development management must include the collection of knowledge from the project, analysis of this knowledge and setting improvements for future projects. Product development is continuously changing and its management needs to change; without change, not only the products but also the whole system become archaic and lost in the past.

This book is intended for people entering or in product development management, at all levels from the project leader to top management. Throughout the book, readers are asked to apply the knowledge in the chapter to their company, so that they can develop their own philosophy and methods for product development. This is not a book on techniques of product development; it is a book to raise the awareness of different aspects of product development and to apply the new or revived knowledge in the practical situation of managing product development in a food company.

Mary Earle
Richard Earle
Allan Anderson

Part I

Introduction

Product development has been a major activity in the food industry for over 40 years, but only gradually has it developed as a strategic business area and also as an advanced technology. For a long time it was essentially a craft, loosely related to the research and engineering areas in the company. The pressures for product development came very strongly from the needs of the growing supermarkets for a constantly changing, extensive mix of products and for continuous price promotions. So there was the drive for product difference, including minor product changes sufficient to distinguish products on the shelves, and for cost reductions. There were also underlying social and technological changes which caused major product development; for example the increasing number of working women which sparked the need for convenience foods and eating out, and the development of spray and freeze drying which was the basis for instant foods.

When one looks at overall success and failure in the food industry during past years, socially there has been success in providing sufficient cheap food in developed countries, but failure through developing such a poor reputation that the food industry became highly regulated; commercially there has been success in developing large multinational companies, but failure with continuously reducing margins on food products. Can the failures be related to narrowly focused business strategies, to lack of innovation strategies and organisation or to lack of knowledge?

There are now compelling social and technological pressures on the whole food system to change rapidly, such as the pressures from the growth of information technology in the more affluent countries, and from the growing economic strength in some of the developing countries. Can the food industry meet this challenge? Has the food industry the knowledge and the people? How can it respond? The aim of Part I of the book is to look at the causes of product success and failure in the past, and to identify the key issues for successful product development in the future.

1

Keys to new product success and failure

The aim of this chapter is to identify the important factors in food product development to be studied in detail in the succeeding chapters. Firstly the different groups of food products are identified as a basis for organising product strategy. Then the published research on the factors in product failure and product success in all types of industries is used to identify the key factors in food product development. This leads into the management of product development at three different levels:

1. Business strategy.
2. Product development programme.
3. Product development project.

Finally specific aspects of food product development are identified as the basis of the book, and the structure of the book is outlined.

1.1   Food products – the basis of innovation

What are food products? What are new food products? Everyone agrees that a food is material eventually consumed by humans to satisfy physiological and psychological needs, but the food company and the consumer can have quite different descriptions of the food product presented for sale. The company defines a basic functional product to which it has added packaging, aesthetics, brand, price and advertising, to give a total company product. The consumer describes the product as a bundle of benefits, relating its tangible and intangible attributes to their needs, wants and behaviour. For a basic food product, for example flour, the description can be simple and pragmatic, but for products such as a meal at a restaurant, it can be complex and emotional. The company defines a new product as having some difference in the basic functions and aesthetic presentation; but consumers compare it with the 'old' product and competing products and if they recognise a difference then it is a new product to them (Schaffner *et al.*, 1998). Product development is all about reconciling these two points of view.

There are many thousands of food products and they can be grouped together into product categories according to:

- food system position;
- market they serve;
- processing technology used to manufacture them;
- basic common characteristics such as nutrition and health;
- product platforms;
- level of innovation.

Grouping products is a useful method of developing new product ideas using techniques such as product platforms, product morphology and gap analysis. One can identify spaces for new developments, methods of product improvement and indeed innovation related to changes in food system or technology.

1.1.1   Food products and the food system

Products interact with every part of the food system from primary production to the consumer as shown in Fig. 1.1. The new cereal, high in protein, may go to the processor to produce a specialised protein product for bakers, or to a food manufacturer to make a high-protein breakfast cereal, or to a vegetarian fast-food outlet as a meat replacer, or to a supermarket as an ingredient for home-prepared muesli or directly to the consumer for use in a home breadmaker. A new product in one part of the food system can cause new products in other parts.

There is a need to distinguish the three groups of products:

1. Primary products from sea and the land.
2. Industrial ingredients from food processors.
3. Consumer products from food manufacturers and food service.

They basically have the same product development process, but there are activities and techniques specific to each area.

There is a need to recognise the total product in each case. There is a formal product with its associations such as service, know-how and image as identified by the company (Crawford, 1997), and then the product concept of the consumer or customer. A McDonald's hamburger may seem a simple product but it has strong associated benefits such as convenience, price, fast service and hygiene, along with a very powerful allure especially for young people of the good things in American life. Food service products usually have a high proportion of services, but so do industrial products and increasingly primary products.

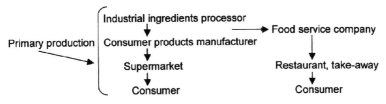

Fig. 1.1   The food system.

The industries upstream from the food manufacturer are important contributors of innovation (Rama, 1996). Both the ingredient suppliers and the equipment suppliers can have a pivotal role in innovation in the food industry. Agricultural and now marine farming are also major sources of innovation both of fresh products, and of materials designed for processing. So the innovation spectrum broadens and deepens.

1.1.2   Food products and the markets

The basic principle of product development is to identify the needs of the buyers and the users, and design the products towards meeting these needs. This means that the market segments for the products are an important basis for grouping products. There are five main market segments:

1. Consumers: mainly branded products.
2. Retailers: branded products, ingredient mixes.
3. Food service: partially prepared meals, meal ingredients.
4. Industrial processors and manufacturers: differentiated ingredients.
5. Primary processors: commodities, undifferentiated raw materials.

It is important to recognise that there are major differences in the development of products for these different segments. If a company moves from differentiated ingredients for food manufacturers to consumer products to be sold through retailers, there is a need for new knowledge and new resources in the company.

Each of these five main segments can be divided into further segments. There are five common consumer market-segmentation categories:

- Geographic.
- Sociocultural.
- Demographic.
- Psychographic.
- User behaviour.

Regions, social classes, ethnic groups, households, age, sex and income are typical groupings for which statistical census data can be found, but consumer targeting can be more accurate if psychographic segments based on life style, behaviour, personality and attitudes are used. User behaviour segmentation on usage rates, brand loyalty status, purchase occasion and benefits sought are useful for targeting product development. In industrial segmentation, two stages can be used: firstly companies are grouped according to location, size and type of processing, and secondly by company factors such as technical expertise, product needs and service needs. It is important that both the product and service needs are recognised in segmentation for industrial product development. Food service is divided into two broad groups – commercial and institutional; but of course there are important internal segments in these such as large chain fast-food companies, fine food restaurants, family restaurants (Schaffner *et al.*, 1998). The segmentation strategy depends on the company's overall business and marketing strategies. But it is important that the market segments are clearly recognised in developing groups of products for product strategy.

1.1.3   Food products and processing technology

Food products in the past have often been grouped according to their preservation technology – frozen foods, canned foods, chilled foods, dried foods, ambient foods. For example milk products are grouped as 'fresh', UHT (ultra-heat treated), canned, dried; fruits as 'fresh', canned, dried, frozen. The main reason for this grouping was that the preservation method was dominant in processing, distribution and retailing; and therefore to change the preservation method was a major undertaking in resources. The first three, freezing, canning and chilling, are thermal processes controlling food quality by temperature and time. Non-thermal processes, controlling water activity, atmospheric gases and packaging, preserve dried and ambient foods. In recent years, there has been increased interest in non-thermal preservation of food for example by irradiation and by high pressures (Knorr, 1999). Both processes have arisen in an effort to avoid damage to food quality in processing, but both have their own difficulties.

Products are also grouped according to processing technologies such as baking, extruding and fermentation, and according to the form of the food such as liquids, emulsions and powders. This is useful because it recognises the basic technologies and the knowledge of them in the company. If the greatest knowledge in the company is in emulsions, then the product groups include, cooking oils, salad dressings, margarines, ice creams, sauces, and new products can be developed from basic emulsion knowledge. Other typical groupings are bread, rolls and cakes; biscuits and crackers; confectionery; sauces and pickles.

A new process technology can start a family of products and indeed several families of products. For example, extrusion technology was the basis for many new snack products from flavoured, puffed snacks to muesli bars. Knowledge of products and processing is important in product development because it can lead to major innovations – the 'new-to-the-world' products.

1.1.4   Nutrition and health

An important grouping is related to the function of the products in nutrition and health. Provision of calories has dominated the food industry for many years: firstly the basic need was to provide calories and then in recent years, the push to reduce calories. Early products in small groceries at the beginning of the 20th century were bread, butter and margarine, sugar, jam, bacon, beef suet – all high-energy foods. In contrast at the end of the century, supermarkets now sell low-fat milks, diet colas, trimmed pork and so on. There will always be 'calorie' foods but the question is what calories they should provide in the next 50 years? Together with calorie foods, came protein foods – legumes, dairy products, meat and fish. It has taken some time to raise the amount of protein in the diet and even in the developed countries there are poor people who are not getting adequate amounts of protein. Legumes and cereals are the cheapest protein foods and these may be stronger areas for protein product development, but of course dairy products, meat and fish will remain major areas for product development for more affluent consumers. There are many more nutrients needed as well as the basic calories and protein, and there have been specific foods designed with fibre, vitamin and mineral enrichments. There is recent re-emphasis on what might be termed the older deficiencies such as calcium, iodine and iron. There will always be foods designed with this supplementation as there have been in the past (Deutsch, 1977).

Recently, the emphasis has shifted from foods supplying the essential nutrients to sustain life and growth to foods for prevention or indeed curing of disease; what have been termed nutriceutical or functional foods (Sloan, 1999). These functional foods have expanded from the health-food stores to the supermarkets, but there is some difficulty in defining what they are. One British definition is 'processed foods containing ingredients that aid specific bodily functions in addition to being nutritious' (Alldrick, 1997) and an American definition is 'foods that encompass potentially healthful products, including any modified food or food ingredient that may provide a health benefit beyond the traditional nutrients it contains' (Platzman, 1999). These definitions are very broad and cover a wide variety of products. If functional foods are to survive in the future they need to be based on scientific evidence and not emotional effects.

1.1.5 Product platforms
A useful method of organising food products is to link them on product platforms (Meyer and Lehnerd, 1997). This is based on the fact that families of products can be grouped together because they have a common architecture or common morphology (Schaffner *et al.*, 1998). Product morphology is the breakdown of a product into the specific characteristics that identify it to the consumers, by analysis of the product family and the individual product. A product platform is formed by a set of linked products, which are distinctive but also have a strong common linkage, such as fresh fruit juices, nutritional breads, cold breakfast cereals for children. The product platform is defined as 'a set of subsystems and interfaces that form a common structure from which a stream of derivative products can be efficiently developed and produced' (Meyer and Lehnerd, 1997). Product platforms are a useful basis for developing a product strategy for the company, and also for creating ideas for new products. If a new platform is started, derivative products can be based on this platform, and then

the next generation of products is started on a new but related platform. They can be considered as Generation 1, Generation 2, Generation 3 of the Product Family. In Case Study 1, Chapter 7, the development over 15 years of a new product platform in the apple industry is described.

The first stage is to identify the present product platforms and to show how they have developed in the past; clearly showing the generations of product platforms and the derivative products on each platform. It would be useful if these were identified as successful and failed product changes, so that a historical picture could be built up as a basis for the future. The second stage relates each product platform with the different market segments to which the products were aimed and to which they were finally related in sales. The third stage identifies the building blocks that were used to achieve these changes in the product platforms: consumer insights, product technologies, manufacturing processes and organisational capabilities. An example of a general systematic grouping is shown in Fig. 1.2.

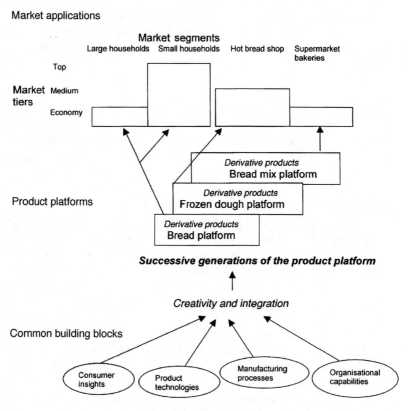

Bread company

Market applications

Market segments
Large households   Small households   Hot bread shop   Supermarket bakeries

Top

Market **Medium**
tiers
Economy

Derivative products
Bread mix platform

Derivative products
Frozen dough platform

Product platforms

Derivative products
Bread platform

Successive generations of the product platform

Creativity and integration

Common building blocks

Consumer insights   Product technologies   Manufacturing processes   Organisational capabilities

Fig. 1.2   Systematic grouping of food products for use in product development
(Source: After Meyer and Lehnerd, 1997).

The company can combine this product knowledge with knowledge of the predicted social and technological changes to identify the changes in the product mix for the near and distant futures.

1.1.6   Level of innovation

In product development, there is a variety of 'new products' and it is necessary to define 'newness' at the beginning of the project since the activities, risks, costs and indeed the product development process vary with the type of new product. The designation of a product as new is used to cover a wide range of product changes from major innovations to cost reduction leading to a lower-priced product (Fuller, 1994). Some of these categories are shown in Table 1.1. Generally the major innovation is followed in time by product improvements as product quality increases with production improvements, then perhaps new packaging, followed perhaps by repositioning in another market segment or a relaunch of the product, and finally ending in price reductions.

New product development provides a wide range of product changes, many of which may not be very marked either technologically or to the consumer. Innovation is most dramatically represented in the 'new-to-the-world' product. Even in cost reduction, however, there can be major innovations in processing to achieve the lower costs. In considering new products, it is necessary to look at the total product mix and to decide how this could be changed over time to maximise growth or return on investment or some other company objective. There is a need to develop a product strategy for the future. The innovation strategy defining the overall new directions for the company, and the product strategy defining the product changes and additions, are the bases for the new product development strategy. Both the product strategy and the innovation strategy need to be embedded in the company's business strategy. In this book, we talk about product development and not new product development, as the company always needs to be aware of the effects of new product development to the product mix.

Table 1.1   New product categories

New-to-the-world	Products are innovations to society.
New product lines	Products are new to the company.
Product line extensions	Additions to company's existing product lines.
Product improvements	Replacement of a present product with an improved version.
Product repositioned	Products are targeted for a new use or application and usually a new market segment.
Product cost reductions	Repositioning as a cheaper product, with similar benefits but cheaper costs and therefore lower price.

Source: After Cooper, 1993 and Crawford, 1994.

Think break

1. 'Innovation is a predictable process.' Do you agree or disagree with this statement? How can a company organise to give 70–80% predictability to product development but allow 20–30% for the unknown?
2. Take one product family in your company, and identify the generations in the product platforms, and then the derivative products on each platform. What have you used as the basis for the family and for each platform – preservation method, other technology, nutrition and health, place in the food system or some other general family characteristic? If you have not used nutrition and health, try building up the platforms on this basis.
3. Identify the market segments that your company targets for this family of products, and relate them to the different platforms and if necessary particular products.
4. What building blocks has the company used to form these platforms?

1.2   Measures of product success and failure

If a company is to build a successful product mix and product strategy, there is a need to study the company's history and current performance and also the history and current performance of the industry and indeed of other industries. The food industry can learn from successes and failures in other industries. The measures for determining success and failure can be for:

- individual new products (financial, market, production, consumer acceptability, targets);
- product development projects (efficiency and effectiveness);
- overall product development programme (success rate, sales and profits from new products, innovation level).

The measures are detailed in Table 1.2.

1.2.1   Individual product success

Individual product success can be measured by financial success, consumer and market success, production success, product/consumer (customer) success.

Financial measures are usually the profits and return on investment. These appear quantitative but they are often fraught with problems. How is the measurement made? Is it the return on investment over one, five or ten years? Does it include the basic research that preceded the product development and perhaps spreads over several present and future products? What is the method of discounting the returns over the 10-year period? There is a great deal written in the product development literature about predicting financial success before launching but not a great deal on financial evaluation after the launch (Crawford, 1997). Obviously at the company level, the annual balance sheet for shareholders is where

Table 1.2   Measures for product development success and failure

---

Individual new product measures
Quantitative targets
Sales volumes and revenues
Market share
Profits
Financial performance

Qualitative targets
Product qualities
Customer acceptance
Competitive position against other companies' products
Extending or completing a product line
Aiding a promotional effort
General company benefits

Product development project measures
Efficiency in time and cost
Effectiveness in achieving product success

Overall product development programme measures
Comparison between old and new products
Number of new products in the last five years
Number of improved products in the last five years
Growth of market due to new product introductions
Proportion of sales related to new and improved products
Profitability of new products compared to old products
Contribution to net margins of new products

The effect on company innovation level
Newness of production technology compared with the industry norm
Newness of marketing technology compared with the industry norm
Newness of markets for the company's products
Innovative advance of company's new products on competing products
Customers' view of the company as innovative

Source: After Earle and Earle, 2000.

the company is judged. But how does this relate specifically to the product development? If it has a product family financial analysis, showing different product families as percentages of the profits, then how the product changes are affecting profitability can be analysed. If the investments in the various product families are recorded then the return on investment in various product families can be determined. But it is seldom possible to track individual products from the annual balance sheet. In product development, a financial benchmark is set which takes into consideration not only the company's own financial needs from a product family but also the financial standards being set by other companies. All the individual products in a product family are often set, for simplicity, the same financial targets, but this may be a false assumption as specific products may have different aims.

Market success, achieving target sales volumes and revenues, is often the measure of success and failure for the overall company, the product families and the individual product. They are usually easy to measure – or are they? Sales are related to time, the marketing effort and the conditions in the market. A simple yearly sales achievement may not relate to the quality or the uniqueness of the product, nor give a true indication of the product development success. Sales over time need to be measured, along with any competing products in the market, and also the other products in the company's product family, together with a breakdown to the different market segments. In industrial marketing to the food manufacturer and the food service company, not only are the actual ingredient sales monitored but also the sales of the resulting consumer products to confirm if the company is achieving its percentage of a growing or static market. The efficiency of the marketing effort to achieve these sales is also a product success measure – the costs of the marketing effort including distribution, advertising and promotion and selling are measured and related to the sales achieved.

Production success is usually analysed by quantity, quality and costs. It has to achieve the product quality in the product specification consistently with only a prescribed variation, to ensure product safety, and also to produce at the correct quantity and time. It has basic production costs that have to be achieved, and the investment capital and the working capital of the process and production development have to be within budget and time. One of the most important measures is the production yield, the ratio of the product output to the raw materials input. The distribution losses and the returns from the retailers are also measured and are very important in the food industry.

Product and consumer (customer) success is measured by the level of consumer or customer acceptance and also by the position of the new product against the competing products. The total product success is determined by how quickly it is bought, how often it is bought and how much is bought, but there needs to be more detailed analysis. The product is set target standards in the product design specifications, and its success is rated according to how it achieves these specifications. It is not just a consumer rating of the product's acceptance but in particular how much it has incorporated the benefits identified as the consumer needs, how much it has achieved uniqueness to the consumer, how much value it is to the consumers. There will also be specific quantitative measures of the product characteristics as identified in the company's product specifications – have they been met? The consumer and the manufacturer's benefits as shown in Fig. 1.3 need to be identified and measured.

In the case of the industrial product, the criteria for success are based on how the product performed in the buyer's process and how it related to the quality of the final consumer product. Very often for industrial ingredients for food manufacturers and for large food service companies, there are strict product specifications and the product quality has to be within a specific range.

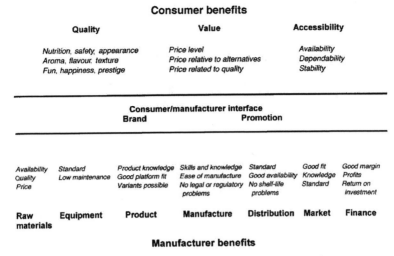

Fig. 1.3   Benefits as seen by the consumer and the manufacturer.

1.2.2   Product development project success

The product development project is also part of the success analysis – its efficiency and effectiveness. The project's efficiency as regards time and costs and use of resources is a basic part of product development. But it is also judged on its effectiveness – the success in developing the product. How often is the product not quite the right quality, does not have the optimum product characteristics, is not what the consumer needs and wants? How near does the new product come to meeting these targets? Companies need to evaluate the success of the product development process (PD Process) at the end of each project, so that they can learn from success and failure, improve their PD Process and achieve better outputs.

1.2.3   Product development programme success

The long-term success is related to the changes in the company's product mix – the structure of the product mix, the sales and profit relationships between old and new products, the growth of the market and the market share. It affects the company value in terms of goodwill, product range depth and potential, brand power, market impact and morale. Product success has also an effect on the innovation level in the company and the technological standard of the company compared with competitors (Campbell, 1999). Weak product development has a long-term effect on the production facilities, which are not renewed or updated regularly, and also on the marketing technology, which tends to become conservative. Most important is the slow growth in company knowledge. With little active product development for a number of years, the knowledge in a company is certainly less than the knowledge in the most innovative companies, and may even be less than in the direct competitors. Griffin (1997), in surveying

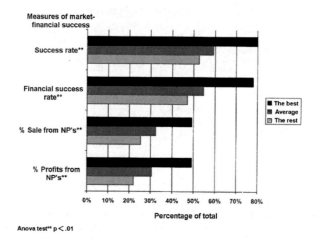

Fig. 1.4   Long-term success of a product development programme.
Success rate: % of products categorised as successes in the last five years.
Financial success rate: % of products categorised as financial successes in the last five years.
% Sales from new products: $ sales of products commercialised in the last five years as % of total sales.
% Profits from new products: $ profits of products commercialised in the last five years as % of total profits.
(Source: From Griffin, 1997 by permission of Product Development and Mangement Association, Moorestown, New Jersey.)

nearly 400 companies in America, used four product programme criteria to measure financial and market success of their product development programmes – product success rate, financial success rate, sales from new products (NPs), profits from new products and the number of new products in the last five years. Figure 1.4 shows a comparison of the best 85 firms and 298 other firms.

The best companies had a higher percentage of successful new products, and also higher percentages of sales and profits from new products. In research with 800 companies in 26 industry sectors in seven countries, the food and drink industry as a whole had 21% of its turnover as new products and services, far below the leader, technology, with 69% (Anon., 2000). This survey showed that a 10% increase in the proportion of turnover generated from new products and services led to a 2.5% increase in revenue growth, year on year.

1.2.4   Selecting success measures for product development
The measures selected are related to the company's business strategy and the level of knowledge and skills in the company. The company must be clear about the measures and if possible choose quantitative measurements (Beaumont, 1996; Hultink and Robben, 1996). The degree of detail in a measure can be very specific, such as the time taken for product design, or can be general such as the percentage of sales that are new products, but it needs to be appropriate, considered, specified and agreed. There is a need to set the measures,

benchmarking and targets before the product development programme and the individual product development project are started, so that everyone involved realises how the final success and failure is to be judged (Zangwill, 1993). Benchmarks set beforehand tend to be less influenced by particular events and circumstances.

Performance measures → Benchmarking → Targets against performance measures

There is a need to look forward, to set up measures for the project's likely success; and also to look backward to assess actual performance against the predicted targets. Once the targets are set, they need to be communicated to all the people and departments involved in product development. At the end of the project, the data from the project are collected and analysed, and improvements identified. For every project, the measures need to be reviewed and set again (Beaumont, 1996). The aim is to have one of the highest success rates for product development in the industry and measures, benchmarks and targets have to be set for the product development programme to achieve this.

The balance of products in the product portfolio, on which the product development programme is based, is also another important measure. The product portfolio is the collection of products manufactured and/or marketed by a company, and it needs to be analysed to give the maximum long-term effects from scarce company resources. The long-term success of the company depends on having some products that generate cash now and other products that use cash to develop the future. All product portfolios include the new product, the growing product, the present breadwinner and the dying product; this succession needs to be preserved for long-term company viability. The product development programme needs to be measured to see it is ensuring the entry of new products and helping the growing product by quality improvements and variety, the mature product by major relaunches, and the dying product by cost reductions.

The techniques used to measure success depend on the knowledge already in the company and the amount of information that can be collected during the project and product launch. Obviously some large companies have detailed databases, extensive staff knowledge and money to collect and summarise the project data. Their measures are more quantitative than with the small company, but for a specific market, the small company can have as accurate a success measure because of close relationships within the market and the company.

Think break

1. Speed to market is the most important performance measure for product development. Do you agree with this or are there other performance measures that you think are important or maybe even more important?
2. For two product families in your company, mark the relative success of their product development programmes on the following scales (from Griffin, 1997):

Programme is a success	1	2	3	4	5	6	7	8	9
Programme meets objectives	1	2	3	4	5	6	7	8	9

Completely disagree   Neutral   Completely agree

Are the scores similar? If not, what has caused the difference?

3. Now studying the individual products in each family, calculate the percentage of:
    (a) financial, market, production, product/consumer successes in each family for the past 5 years,
    (b) the sales, and the profits, from new products in each family for the past year.

4. What are your conclusions on these two product families? How would you develop measures to guide the future product development in each product family?

1.3   Key factors in product success

In the past 30 years, there have been many studies on the factors causing success and failure in product development (some reviews are Ali, 1994; Balachandra and Friar, 1997; Cooper, 1996a). Balanchandra and Friar's review in 1997 of R&D and new product development studies found:

• there were many different factors identified;
• the magnitude of the effect and even sometimes the direction varied in different studies;
• the meaning of similar factors in different studies varied.

They did state there were three common contextual variables, which need to be considered when identifying important factors for product success:

1.   Nature of the innovation.
2.   Nature of the market.
3.   Nature of the technology.

The importance of the market, technology and organisation factors varies according to whether the product is an incremental or radical innovation, the technology is low or high and it is a new market or an existing market as shown in Table 1.3. The market factors are more important in the incremental innovations than in the radical innovations. The technology factors are important in products that have high technology, and the organisational factors in products with low technology in existing and new markets, and high technology in new markets. Balachandra and Friar noted that this was their best guess in 1997, but certainly it is a good basis for starting analysis of product successes and failure in the company.

Many of the factors either identified as leading to success or differentiating between success and failure are under the control of the company. Some *overall company factors* in the product development programme are:

Table 1.3 Important factors for different levels of innovation, technology, market

Technology	Market	Factors		
		Market	Technology	Organisation
Incremental innovations				
Low	Existing	Very important	Less important	Very important
Low	New	Very important	Less important	Very important
High	Existing	Very important	Very important	Important
High	New	Important	Very important	Important
Radical innovations				
Low	Existing	Important	Important	Important
Low	New	Less important	Important	Important
High	Existing	Important	Very important	Important
High	New	Less important	Very important	Very important

Source: After Balachandra and Friar, 1997.

- product development integrated with a clear business strategy;
- systematic PD Process;
- relating the product to the consumer and the marketing;
- knowledge and skills of people;
- regular evaluation.

Many studies have shown that the product development programme needs to be built from the business strategy of the company, and detailed in the innovation and product strategies. If this is not done there is a lack of direction and focus in product development that leads to failure. Over the last 20–30 years, a recognised product development process has developed which is the basis for successful product development; specific activities may vary from company to company but the overall structure is the same. A basic factor that the company needs to recognise is that the product is being designed for specific consumers or industrial customers, and success will be realised if a strong consumer relationship is built up with the product. Several studies have confirmed that the product qualities are important to success, and that there is indeed demand for a superior product that delivers unique benefits to the user (Cooper, 1993). Knowledge of the people in the company is important, with factors identified such as 'marketing and technology are strengths', 'training and experience of own people', 'commitment of project staff'. The product development programme is a complex mixture of specific product development projects that need to be integrated for overall success of the programme and the programme needs to be evaluated regularly for overall success of products as well as efficiency in the running of the programme.

There are also important factors for successful products in the successive stages of the PD Process:

- Stage 1: Product strategy development – integration of the product development programme with the business strategy, clear description of the market and consumers, identification of market and consumer needs.
- Stage 2: Product design and process development – quantitative design specifications, multidiscipline integration, use of new techniques, feasibility analysis.
- Stage 3: Product commercialisation – multifunctional integration, planning and scheduling, market testing, business analysis.
- Stage 4: Product launch and evaluation – organisation and control, fast problem solving, evaluation of launch, production, distribution and marketing, evaluation of outcomes.

Throughout the PD Process there is a need for clear direction at the beginning of each stage, for example at the beginning of Stage 2: Product design and process development, the product design specifications state the consumer's product concept, the quantitative targets for the product qualities, processing parameters and marketing needs. There is also a need for integration of people with different skills and knowledge from different departments. Most important there is a need for constant evaluation throughout the project in feasibility studies, business analysis and post-launch studies.

Fundamental factors in the planning and organisation of the product development project are:

- on-going communication;
- clear aims, objectives and constraints;
- quality assurance of the development;
- final evaluation of the project.

The people in the project need to know what is to be achieved and what other people are doing; this gives an integrated focus to the project, which will lead to success. Studying the quality of the project in its execution and in the end results will increase the chance of success in the future.

To summarise, there are many company-controllable factors related to product development success; the importance of each can vary from project to project. Some important factors, shown in Table 1.4, are common to many projects. Although the company's capability factors are the dominant factors related to product success, they need to be combined with the environmental/situational variables, such as the market characteristics, in selecting products for development (Cooper and Kleinschmidt, 1987). The interrelationships of the product development with environmental factors such as society, consumers and technology need to be considered not only in project selection but also throughout the project and particularly before the product launch. Environmental factors are more important with pioneering innovation than with incremental product development, because often the environment is unknown (Ali, 1994).

Table 1.4   Company-controllable factors in product success and failure

Consumers and markets
Consumers
Closeness to the customer/consumer in product development
The product designed for the consumer's needs, wants and value
Marketing
A strong market orientation

Product
Product
The product superior to competitors
The product has different, unique benefits

Project development process
PD Process
Multistage, multifunctional disciplined process with clear decision points
Integration of product, marketing, production, testing and evaluation
Stage 1. Product strategy development
Product strategy related to business and market strategies
Clear and early product definition
More predevelopment work before product design
Product evaluation and screening to give sharper project selection decisions
Stage 2. Product design and process development
Clear product design specifications
Creativity in design
Integration of product design and process development
Consumer/customer involvement in design
Stage 3. Product commercialisation
Pre-commercialisation business analysis
The new product marketed by the design team to the production and marketing personnel
Integration of production, distribution and marketing planning
Costs definition and reduction
Product quality sustained
Stage 4. Product launching and evaluation
A well-conceived, properly executed launch with a solid marketing plan
Evaluation measures set before launch
Timing of launch optimised
Good control methods
Post-launch evaluation and follow-on.

Product development management
Good technical/manufacturing/marketing interfaces
The right organisational structure and environment
Project evaluation and decision-making procedures
Completeness, consistency and quality of execution of project
Good project leaders and a core group
Time and cost control; continuous evaluation of project and process

Company
Company management
Top management support
Product development in business strategy
Resources in place – time, money, people
Top management in major decision making
Company knowledge
PD project synergy with company's resources/skills/knowledge
Technological synergy and market synergy with company resources/skills

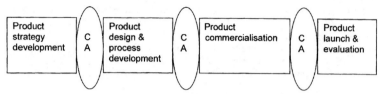

CA is critical analysis and 'Go/Recycle/No Go' decisions by top management

Fig. 1.5   The fundamental PD Process.

1.4   Product development process: the basis for success

The PD Process is important in the food industry, as in all other industries. There is a need for a multidiscipline, multifunctional, well-defined process frame on which the company's different projects can be planned (Rosenau, 1996). In an American survey (Griffin, 1997), 60% of the firms surveyed used formal stages with inter-stage reviews and recycling such as the *Stage-Gate*^{TM} process (Cooper 1990, 1996b), but most companies had developed relatively flexible gates and stage structures. Earle and Earle (1999) suggested a simpler version of four stages – product strategy development, product design and process development, product commercialisation and product launching and evaluation which more clearly delineated the skill and knowledge areas required as shown in Fig. 1.5.

Decisions between the stages are identified and then the necessary knowledge *outcomes* from the stages for these decisions are identified. Then the *activities* within the stages are identified and finally the *techniques* for these activities chosen. This gives a clear basis for planning the project:

Decisions → Outcomes → Activities → Techniques

which will give the sequence leading to the critical analysis and decision making:

Results → Analysis → Reporting → Decisions

In other words, at the start decide on where the project is going, identify the resources available in people, equipment, time and money, then decide how to get to the successful outcome and have an efficient PD Process.

1.4.1   Stages in the PD Process

PD Processes vary according to the level of product innovation, the company's knowledge and resources, the time constraints and the level of risk taking in the company; but there are basic, necessary activities in every stage of the process.

Stage 1: Product strategy development

This has received a great deal of attention in recent years – as one might say taking the 'fuzziness' out of it has been the aim. There is a need for a clear focus; a business definition and a product definition have to be developed in the early stages (Kmetovicz, 1992). The activities of product, market and technology research in this stage are often recognised as vital for successful product development. The product concept and if possible the product design specifications plus a report on the feasibility of the project are the outcomes of this stage. Many times it has been stressed that there should be early definition of what the new product should offer the consumer/customer, that is the benefits, desired product characteristics, uses, safety, value. With the consumers/customers, a product concept is developed describing the product as the consumer sees it and wants it. This product concept is developed into more quantitative descriptions by relating the product concept to both the product metrics, which can be measured by physical, chemical, microbiological or sensory tests, and also the processing, production and marketing methods. The resulting product design specifications give clear directions to the designer. There is an interaction often between the building of the product concept and the product design specification and the initial design of the product.

Stage 2: Product design and process development
This stage is important because a unique product is a key element in product development success. It all depends how one defines uniqueness, but it is mostly a noticeable change that is being achieved rather than a completely new product, as illustrated by Griffin's (1997) comments from the Product Development and Management Association (PDMA) survey in Box 1.1. In the food industry, where thousands of new products are placed on the shelves in a year, the newness can be quite minor such as a different flavour, a package face change, a different slice thickness. It may be opportune to look at the degrees of product

---

Box 1.1   Efforts target product improvements

Over the last 13 years, firms have increased slightly the percentage of projects which improve the performance of already-commercialised products at the expense of projects which merely reposition current products, extend product lines, or reduce product costs, although these trends are not statistically significant. Over half of all NPD [new product development] projects undertaken today represent significant efforts rather than incremental ones. The goal of over $\frac{1}{3}$ of all NPD projects is to add performance capability to current products. There has been no change in the proportion of the new-to-the-world projects, which has been constant at 10% of the total or in new-to-the-firm (but not the world) projects, which has been constant at 20%. The consistently small proportion of new-to-the-world projects may reflect the difficulty of uncovering and delivering radically new solutions for unmet needs, or a bias in firms against very high risk projects. This research was not designed to resolve which alternative is more likely.

Source: From Griffin, 1997 by permission of Project Development and Management Association, Moorestown, New Jersey.

difference being achieved in food design and to decide if the expensive launches are worthwhile for such minor product changes.

It has been noted many times that a unique, superior product is a key issue, but seldom have the methods to achieve it been identified. Industrial design in general is moving away from the traditional belief that it is the creative workings of one person, to the concept of a design team. Industrial designers are also moving into the area of food design, bringing their more aesthetic attitudes in design. The food designers (often called food developers) are much further ahead than other designers in bringing consumers, or at least consumer needs, into the design process. Also because of the close connection between the process and the product qualities, they have often a closer interrelationship between product design and process development, using computer-based experimental designs and analysis. For successful product design there need to be multidisciplinary skills closely integrated, consumer involvement and creativity, combined with the functional areas of marketing and production. In the food industry, the term product design is seldom used, it is more commonly called food product development. Increasingly, there is a need to recognise the principles of design, as the industrial designers become more involved in food design. 'Design and food have decidedly embarked on a union that will forever change the course of both' (Pearlman, 1998). As an example 'Snackitecture', with shapes and colours such as Trix wildberry corn puffs, Heinz Barbie Pasta Shapes in tomato sauce, or pizza-flavoured goldfish, is recognised in design journals (Kalman, 1998).

Product design is the central, creative part of product development and it is important that the different factors influencing it are recognised and integrated into the design process as shown in Fig. 1.6. Product design is based on the tacit knowledge of the designers but it has input from many disciplines and functional areas. It is a blend of creativity, research and testing.

Stage 3: Product commercialisation
This has two activities identified as related to product development success – business analysis and marketing the product to the people in the functional areas (Cooper, 1993; Crawford, 1997), but there are many other aspects of product commercialisation that lead to product success. Business analysis is essential for the decision making at this stage but amazingly there are still companies who never do this before spending large amounts of money in commercialisation and launching. Communication between the design team and the functional groups that will carry the product into production and marketing is essential – this is often called technology transfer, but is better called technology cooperation or technology integration. There are two important facts to recognise in commercialisation – firstly it is still a design process and secondly integration of the functional areas is vital.

The key issues are:

• maintain the product qualities at the same standard as in the design through the process and the distribution;

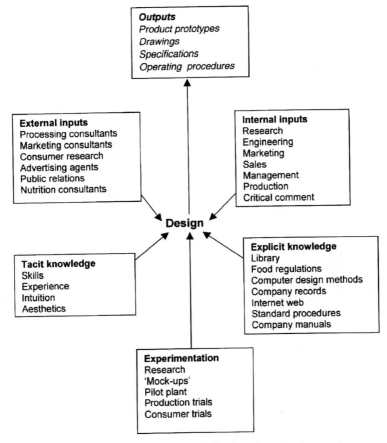

Fig. 1.6   Design components in food product development.

- produce and distribute at the quantities needed;
- develop a total product concept for marketing that agrees with the consumer needs and wants and creates unique value for the chosen target market;
- organise a distribution channel which ensures quality, quantity and costs;
- reduce uncertainty and risk in the launching;
- reach the predicted sales and profits.

In product commercialisation, the product prototype and the preliminary product specifications and marketing strategy have to be developed into a commercial product and production and marketing plans.

Stage 4: Product launch and evaluation
This is the most expensive and risky part of product development. The key issue is to have a fast and effective launch as this can generate the same or more incremental profits as reducing the time for the early stages of the PD process (Ottum, 1996; Stryker, 1996). The targets for the launch should be clearly set to

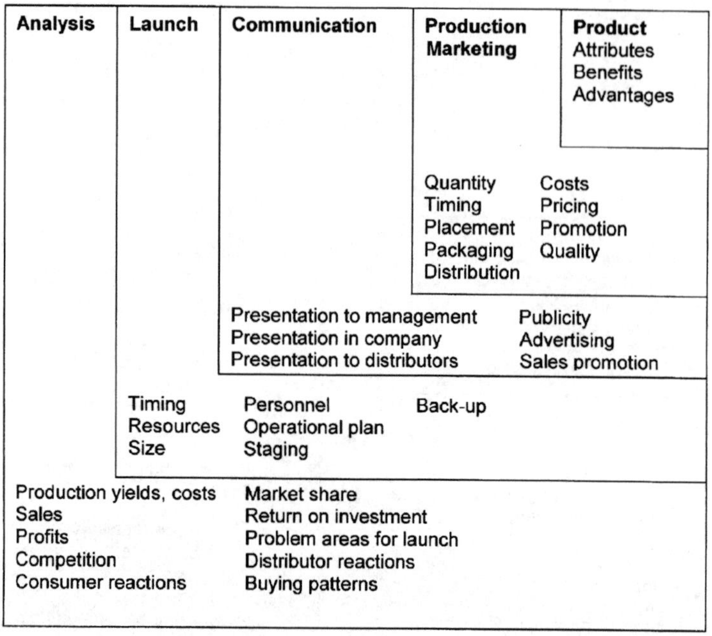

Analysis	Launch	Communication	Production Marketing	Product Attributes Benefits Advantages
			Quantity Timing Placement Packaging Distribution	Costs Pricing Promotion Quality
		Presentation to management Presentation in company Presentation to distributors	Publicity Advertising Sales promotion	
	Timing Resources Size	Personnel Operational plan Staging	Back-up	
Production yields, costs Sales Profits Competition Consumer reactions		Market share Return on investment Problem areas for launch Distributor reactions Buying patterns		

Fig. 1.7 Key factors for launch

provide the basis for the evaluation. It is important for the company to decide on the measures for the launch so that success and failure are clearly defined and measured. The goals might be selected revenues over time, or the market share over time, or profits over time, or for the product to be long term in the product mix. The environment is also important; competition and possibly social, political or economic changes should be monitored. The operational plan for the launch is a key for success but it needs to be sensitive to the situational and operational conditions. Another key factor is the launching to the company and to the trade (distributors). People can easily slow or divert a launch and on the other hand can quickly and successfully overcome any problems that may arise. In industrial product development, it is important to have different departments in the buying company knowledgeable and enthusiastic for the product, and in consumer product development, the retailer as well as the consumer must not be fearful or bored about the product, but encouraged to buy it. Evaluation is continuous so that improvements can be made quickly. The key factors for the launch are shown in Fig. 1.7.

Think break

1.   'The launch is designed for the end user, but considers also the company personnel, the distributors and the society.' Discuss how these people may have

different aspirations for the launch. How can the launch be planned and organised so that the aspirations are integrated to give a successful launch?

2. Choose two of your company's recent new products that were successes. What were the key issues in leading to these successes?

	Not important	Very important
Market orientation		
Consumer focused		
Product design specifications		
Product design creativity		
Product commercialisation integration		
Product launching organisation		
Business strategy		
Knowledge of product technology		
Other		
Other		

3. What do you identify in your company as product design? How do you think food product design will change in your company in the future?

4. What do you think is the quality of the execution of the product commercialisation in your company? Do critical activities happen as they should, or are they slow and late or sometimes do not happen at all? Do you see any changes that could be improved in the product commercialisation to improve the outcomes?

1.4.2 Product development processes for different products

Products can be classified as business-to-business (or industrial), service and consumer. All have the same basic product development but there are significant differences, and the food industry has added differences to each type of product development.

In business-to-business product development there are usually only a few customers, but each customer is a company with a group of people interested in the product, who usually have technical knowledge of the product and its relationship to their processing and product. Not only has the immediate customer to be considered but also the ultimate customer/consumer of the buyer's product. Product specifications are very important and the customer may detail the product design specifications that will then limit the area for product design. The total product of quality, quantity, delivery, price, service has to be recognised in the product design, but the product quality and price are usually the dominant factors. The PD Process has four important features: detailed, quantitative, product design specifications; customer testing of product prototypes; pilot plant and then production testing with the customer; final contracts for supply (Earle and Earle, 2000; Fuller, 1994; Haas, 1995).

Food service can be divided into two parts – marketing of the ingredients or part meals to the food outlet and of the meal/snack to the consumer. Both types of marketing are a mixture of product and service; but in the restaurant, the service can be the major component. Eating meals in a McDonald's or a fancy restaurant is an experience for the individuals and includes intangible experiences such as fun and sophistication. The service is produced and delivered in close proximity, and there are direct reactions between the supplier and the consumer. In both types of food service development, there is participation of the supplier and the buyer in the product concept development and in the product design, because the interaction with the product and each other need to be part of the design (Johne and Storey, 1998; Shekar and Earle, 1997; Terrell and Middlebrooks, 1996). There is a degree of heterogeneity in food service development because people are different but as can be seen in McDonald's product design this can be reduced if the complete product plus service is developed. In developing ingredients for the food service customer, there are fundamental product qualities such as price, quality and safety, but also one must consider the buyer's food preparation and storage facilities and use of labour and energy. The consumer buying the meal or snack has the needs shown in other consumer products but also wants a social eating occasion. 'Food must please; food must entertain; food must satisfy; food must comfort' (Fuller, 1994).

Developing consumer products for marketing in supermarkets is the product development usually described in textbooks (Stinson, 1996). The development of the consumer products is mostly concerned with the design of the physical product, and there are minimal services for example information on the package; so that there is concentration on the consumer/product relationship in the design.

Management has to recognise the difference between the different types of product development and develop the product strategy and the company organisation to work efficiently in each area.

1.5   Managing for product success

There are two very important management inputs for product development success (Cooper, 1993; Crawford, 1997):

1.   Company management's involvement in product development.
2.   Direct management of the product development programme and project.

General management needs to provide direction from the business strategy, resources and major decision making on the programme and the individual project. Product development management needs to provide the plan, the multidisciplinary cooperation and the quality assurance to product development that will result in an effective outcome and efficient procedures.

Many textbooks have been written about PD management in general (Twiss, 1986; Cooper, 1993; Zangwill, 1993; Crawford, 1994, 1997; Urban and Hauser, 1993; Jackson and Frigon, 1996); a list up to 1995 is in the *PDMA Handbook of Product Development* (Rosenau, 1996). Some key elements identified by three different authors are shown in Box 1.2. From all of these books and research, and also from Griffin's 1997 summary of the PDMA survey on American

Box 1.2   Some key elements in product development management

After Vrakking & Cozijnsen (1997)

Failed scenario	*Success scenario*
Lack of strategic control	Enhancing strategic control
Vague innovation objectives	Clear innovation objectives
Competence conflicts between disciplines	Cooperation between disciplines
Defective decision making	Effective decision making
Company culture: survival of the fittest	Company culture: consensus about objectives
Weak role of marketing	Initiating role of marketing

After Zangwill (1993)

Expertise and technological foundations, cultural foundations, managerial foundations, planning foundation and risk management
Eradicate fumbles, place customers first
Develop a business strategy, design the product, improve continuously

After Kmetovicz (1992)

Make new product development a controlled process
Keep an eye on the world
Involve all relevant people from the start
Collect information together
Use information cooperatively
Have a representative object of the end product in view
Learn how to make decisions quickly
Work with competitive tools and methods
Entrust execution to competent people
In the event of problems, adjust only the affected areas
Maintain the 'can do' vitality in the organisation

companies, there are some fundamental needs in product development management in general which are also applicable in the food industry. Firstly there is involvement of the general management, which is crucial and then there is the management of the product development programme and projects.

1.5.1   Company management and product development

Top management leads product development by defining the focus, setting up a system and organisation, ensuring resources and making the critical decisions. It

Table 1.5   Product analysis in developing the product development programme

Demand for products and services
Long-run growth or decline
Stability of demand for products
Stage in product life cycle
Supply of products and services by the industry
Capacity of the industry
Availability of processing technology
Availability of raw materials
Food regulations and other industry constraints
Social constraints
Competitive conditions of the industry
Structure of the industry
Market shares
Finance availability
Relative technologies
Analysis
Probabilities of sales volume and revenues
Key factors for success of the product development

needs to include the innovation strategy and the product strategy in the business strategy, and also to identify any deficiencies in innovation culture, knowledge, management and resources.

In company planning and in the building of the business strategy, the changes that drive innovation are predicted so as to give a long-term direction for the business strategy (Earle and Earle, 2000). It is important to study technological changes and social changes, including political and economic changes, and these give the basis for the innovation strategy. A checklist for analysis of the company and the industry, as shown in Table 1.5, is the basis for the changes to be made in the product mix and the associated services. These two studies – innovation and products – lay the basis for the product development strategy within the business strategy. This analysis is balanced by the company's capabilities – the market position, the supply position, the product, production and marketing knowledge and skills, the plant, marketing and financial resources. Then the product development strategy and the plan for the future can be developed. Planning in this way ensures that the strategic management of product development is connected with the company's overall business strategy. The product development strategy is then an amalgam of the company's innovation, product and technology strategies and gives direction to the whole product development programme.

From the strategy is developed a product development programme and top management sets the budget and the other resources for this programme. The question is how to set this budget. For example on past sales and profits, on the predicted sales and profits if the product development is successful or on the predicted costs? This is an important decision as too tight a budget restricts the project and also the breadth of changes in the final product; too expansive a

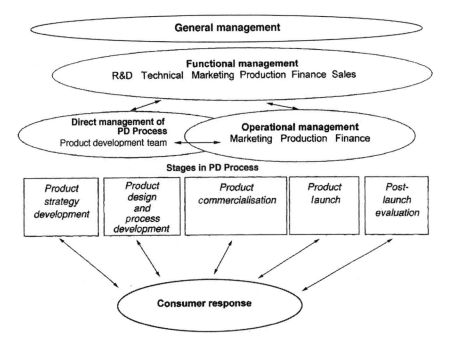

Fig. 1.8   Management for product development.

budget can lead to a wandering programme and too much research for research sake and not for development of the product.

The top management has to be ultimately responsible for setting up an organisation for the product development programme, a diverse activity involving general management, production, marketing and finance, with the product and process development team. Today product development is recognised as an integrated multifunctional process, which combines and coordinates the work in all departments and groups as shown in Fig. 1.8. The organisation has to be structured to give an efficient process with an effective outcome. Companies often place the incremental improvement projects in functional areas or in a strategic business unit, and the major innovation projects in joint committees or separate venture groups (Griffin, 1997). The top management introduces integrated procedures, aims, attitudes and methods into product development. It also indicates the decisions that it will take at specified times and the knowledge it needs for these decisions.

1.5.2   Product development programme management

There are two areas of product development management – of the whole programme and of the individual project. If there is a focused product development strategy, the projects in the programme are more clearly defined and there is less need in the initial stages to do a great deal of research to cull the projects that will not be successful. There is an interaction between the product strategy and product development strategy in which there is product, market and production analysis, and this eliminates projects before they are absorbed into the product development programme.

Management of the product development programme involves the effective use of the available skills, knowledge and other resources between the different projects. Timing is vital not only for the efficient use of resources but to ensure that the outcome – the product – is launched at the optimum time. Planning and control and costs within the programme budget are not always easy when there is a variety of projects whose progress cannot be exactly predicted. But the most important facet of programme management is to ensure the quality of the development and in particular the total outcome of product, production, marketing and finance.

1.5.3   Product development project management

Management of the individual project has the same overall parts of the programme management, but it is very much 'hands-on' management with day-to-day super-vision of the development as regards the activities, the techniques and the quality of the results. This is the basis for the quality of the outcomes and the project management's responsibility is to organise the team and its work to ensure this quality. It is a skilled task, as the project manager has to balance the quality of the outcomes against the costs and time. Also personnel management is very import-ant, as the creative designers have to be integrated to work harmoniously with the more pragmatic functional groups. The leader is usually a formal project manager, but sometimes a project champion with a less formal position leads the team.

To summarise, there are some key questions to be answered by management especially when organising the product development programme, but also at the beginning of major projects, as shown in Table 1.6.

Think break

1. 'Product development is a top management responsibility.' Discuss this statement and delineate top management's areas of responsibility and how it can coordinate this responsibility with other people in the company.
2. In your company, identify the management roles in product development and the various people in these roles – their positions in the company, their expertise and knowledge.
3. Draw a diagram to show how these people interact in the overall product development programme and in the individual product development projects, to ensure project coordination and staff coordination.
4. What methods do top management, product development programme managers and product development project managers use to ensure the quality of the out-comes from product development and the efficiency of the product development?

Table 1.6   Key questions in product development management

Business strategy: does it focus on product strategy and innovation strategy?

Product strategy: is it a predicted, continuous development of the product mix? Does it show the product improvements and the major product innovations, which will be the basis for the product development programme?

Product development programme: is it based on the business strategy and on predicted social and technological changes? Does it specify outcomes needed, time and costs? Are there clear objectives?

Product development organisation: is there a multifunctional, integrated organisation uniting teams and functional groups? Are there identified organisations for incremental product improvements and for major innovations?

Top management control: has top management agreed to the programme and the individual projects? Has top management set the decisions it will make throughout the project and indicated the information it needs for these decisions?
Has top management identified the resources needed for the programme?

Knowledge: is there the level of product, processing and marketing technologies for the planned product development? Is there product design knowledge and creative abilities to create unique products?

Consumer/product relationship: does the company recognise this relationship as a major factor in product development success? Are the consumers integrated into the product development process?

Systematic product development process: has the company recognised the important stages in its planned product development and designed a suitable basic PD Process, and identified variations for different products?

Product design and process development: are there clear definitions of the product concept and the product design specifications? Is there integration of the product design and process development?

1.6   Relating to consumers and markets: the key to product success

There is no doubt that in the food industry the consumer's concept of the product and the relationship of the product to the consumer's needs, wants and behaviour, are critical to success of the product (Saguy and Moskowitz, 1999). There needs to be a clear target market segment(s) identified early in the project. The interaction of the consumer or customer with the product must be identified in the early stages of the product development project, and then followed through each stage of the product development process and finally evaluated after the launch. Knowledge from the evaluation needs to be built up as a knowledge base for incremental product changes, so that a great deal of knowledge on the product/consumer relationship is known at the start of the next project. With a major innovation, there is a need for a great deal of research on the consumer/product relationship in the early stages and constant consumer testing throughout the project.

Relationships between the manufacturer/supplier/user vary a great deal between industrial selling of ingredients, the food service selling of meals and snacks, or the retailing of consumer products in the supermarket. In each there is a blend of product and services, but the proportion of each varies. In industrial and food service product development, there is a need to combine product development and service development, as both are related to the product success. In marketing meals in a restaurant and ingredients to a food manufacturer, there is usually personal involvement of the supplier with the user. Food manufacturers may not have much direct contact with the consumer of their products who experience the product in a supermarket or other retail setting. Therefore the product/consumer relationship is all-important in product development. The level of manufacturer/user involvement is related to the ratio of product quality to services in the total product, and to the blend of product and service development.

Grunert *et al.* (1996) showed, by extensive studies in the European food industry, that a strong marketing orientation is vital for successful product development. Grunert *et al.* (1997) found in examining a number of case studies in the food industry that sometimes innovation was driven by process development and sometimes product development, but the common need was for a marketing focus.

1.7   Knowledge of society, industry and technology

Product development practice is surrounded by a complex environment, which is constantly changing (Earle and Earle, 2000). A product development practitioner must have knowledge of the changes that are taking place in society as a whole, and in the industrial environment including technology and market environment, as well as the changes that are occurring in the company.

1.7.1   Knowledge of societies and their changes

Societies with their social and political systems and their economic, environmental and future needs affect the consumers' behaviours and attitudes in addition to the legal controls on foods and so need to be integrated into product development. To ignore these in product development can lead to product failure and indeed sometimes to violent, anti-product reactions by the society. Innovation is related to change in society; there may be:

- change in economic status of the society so that there is first increased calorie and then increased protein consumption;
- decrease in the size of households causing a shift from bulk foods and jumbo packs to specialised foods and small packs;
- change in knowledge of the consumers causing change in the nutritional and aesthetic qualities of foods (Earle and Earle, 2000).

It is important to recognise changes that are occurring in economic status, society's behaviours and attitudes, so that products can be designed to fit into

Box 1.3   Political effects on product development

Over a number of years, the New Zealand dairy industry developed a truly spreadable butter. Significant time and effort was put into the technical development of this product and into consumer testing in the target market, the United Kingdom. The launch of the product was a great success with good market up-take and repeat purchases. The signs for on-going market growth were very good. Then, with only little warning, the UK Customs banned the import of spreadable butter, based on a claimed non-compliance with regulatory requirements for butter. Subsequently, after a lengthy court battle and significant loss of market share, the decision was overturned.

these changes (Earle and Earle, 2000). Specific attitudes can rapidly develop into political action, which can impose new regulations or indeed ban the new product. An example of the effect of political country barriers on new product development is described in Box 1.3, showing the British attempts to stop a new product – spreadable butter from New Zealand.

1.7.2   Knowledge of industry and technology

Knowledge of the food system, industry and markets with their technologies and organisation is basic to successful product development. How the new product and the product development project relate to the total food system from the producer to the consumer is a key issue in food product development. What does the project want from the system? What is it putting into the system? Obviously a minor product improvement such as a flavour change affects the company's place in the system very little – it is just a case of sourcing new flavouring material, and adjusting the process a little. But if the aim is to launch a new line of frozen fish snacks, then supply becomes a major problem:

- How to find the right species in the right quantity.
- How to have controlled temperature distribution to supply the factory and then for the distribution to the retailer.
- How to organise freezer space with the retailer.
- Has the target consumer suitable freezer space?

So the whole system from producer to consumer is part of the product development project. It is even more complicated if the new product is live crayfish! Today the food system is an interwoven, international network with raw materials and ingredients going from country to country, so that a formulated product may have ingredients from at least five or six countries and maybe every ingredient except for the basic raw material is from a different country. This makes product development very complex, as materials have to be found from many sources. The food system has regulations and controls which may be different, in fact probably are, from country to country. For international marketing, this means product development has either to be aimed at one country or to work in an amalgam of regulations from different countries.

The sizes of companies in an industry and in particular their related market shares lead to a structure that defines the types of innovation (Ali, 1994). The food industry very often has developed from monopolistic to national oligopoly to international oligopoly with large multinational companies dominating the food system in ingredients processing, food manufacturing and retailing. The small company may therefore be forced into pioneering new products, targeting niche markets, or considering late entry when markets are too small for the large companies. The large company has the resources for innovative product development but its total system may be too slow, so that it often gains innovation quicker by buying the small company. Also it may think it should wring everything from the present products before seeking the innovation – but it may then be killed by the innovation from another company.

The infrastructure of suppliers and distributors also has an important effect on product development; the quantity and quality availability of raw materials limits or enhances product development, as do the buying capabilities and product selection of supermarkets and food service outlets.

The food industry, typical of non-durable products, has fewer pioneering innovations than some other industries such as communications and electronics, and mostly pursues incremental product development. The technology of improvement is comparatively simple but can be easily copied. New technology can be difficult, needs increased knowledge and resources, and can be risky because of the intimate relationship of the products with the consumer (Galizzi and Venturini, 1996). Even large food manufacturing companies with large resources tend to veer away from the pioneering innovation, and take the easier path of competing on simple product differentiation. Many small companies are still near the craft stage in their development and have insufficient knowledge of recent food technology for major innovation. So the technology has only changed slowly in the past, but one wonders what will happen in the future with the developments in genetically modified raw materials and in information technology.

It is important to recognise in product development that there is also an internal company environment for product development practice that sets the atmosphere for product development – management, resources, philosophy, beliefs, skills, knowledge and behaviour. The product developer ignores this at their peril.

1.7.3 Creating knowledge for product development

Knowledge is the basis for success in a company and is fundamental to product development. Although knowledge and information are often used interchangeably, there is a clear distinction. Information is a flow of messages, while knowledge is created from information received. There is basic knowledge in the company – both tacit knowledge in people's heads, and explicit or codified knowledge that is transmitted in formal, systematic language. Knowledge can be a major part of the capital structure and is recognised as such in the successful modern company.

Knowledge is becoming more highly regarded in the food industry, and from this can develop new synergy with developing technology and markets both inside and outside the industry. There needs to be inside the company:

- synergy between the knowledge in the differing functions, such as marketing, R&D and production;
- synergy between the functional knowledge and the knowledge needed for the product development.

In recent years, there has been increasing recognition that knowledge creation and management is important in product development (Clarke, 1998).

Nonaka *et al.* (1996) developed the theory that there is knowledge stored and knowledge created in individuals; and the organisation can amplify this and then crystallise it as company knowledge. This bank of individual knowledge in the company, often called tacit knowledge because it is within individual minds and not recorded, is the basis for innovation in the company. In product development, this bank of company knowledge, together with sources of information, is used to create the ideas throughout the PD Process. There are different types of knowledge, some from general experience and education, some from experience in a specific area of the company's technology and some from working in the company and the industry's organisation and environment. Therefore companies have different levels of company knowledge and this is directly related to the level of product development that they can undertake (Court, 1997). Tacit knowledge is important in product development because much of the experience in projects is not recorded; so experienced personnel are important in the project teams as well as in the management decision makers.

Think break

1. 'The consumer is paramount to product success.' Is this statement true for your company? Can you identify successful projects that were based on ensuring an optimum product/consumer relationship? Successful projects that took little consideration of the consumers' needs, wants and behaviour?
2. Take two PD projects that you would identify as failures and two as successes. Place the four projects on the following scales:

	Weak		Strong
Marketing orientation	_____
Product/consumer relationship	_____
Knowledge in company	_____
Information sources	_____
Marketing capabilities	_____
Production capabilities	_____
Product differentiation	_____
PD Process	_____
Product development management	_____
General management	_____

3. What are the differences between the successful products and the failed products? Can you identify places that the product development could be improved?

4. What is the level of tacit knowledge (knowledge within the individuals in the company) in your company of product, production, marketing technologies and of markets and consumers or customers? How does this restrict or widen what your company can do in product development?

1.8   Product development management in the food industry

The food industry has its own specific problems in managing product development:

- Biological raw materials.
- Seasonality of raw materials.
- Complex interactions in the food system.
- Interrelationship of processing conditions and product qualities.
- Direct relationship between the product and the nutrition of the consumers.
- Complex relationships between products and health for different groups of people.
- Instability of food products.
- Continuous supply and buying of food products.

There are problems in designing new fresh fruit and vegetable products because it takes time to develop new types, new varieties – by the time they are developed the consumer may have other needs and desires, or food retailing may have changed. Predictions have to be long term and there is usually the need to develop a range so that changes can be accommodated. There is usually only one season a year, or two by incorporating northern and southern hemispheres, so that development and testing is difficult. Fresh fruit and vegetables and whole chilled fish and now live fish, are growing markets and there is a need to study the management of product development for these fresh products.

Because of the product/process interaction, product design is integrated with processing development. Often in other industries product design and production design are two separate activities, done in series; but this is not possible in the food industry. Hence the second stage is called product design and process development. The consumer nutritional and health relations with the product are a vital part of food product development, especially in the future when products are going to be designed for specific effects in the human body instead of general nutrition. This will mean an ethical responsibility and will lead to product development similar to the pharmaceutical industry with a great deal of ethical testing. Because of food instability, the research on distribution is vital in food product development. The shelf-life in storage and the change in quality during transport are important parts of the product development process. Many products have failed because of a rush to the market without shelf-life trials, with disastrous quality results – better to spend time on testing than money on removing from the supermarket shelves.

Finally there is the continuous buying of food which leads to placing of more new products on the market. There usually is need for a continuous product development programme, which leads to more forward product planning and a rolling product mix. This needs efficient and effective planning and control in product development (Stinson, 1996).

1.9   Basis and structure of the book

Managing innovation is a necessary skill for senior management of all food companies producing new raw materials, new ingredients or new consumer products. Company growth and even survival depends on the continuing introduction of successful new products into old and new markets. Product success or failure depends on many factors, but the most important are the product, the skills and resources of the company, the market and the marketing proficiency, and an organised product development process. There is a vital need to understand the consumers' behaviour and attitudes and to be able to design a product to meet the users' needs. But it is also necessary to have the technological knowledge, and the skills, and the organisational ability to bring a product to a successful commercial conclusion in the marketplace. This book studies these key issues in product development and outlines the methods of managing them. It differs from other books on product development because it recognises:

- different approaches to product development at different stages in the food system;
- supply of biological raw materials that affects food product development;
- central place of the consumer in all aspects of food product development because of the close, daily relationships between the consumer and food;
- very fast turnover of food products;
- the political effects on food products and their marketing.

Food product development aims to develop:

- understanding of the place of product development in the company's business strategy and how this is related to the technological, political, societal and economic changes occurring in the environment;
- ability to analyse the complex food system as the basis of delivering food products to the final consumer;
- understanding of the customers' needs for food commodities and for industrial ingredients as a basis for production research and for product development of ingredients;
- understanding of the food consumer's needs (nutrition, safety, sensory, social and psychological) in different cultures and societies as a basis for consumer product design;
- knowledge of the product development process and the ability to select decisions and outcomes for the various stages;
- knowledge of the activities in product development and the techniques related to these activities;
- ability to plan and manage a product development programme and specific product development projects;
- ability to evaluate the outcomes of product development projects and design improvements to the PD process so as to raise the level of success.

The material in the book is divided into three sections:

- Part I Introduction
- Part II Key requirements for successful product development
- Part III Managing and improving product development

Part II explores four basic aspects of product development – developing an innovation strategy, the product development process, the knowledge base for product development, the consumer in product development. Part III studies the management of product development in general and in different parts of the food system and in different types of food companies. It also discusses the evaluation of the launch of a new product and also the outcomes of a complete product development programme and how changes can be implemented to improve the outcomes and the efficiency of product development in the company.

1.10   References

ALI, A. (1994) Pioneering versus incremental innovation: review and research propositions. *Journal of Product Innovation Management*, 11, 46–61.

ALLDRICK, A.J. (1997) Functional foods: assuring quality, in *Functional foods: the Consumers, the Products and the Evidence*, Sadler, M.J. & Saltmarsh, M. (Eds) (Cambridge: Royal Society of Chemistry).

ANON. (2000) Innovate or die, industry warned. *Chemistry & Industry* (1) 3.

BALACHANDRA, R. & FRIAR, J. (1997) Factors for success in R&D and new product innovation: a contextual framework. *IEEE Transactions on Engineering Management* 44(3), 276–287.

BEAUMONT, L.R. (1996) Metrics: a practical example, in *PDMA Handbook of New Product Development*, Rosenau, M.D. (Ed.) (New York: John Wiley & Sons).

CAMPBELL, H. (1999) *Knowledge Creation in New Zealand Manufacturing*. Masterate Thesis, Massey University, Palmerston North, New Zealand.

CLARKE, P. (1998) Implementing a knowledge strategy for your firm. *Industrial Research Institute*, July/August, 28–31.

COOPER, R.G. (1990) Stage-Gate System: a new tool for managing products. *Business Horizons*, 44–54 (May–June).

COOPER, R.G. (1993) *Winning at New Products* (Reading, MA: Addison-Wesley).

COOPER, R.G. (1996a) New products: what separates the winners from the losers, in *The PDMA Handbook of New Product Development*, Rosenau, M.D. (Ed.) (New York: John Wiley & Sons).

COOPER, R.G. (1996b) Overhauling the new product process. *Industrial Marketing Management*, 25, 465–482.

COOPER, R.G. & KLEINSCHMIDT, E.G. (1987) What separates winners and losers?, *Journal of Product Innovation Management*, 4, 169–184.

COURT, A.W. (1997) The relationship between information and personal knowledge in new product development. *International Journal of Information Management*, 17(2), 123–138.

CRAWFORD, C.M. (1994) *New Products Management* 4th Edn (Burr Ridge, IL: Irwin).

CRAWFORD, C.M. (1997) *New Products Management* 5th Edn (Burr Ridge, IL: Irwin/McGraw-Hill).

DEUTSCH, R.M. (1977) *The New Nuts Amongst the Berries* (Palo Alto, CA: Bull Pub. Co.).

EARLE, M.D. & EARLE, R.L. (1999) *Creating New Foods: The Product Developer's Guide* (London: Chadwick House Group).

EARLE, M.D. & EARLE, R.L. (2000) *Building the Future on New Products* (Leatherhead: Leatherhead Food RA).

FULLER, G.W. (1994) *New Food Product Development: From Concept to Marketplace* (Boca Raton: CRC).

GALIZZI, G. and VENTURINI, L. (1996) Product innovation in the food industry: nature characteristics and determinants, in *Economics of Innovation: the Case of the Food Industry*, Galizzi, G. & Venturini, L. (Eds) (Heidelberg: Physica-Verlag).

GRIFFIN, A. (1997) *Drivers of NPD Success: The 1997 PDMA Report* (Chicago: Product Development & Management Association).

GRUNERT, K.G., LARSEN, H.H., MADSEN, T.K. & BAADSGAARD, A. (1996) *Market Orientation in Food and Agriculture* (Boston: Kluwer).

GRUNERT, K.G., HARMSEN, H., MEULENBERG, M. & TRAILL, B. (1997) Innovation in the food sector: a revised framework, in *Product and Process Development in the Food Industry*, Traill, B. & Grunert, K.G. (Eds) (London: Blackie Academic & Professional).

HAAS, R.W. (1995) *Business Marketing: A Managerial Approach* (Cincinnati, OH: South Western College Publications).

HULTINK, E.J. & ROBBEN, H.S. (1996) A framework defining success and failure, in *PDMA Handbook of New Product Development*, Rosenau, M.D. (Ed.) (New York: John Wiley & Sons).

JACKSON, H.K. & FRIGON, N.L. (1996) *Achieving the Competitive Edge* (New York: John Wiley & Sons).

JOHNE, A. & STOREY, C. (1998) New service development: a review of the literature and annotated bibliography. *European Journal of Marketing*, 32(3/4), 184–251.

KALMAN, T. (1998) Snackitecture. *International Design Magazine*, 45(6), 60–63.

KMETOVICZ, R.E. (1992) *New Product Development: Design and Analysis* (New York: Wiley).

KNORR, D. (1999) Process assessment of high-pressure processing of foods: an overview, in *Processing Foods: Quality Optimization and Process Assessment*, Oliveira, F.A.R. & Oliveira, J.G. (Eds) (Boca Raton: CRC Press).

MEYER, M.H. & LEHNERD, A.P. (1997) *The Power of Product Platforms* (New York: The Free Press).

NONAKA, I., TAKEUCHI, H. & UMEMOTO, K. (1996) A theory of organisational knowledge creation. *International Journal of Technology Management. Special Publication on Unlearning and Learning*, 11(7/8), 833–845.

OTTUM, B.D. (1996) Launching a new consumer product, in *PDMA Handbook of New Product Development*, Rosenau, M.D. (Ed.) (New York: John Wiley & Sons).

PEARLMAN, C. (1998) Food for thought. *International Design Magazine*, 45(6), 47.

PLATZMAN, A. (1999) Functional foods: figuring out the facts. *Food Product Design*, 9(8), 32–62.

RAMA, R. (1996) Empirical study on sources of innovation in international food and beverage industry. *Agribusiness*, 12(2), 123–134.

ROSENAU, M.D. (1996) Choosing a development process that is right for you, in *PDMA Handbook of New Product Development*, Rosenau, M.D. (Ed.) (New York: John Wiley & Sons).

SAGUY, I.S. & MOSKOWITZ, H.R. (1999) Integrating the consumer into new product development. *Food Technology*, 53(8), 68–73.

SCHAFFNER, D.J., SHRODER, W.R. & EARLE, M.D. (1998) *Food Marketing: An International Perspective* (New York: WCB McGraw-Hill).

SHEKAR, A. & EARLE, M.D. (1997) Challenges facing consumer research in new service development. ANZMEC Conference Proceedings Volume II, Melbourne, 1–3 December.

SLOAN, A.E. (1999) The new market: foods for the not-so-healthy. *Food Technology*, 53(2), 54–60.

STINSON, W.S. (1996) Consumer packaged goods (branded food products), in *PDMA Handbook of New Product Development*, Rosenau, M.D. (Ed.) (New York: John Wiley & Sons).

STRYKER, J.D. (1996) Launching a new business-to-business product, in *PDMA Handbook of New Product Development*, Rosenau, M.D. (Ed.) (New York: John Wiley & Sons).

TERRELL, C.A. & MIDDLEBROOKS, A.G. (1996) Service development, in *PDMA Handbook of New Product Development*, Rosenau, M.D. (Ed.) (New York: John Wiley & Sons).

TWISS, B.C. (1986) *Managing Technological Innovation*, 3rd Edn (New York: Longman).

URBAN, G. & HAUSER, J.R. (1993) *Design and Marketing of New Products*, 2nd Edn (Englewood Cliffs, NJ: Prentice-Hall).

VRAKKING, W.J. & COZIJNSEN, A.J. (1997) Monitoring the quality of innovation processes and innovation successes, in *The Innovation Challenge*, Hussey, D.E. (Ed.) (Chichester: Wiley).

ZANGWILL, W.I. (1993) Managerial foundations: competitive benchmarking, in *Lightning Strategies for Innovation* (New York: Lexington Books).

Part II

Key requirements for successful product development

Product development is the key to the future in the business strategy. In the business strategy, the top company management signifies the changes necessary for the company's future survival and growth and from this can be identified the basis for product development. The future is what product development can provide, delivering new products by the least risky, most efficient process to carry the enterprise forward. It must harmonise with the existing company activities as it delivers the new. Therefore it has to be an integral part of the business strategy with full support of top management and cooperation from management in all areas of the company.

Product development builds systematically from the resources and within the constraints of the business, progressing through four key stages towards a satisfied market. The product development process is the system that integrates the activities in the four stages and its efficient and effective organisation is one of the major factors ensuring success in product development.

All of product development is underpinned by knowledge – of the product, market, production, distribution, consumer and society. The more extensive, complete and accurate the knowledge, the greater the probability of a good fit of the new product to the business and the market. So the fullest, practicable, exploration of existing knowledge is important early in the product development process. This knowledge is extended by information from outside the company, and by creation of new knowledge inside the company.

The consumers are an integral part of all product development projects, even in industrial marketing where the immediate customer is the food manufacturer or food service outlet. The product is built around what the consumers need and want, and their behaviour in buying, using and eating the product.

The four important facets of understanding product development are the place of product development in the business strategy, the product development process, the knowledge in product development, and the consumer/product relationship. Thus the new product can be developed to fit most comfortably to the business that produces it and to the customer who consumes it.

2

Developing an innovation strategy

Product development does not occur in isolation as a separate functional activity. It is a company philosophy, a basic company strategy and a multifunctional company activity. In recent years to show this all-encompassing basis, bringing together product, process, marketing and organisational innovations, there has been development of an overall innovation strategy. This innovation strategy is related to the company's overall business aims and strategy, as well as the social, economic and technological environment, and the company's own knowledge and skills. The business strategy also includes a product strategy outlining the products of the future. The combination of the innovation and product strategies is the basis for the product development strategy, and from this can be developed, with the company's technology strategy, the product development programme as shown in Fig. 2.1. In building business and innovation strategies, it is important to recognise that from them comes a product development programme both for many years ahead and for the immediate year.

The innovation strategy is built up in the business strategy from the innovation possibilities, but only after thorough coordination with the product, marketing and technology strategies. The product development strategy is then built from the innovation strategy, together with other parts of the business strategy such as product mix planning and marketing strategy. Finally from the new product portfolio and the product development strategy is built the product development programme. In this way the product development programme sits harmoniously with the strategic direction of the company, the company's technical and marketing capabilities, and the customers in its ultimate market.

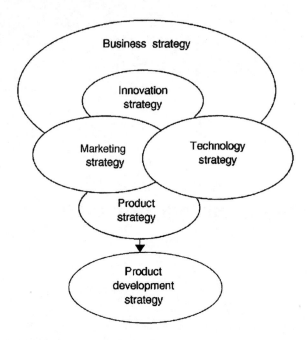

Fig. 2.1   Product development strategy generator.

2.1   Possibilities for innovation

Innovation is an integral part of society, and therefore an integral part of an industry and a company. There are three basic principles of innovation:

1. An innovation is an idea perceived as new by the individual (Rogers, 1962).
2. An innovation causes change, which can be technological or sociological but is probably a combination of both (Earle, 1997).
3. An innovation involves a wide range of people, in the company, the company's environment and the society (Earle, 1997).

Innovation is seen as the state of mind in the company (Kuczmarski, 1996). The traditional definition of innovation in companies as product development and process development has expanded to include all the other changes that can occur (Voss, 1994). Innovation can include ideas for different changes – philosophy, technology, methods, organisation, market, people. But it is important for the company to recognise that any of these changes will affect not only the company but also the other organisations in the food system, the consumers and the society. Innovations outside the company also cause changes inside the company; for example, the technological innovation of the supermarket changed food manufacturing and marketing, the social change of more women working caused an increase in convenience foods. So innovation is related to the climate within

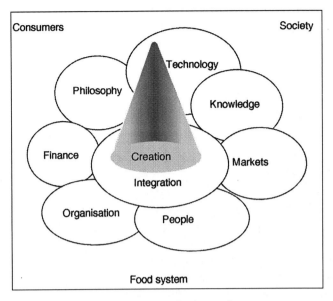

Fig. 2.2   Climate for innovation.

the company and also that surrounding it in the food system and the society as shown in Fig. 2.2. It is important to observe the changes already occurring outside and inside the company, and to predict the possible changes that can achieve the aims of the company to survive and grow. One of the great difficulties is to differentiate between the true, long-term changes and 'fashions' which die quickly. Judging wrongly may adversely affect the company.

The rate of innovation in a company depends on its ability:

- to sense possibilities and to perceive and assess the likely outcomes of feasible changes;
- to evaluate and rank such outcomes strategically and operationally, in relation to company objectives;
- to make decisions on the basis of such information and prepare appropriate strategies;
- to implement plans and changes in managerial and technical terms (Frater *et al.*, 1995).

These steps are shown in Fig. 2.3.

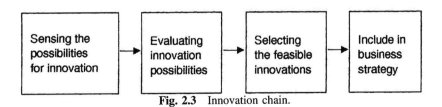

Fig. 2.3   Innovation chain.

Table 2.1   Changes in society leading to food innovations

Major long-term living patterns: urbanisation, suburban and in-city living
Working patterns: increase in office workers and decrease in blue collar workers
Sex roles: women working, women in former male-dominated positions, women in senior positions
Economic status: increasing incomes, more equal distribution or more unequal distribution of incomes
Educational status: knowledge growth from education and the media
Age structure: increasing percentage of old people in Europe and of young people in South America

Source: After Earle, 1997.

2.1.1   Sensing the possibilities for innovations

In sensing the possibilities, it is important to study the major changes that are taking place or predicted in society, in technology, food system, the marketplace and the consumers. Only then can the possible company initiatives be created.

Social and political changes cause changes in the food industry or may even prevent innovations in the food industry. Eating food is a universal activity and therefore the food industry perhaps more than any other industry is enmeshed in the social and political systems in every country. Society changes in many ways as shown in Table 2.1.

The political systems and their attitudes to the food industry also change with societal changes. In 1982, Throdahl suggested that the most important governmental method of encouraging innovation in the food industry was to reduce the adverse impact of regulations on innovation but did add 'without sacrifice of social objectives'. This has been the food industry's dilemma for the past 100 years and even earlier – innovation with or without consideration of society. The political system itself can encourage or discourage innovation, by placing trade barriers or subsidies which encourage local food production and discourage imports. National policies, based on societal concerns, needs and wills, can create a reactive environ-

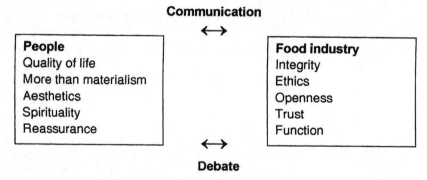

Fig. 2.4   Human values and the food industry
(Source: From Earle and Earle, *Building the Future on New Products*, © LFRA Ltd, 2000, by permission of Leatherhead Food RA, Leatherhead, UK).

ment for innovation in the food industry. In looking for innovation possibilities, food companies need to be aware of changes in societal attitudes that fuel political changes as well as food changes. The social and related political changes have caused food innovation in the past and will continue to do so in the future. Food companies need to have methods of monitoring social changes and predicting future changes (Earle and Earle, 2000). There needs to be greater recognition of human values in developing innovation strategies as outlined in Fig. 2.4.

Think break

1. Identify important social and political changes occurring in your company's external environment.
2. What changes could be made in the company to relate to these changes so that the company not only survives but also grows?

Technological innovation spans a broad spectrum of areas from the new crop and the newly farmed fish, through new refining methods, new preservation methods, new manufacturing methods, new distribution methods, new retailing methods, new cooking and preparation methods. But it also includes changes in technologies of other industries, particularly in those related to the food industry such as the processing technologies in the pharmaceutical and chemical industries, in the home appliance technologies and in the electronic and information technologies. There also needs to be consideration of new scientific knowledge that may be the basis for new technologies in the future. Companies, even very large food companies, are often based on one technology; for example emulsion technology may be the main emphasis and this covers a very wide spectrum of foods from margarine to mayonnaise to ice cream to sausages. Their knowledge is extensive in this one technology and it is often more successful to seek innovation from this basis. When going to a new technology, a great deal of knowledge has to be found as quickly as possible; this means building up resources either by learning or by buying a company already using the technology. It is important for companies to select a basic technology that can lead to many different types of products to satisfy different markets. Some of the technological areas for innovation (Rizvi *et al.*, 1993) are shown in Fig. 2.5.

Think break

1. What would you identify as major technological developments in the processing and distribution of food products in your company's present technological system?
2. What new technological developments in other parts of the food industry or in other industries at the present time might cause changes in your company's technology?

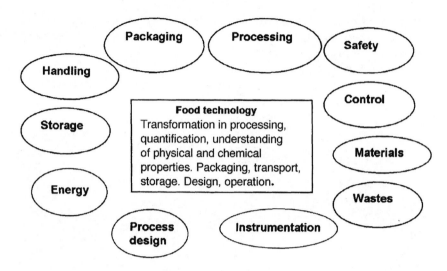

Fig. 2.5 Technological areas for innovation.

Changes in the structure of the food system are also an important source of ideas for innovation possibilities. There are often changes in the importance of the various parts – production, ingredient processors, food manufacturers, retailers, food service – and the pressure for innovation moves from one section to another. Recently there has been increased new product activity in the ingredients industry, which is being transmitted to both producers and food manufacturers. Both vertical and horizontal integration have occurred in the last 50 years, and caused major innovations. For example vertical integration in the chicken industry led to chicken as major meat, and to the development of many new products; horizontal integration led to many new products in the baking industry.

A change in one part of the food system leads to new products in other parts. In particular, innovations in the primary producing industries produce new ingredients, which then advance to new consumer products. Canola seed was developed with low erucic acid, and these seeds were used to produce oil with high polyunsaturated/low saturated fatty acids, and the oil was then used to develop oil-based consumer products which were more attractive nutritionally.

The food system changes slowly. During the last century changes were incremental with some major changes, and the radical changes were well spaced. This may be caused by:

- maturity of the industry – compared with the innovative industries such as electronics, it is more difficult to invent new products;
- consumers – many are cautious and suspicious in judging new foods; food consumers change slowly unless they recognise marked benefits in the new product, and new benefits are more difficult to design than in other industries, for example information technology;

- biological product development – it takes time to develop a new plant, a new animal, a new fish, and even a new safe process, and these are the basis of the radical changes. Because of the expense of these developments, they were mainly funded by governments in the past (Earle and Earle, 1997), which also led to slowness;
- marginal returns on new food products – compared with other industries, for example the pharmaceutical industry, the profits on new food products are small.

So it may be that this has been the most suitable innovation method – mostly incremental product changes, some major changes and a few radical changes. But with today's high rate of technological change in other industries, there may be change in the rate of innovation in the future.

The knowledge and understanding of technology in the total food system is continuing to grow rapidly and, if this is recognised by the industry, it will impact on new products, new processes and new manufacturing systems but more important on the consumers' and society's attitudes and behaviour towards food. However, if acceptance by the consumers is to be widespread and willing, then they must see obvious overall benefits to themselves; this needs, among other things, full and clear information. A striking lesson in the difficulties that may arise is to be seen in the introduction of genetically engineered foods. The bundle of products and services that the food company calls an innovation is now in the eyes of the consumers an experience, which they hope is safe and enjoyable (Pine and Gilmore, 1998). The food industry has gone through:

Commodities $\rightarrow$ Products $\rightarrow$ Services $\rightarrow$ Experience

Innovation occurs today at all these levels in the various parts of the food system.

Think break

When searching the food system for innovations, some leading questions are:

- What are the changes in the relative importance of the various sections in the food system?
- How is the capacity of the industry changing?
- What are the changes in the ownership structure?
- Are there predicted take-overs in the industry?
- Are there predicted take-overs from outside? Hostile? Friendly?
- Are there predicted investment changes?
- Are there new companies entering the industry?
- Who are the innovators in the industry?

1. Try to answer these questions.

2.   Identify possibilities for new products in your company based on this
     information.
3.   What new technological developments are predicted in the food system that
     could affect your company?

Marketplace changes provide a rich source of innovation possibilities
(Earle, R.L. and Earle, M.D., 1999). There needs to be searching for long-term
possibilities, as well as tactical thinking for the immediate marketing plans. Four
areas to consider when looking for long-term marketing possibilities are:

• international comparisons;
• product and service developments;
• market specialisation;
• new distribution methods.

Looking internationally, it is important to take a broad look over many
markets and compare them. The home market in the USA or Europe may be
static, but markets in Asia are increasing rapidly. Alternative possibilities are
either in the home market to increase a market share or to have higher value
products, or in the new market to relaunch the old basic products. As can be seen
with McDonald's and Coca-Cola, relaunching on a new market is successful in
the long term, but there is a need to keep the home market viable as the basis for
the new venture. The reverse also occurs: products on an overseas market can
produce ideas for the new product in the home market.

Changing the ratio of product to services is another way of identifying
innovations. Once the new product was the main innovation in consumer
marketing but increasingly service has become important. How far does the food
manufacturing company go in providing services for the consumer; how far does
the ingredients company go in providing services for the food manufacturer?
Certainly innovation can be found; for example in food service providing the
materials and the recipe for the dish opens up a whole range of new products to
be supplied to small restaurants; providing complete chilled meals, ready-to-
heat, in supermarkets again leads to many new products.

Market specialisation has gained increasing recognition in searching for
innovations. In the past the food companies tried to provide a wide range of
foods, and their innovation growth was often achieved by buying or
amalgamating with other companies. Today should marketing be more focused
and the innovations aimed at specific target markets? In other words, should
variations of a product be developed for different market segments, so that the
new products are more focused on the people in that market?

Distribution has always been an area for major new developments – from the
grocer's shop to the supermarket to the mega-market – and one would predict
that there are going to be major changes in the next ten years with the
introduction of e-commerce and other uses of the Internet. The information age

Box 2.1   Some future consumer needs predicted in 1969

- Increasing importance of smell in foods
- Foods light in substance but strong in flavour
- Texture a more important featured characteristic
- Packaged goods accepted as norms
- Dieting and slimming will become an increasing occupation
- The family mealtime will break down
- Strong conservatism in food taste progressively breaks down
- Better nutritional standards eliminate the danger of between meal hunger
- Meal nibbling for social/psychological reasons increases
- Increased public sophistication in dietary and nutritional matters

Source: After Hedges, 1969.

is certainly having a strong effect on all aspects of marketing technology – the distribution system, the places for selling food, the communications, the promotion, the sales methods. New food products will certainly come from the four consumer trends: using the Internet to buy food, food shopping as entertainment, food shopping for freshness and food shopping for health, all of which will affect the distribution system (Earle and Earle, 2000).

In searching for the long-term market possibilities, the basic research is to study the consumers and in the case of the food ingredients company also their immediate customers. There is a need to take a broad look at the possible consumers and their future needs, wants and behaviour (Earle and Earle, 2000). The research is about people – how they think, feel and behave, and why they think, feel and behave in these ways, and then to relate this to their needs in future products as shown in Box 2.1 (Hedges, 1969). It is interesting that this paper was published over 30 years ago, and how many of the predictions have become reality. An interesting question today is: are the consumers' knowledge and attitudes pushing the food industry towards the product quality standards of the pharmaceutical industry, guaranteeing the safety and the effectiveness of the food products? What innovation possibilities does this uncover?

Think break

1. Identify eight innovation possibilities for your company, two under each of the following areas:
   (a) society changes,
   (b) political changes,
   (c) technological changes,
   (d) food system changes.

2. Select an important product area for your company and study it for:
    (a) international comparisons,
    (b) product and service developments,
    (c) market specialisation,
    (d) new distribution methods.

From this identify eight innovation possibilities.

2.1.2   Evaluating the innovation possibilities for the company

The innovation possibilities may be market related, e.g. a new market niche, a growing market area; technology related, e.g. a new process, increased automation; resource related, e.g. a new crop, a new ingredient; society related, e.g. increased income, poorer health; consumer related, e.g. single complete meals, children-friendly meals. These innovation possibilities need to be analysed against the company's capabilities and the company's objectives. The company evaluates from 'might do' to 'can do' to 'should do'.

The company's climate and capabilities are a major evaluation factor in studying innovation possibilities. One company may be very conservative, and not want change, so it chooses a low level of innovation as the company climate and therefore in its business strategy. Another company may want to be at the forefront of change, so it has a company climate of innovation, and includes innovation as a major part of its business strategy. This incorporation of innovation into the company philosophy sets the basis for the product development. If the company has low-level innovation, product development consists of cost cutting and minor product improvements; at high-level innovation, product development is searching for a unique product that will cause a major change to industry, market and consumers. Many companies have a mixture of innovation and conservatism.

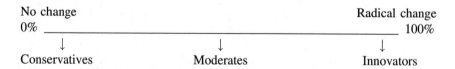

The company may think of change as technical, but it is the commercial change, particularly as related to the consumer, that is the important change. This spectrum is also related to risk-taking: companies can vary from aversion to risk to seeking risk. It is important to recognise the present level of innovation in the business strategy and also the philosophy for risk-taking in the company.

Companies cannot quickly change from one level of innovation to another. Before viewing the innovation possibilities for the company, it is often interesting for the company to take a look at itself:

- Is it blinded by the glare of the oncoming future, trying to muddle along in its present markets and technology?
- Is it searching fearlessly and widely for new opportunities?
- Is it moving in a focused direction with a strong sense of purpose?

There are basic company qualities that affect evaluation of possibilities such as size of the company, financial status, type of product mix, place in the market, standard of production and marketing. But when judging the innovation possibilities, it is more important to study the company's experience, expertise and knowledge in innovation. It is important to make a quantitative analysis of the company's rating in innovation, and it is helpful to use a set of innovation indices and compare these, if possible, with the ratings of other companies or the industry in general. Various suggestions have been made for innovation indices, including the success of new products, new product development effectiveness and the innovation level of the company as shown by Kuczmarski (1996). He suggested that the following indices should be determined over a three-year period.

1. Success rate of new products:
    (a) survival rate: new products still on market/total number of products commercialised,
    (b) success rate: new products exceeding revenue forecasts/total number of products commercialised,
    (c) innovation sales ratio: cumulative annual revenues from new products/total annual revenues.

2. New product development effectiveness:
    (a) R&D innovation effectiveness ratio: gross profits from commercialised new products/R&D expenditures to new products,
    (b) return on innovation: cumulative net profits from new products/cumulative new product total expenditures for all commercialised, killed and failed new products,
    (c) process pipeline flow: number of new product concepts in each stage of the development process at year-end,
    (d) innovation revenues per employee: total revenue from new products/number of employees devoted to innovation initiatives.

3. Innovation level:
    (a) R&D innovation emphasis ratio: R&D expenditure to new products/total R&D expenditure,
    (b) newness investment ratio: expenditure to new-to-world products/new products total expenditures,
    (c) innovation portfolio mix: percentage of products new-to-the-world, line extension, repositioning, new-to-company, product line improvements.

Table 2.2   Company innovation indices in New Zealand manufacturing over five years

Innovation indices	Highly innovative	Moderately innovative	Least innovative
Number of new products	26	9	8
Number of improved products	48	22	5
New products, % of total sales	42	19	18
Improved products, % of total sales	32	25	15
Success of new products*	4.0	4.6	3.5
Change production processes[†]	2.6	2.5	1.9
Change management, marketing, support systems[†]	3.5	2.7	2.4
Comparative status of plant equipment[‡]	2.9	2.9	2.1

* Scores, 1 (most failed) to 5 (highly successful).
[†] Scores, 1 (not at all) to 4 (completely).
[‡] Scores, 1 (more than 10 years behind) to 4 (fully up to date).

Source: From Campbell, 1999.

These are quantitative measures (metrics) of new product development success and effectiveness, and of the innovation level of the company, and these can be used to compare the company's performance with that of other companies. Campbell (1999) studied innovation in manufacturing companies in New Zealand over a five-year period using a simple comparison of product success:

• number of new products;
• number of improved products;
• new and improved products as percentage of total sales;

and asked the companies to state their level of success in new products and their level of change in technology as shown in Table 2.2. These are mean scores for New Zealand manufacturing companies in a variety of industries so are not typical scores for the food industry. But they show the differences that can be found between the most innovative and the least innovative companies. The innovative companies tended to be innovative in all parts of their business, as can be seen from their much higher scores, than least innovative companies, for change in production, plant equipment, marketing and support systems. It is interesting to note that the highly innovative companies launched more products but had a slightly lower success score than the moderately innovative companies. It was found that these highly innovative companies tended to have a truncated product development process and missed some of the evaluation steps, while the moderately innovative tended to have more stages and more analysis.

Think break

1. Compare the innovation scores either between your company and other companies in the industry, or if this is not possible between different product areas in your company.
    For the last five years, collect the following information:

Sales growth over last 5 years   _____
Number of new products   _____
Number of improved products   _____
New products – proportion of sales   _____
Improved products – proportion of sales   _____
Success of new products
All failed _____ 100% success
Changed production processes
Not at all_____ Completely
Changed marketing methods
Not at all_____ Completely
Changed company organisation
Not at all_____ Completely
Age of technology
More than_____ Fully
10 years behind   up-to-date

From these results, how do you rate your company – highly innovative, moderately innovative, not innovative?

2. In what areas do you think your company has the knowledge and skills for innovation in the future – raw materials, processing, products, distribution, marketing, communications, consumer experience?
3. In what areas do you think your company has the financial resources for innovation in the future – raw materials, processing, products, distribution, marketing, communications, consumer experience?
4. What do you see as your company's barriers to innovation?

The **company objectives and goals** are also important in studying innovation possibilities. What is the company wishing to achieve, where and when? The innovation possibilities need to be ranked against these objectives – in particular innovation possibilities need to fit into the general direction of the company and not involve technologies, markets and finances, which are well outside the objectives of the company.

The innovation possibilities are screened to choose the most suitable for further study. In selecting the innovation paths, it is important to retain contact with the twin areas of business and society, as shown in Fig. 2.6. The factors used for screening vary with the company and the types of innovations, but

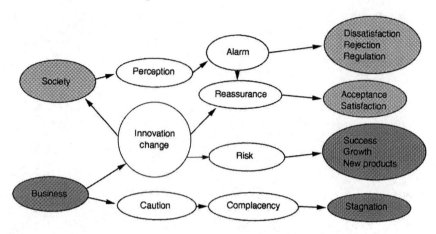

Fig. 2.6　The business and societal decisions for innovations.

important factors are related to the company, market, technology, society, predicted outcomes, project needs and company resources. Some important evaluation factors for innovation possibilities (Kuczmarski, 1996) are shown in Table 2.3. Major factors are those that are important in evaluation while critical factors are those that are directly related to product success and must be

Table 2.3　Evaluation factors for innovation possibilities

Major factors	Critical factors
Company	
Fit with strategic objectives	Exploits internal strengths
Impact on existing business	
Market	
Consumer need intensity	Product/service uniqueness/differentiation
Source of competitive advantage	
Technology	
Company competence in technology	Relation to present technologies in company
Society	
Impact on ethical constraints	Agreement with religious rules
Impact on political constraints	Agreement with government regulations
Predicted outcomes	
Sales and profits potentials	Return on investments
Degree of risk	
Needs and resources	
Financial needs	Financial resources
Knowledge needs	Knowledge resources

Source: After Kuczmarski, 1996.

evaluated. The remaining innovation possibilities after screening are incorporated into the building of the business strategy.

2.2   Incorporating innovation into the business strategy

The innovation strategy/strategies are formed within the business strategy, along with other strategies such as product, technology and marketing, as shown in Fig. 2.1. The formulation of company goals and strategies is very much an iterative process, integrating the various strategies in the direction of the business goals, and using forecasts and analysis of possible outcomes, with an understanding of the company's capabilities.

Top management develops an **innovation blueprint** – a vision that defines the future role that innovation plays in the long-term goals of the company (Kuczmarski, 1996). The basis of this is an understanding of how innovation affects the company's main stakeholders – consumers, staff and shareholders; how it is related to the value of their company – capital value, share price; and how it is related to the value of their brand(s). This blueprint is the standard for accepting an innovation possibility into the business strategy.

The top innovation possibilities are combined with the blueprint to develop an **innovation summary**, which is built up with the product, marketing and technology strategies into an **innovation strategy**.

2.2.1   Combining strategies – product and innovation

The product strategy develops a balanced and rolling programme for the product mix during at least the next five years, with an outline product mix for later years. The forward planning of the product mix depends on the culture and size of the company, volatility of the market and the rate of technological development. In a large company with a reasonably stable market and slow change of technology, planning can be ten or more years; in a small company with few resources it can be one or two years.

In the product mix planning, there is recognition of today's breadwinners and also of the future breadwinners, the place in the product life cycle of the product areas, the competitive status of the products now and in the future. This identifies the areas for product improvements, line extensions, repositioning, new innovations in the present product system and radical new products outside the present system. There needs to be a constant interplay between the innovation summary and the development of the product mix, so that out of it will come the product development strategy that will be the basis for the product development. It is also important at this time to predict the effect on products that other innovations will have – for example, a new processing line, or a restructuring of part of the company. Often these are analysed separately, especially the company reorganisation, with little thought of how this would affect the product strategy and therefore the financial outcomes of the company in the future. There can be very adverse outcomes. They may be cost saving at the time, but may have severe impacts on the new product planning and the future returns.

Think break

1. What are the basic product areas in your company's product mix? In the product mix, identify today's breadwinners – the products providing the main part of the sales revenue, and tomorrow's breadwinners. What is the place of the other products in the product mix?
2. From your study of the product mix, what types of product innovation do you predict for the next few years?

2.2.2   Combining strategies – technology and innovation

The technology strategy for the company is also interwoven with the innovation strategy. In building the technology strategy it is essential first to identify the competence of the company with the present technologies and the ability to develop new technologies. A systematic method is used, comparing the technological competence of the company against other companies. This gives a truer indication of how technologically skilled the company is, rather than using subjective statements of company staff who may have vested interests in the present technology. It is difficult for outside consultants to assess the company's abilities for new technological areas. A combined project team with company staff and consultants using quantitative analysis is probably best for analysing technology competence throughout the company – raw materials, processing, distribution, marketing, and products. There is a need to study:

*   base technologies that are necessary for the chosen product–market mix;
*   key technologies which provide competitive advantage;
*   new technologies, which could become tomorrow's key technologies.

A technology mix needs to be developed for the future incorporating all of these. The technology strategy is related to the innovation possibilities that have been selected in the innovation summary, the product mix and the technology mix as well as the company's technological capabilities as shown in Fig. 2.7.

A technology strategy can identify:

*   new base or core technology that may lead to a range of new products;
*   base or core technology that is needed for an original new product;
*   key technology change that will be a unique competitive chance for the company;
*   improved technology that will lead to higher product quality, more varieties of products or cost reductions.

In developing a technology strategy, it is important to relate it to the products, consumers and markets. Sometimes a new processing or production technology

Information ⟶	Evaluation ⟶	Strategy
Innovation summary	Technological capability Organisational capability Rate of technological change Time for development	Strategy 1
Base technologies Key technologies New technologies	Product mix needs Distribution needs Relation to present technology Production capacity	Strategy 2
Overall company aims Business strategy	Financial resources Risk Government controls/regulations	Strategy 3

Fig. 2.7 Building the technology strategy.

may appear an attractive advance but may give product changes that are not recognised by the consumers or may even be unattractive to the consumer. Irradiation is a long-time technological innovation that has not come to be used because of consumer resistance to it. Genetically engineered crops are another instance today. It is also important to study the technological need, possession and lack of technology in the company and outside sources of technology, in developing the technology strategy:

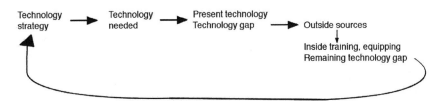

In looking at new raw materials, some factors to study are shown in Table 2.4. Raw materials are an important technological area for innovation in the food

Table 2.4 Factors in raw material innovation

Materials
New agricultural and marine resources/materials
New processed raw materials, ingredients
New packaging materials

Constraints and new freedoms
Social constraints on raw materials changing
Availability/costs of raw materials changing
Government controls on raw materials changing
Effects of economic/political changes on materials
Changes in company standards for materials

industry, but consumer and political pressures today indicate that more care must be taken in sourcing them, so that the development and the production environment and methods are visible. Saying that the raw material pathways in international trade are too complex will not be an answer in the future – this may see more joint ventures in raw materials innovations.

In the food industry, it is critical to combine the necessary production, processing and marketing technologies in technology innovation to ensure a successful innovation.

Think break

1. What are the basic technologies in raw materials, processing, distribution, marketing in your company?
2. What are the key technologies in each area?
3. What new technologies do you see developing in each area in the future?
4. Does your company have the competencies for these new technologies?
5. If not how could the company acquire them?

2.2.3   Combining strategies – market and innovation

The important first step in market innovation is to identify the target market segment (Schaffner *et al.*, 1998). The company must group consumers, industrial customers, retailers or food service organisations into coherent groups which have similar behaviour, attitudes, needs and wants, so that the same marketing method can be used for all members of the group. Once the target market is identified, then the information on which to build the market innovation strategy can be collected. The innovation strategy can stay with the present target market, expand this into similar market segments, or look for new market segments in the national or international markets. In a true innovation, it can be creating a new market.

The marketing orientation in the business strategy is guided by the company's interpretation of consumer or customer needs and wants. This is implemented in the marketing strategy. The overall marketing strategy is based on developing the consumers/customers' concept of the company and also their concept of different product areas. As consumers' behaviour, needs and wants change, the company needs to adapt and develop an innovation strategy in parallel with these changes. The company needs to be flexible and change its relationship with the consumers as consumers change. The innovation strategy is based on the company/consumer relationship and also the product/consumer relationship:

Company $\leftrightarrows$ Consumer $\leftrightarrows$ Tangible product

In the case of the tangible product in the supermarket, the consumer is relating to the brand, the company and the product.

In the case of the food service, there is also the service relationship where seller, buyer and product interact:

Consumer ⇄ Food service provider

Product

In the take-away or restaurant or institution, the consumer is reacting to the service provider and the service as well as the food. In industrial marketing, the customer is also reacting to the seller as well as the product and the services provided.

In developing an innovation strategy in marketing, the company can be changing the consumers' concept of the company, the brand, the products and the services.

The marketing strategy is strongly related to the product strategy. It is influenced by the stage of the product life cycle the product has reached. Is the market innovation to launch a new product in a new market, to launch an old product in a new market, an improved product in the present market, relaunch a product in the present market? Or even to drop a product – also an innovation but with the aim of death instead of life?

Another important aspect of developing the market innovation is the competition and the company's competitive position. Is the innovation reacting to the competitors' actions or is it proactive, acting before the competitors? The positioning of the products relative to the competitors is important in building the market. The market segment may be the same but the positioning of the product to the segment may be the innovation. For example, a tin of baked beans is an everyday commodity product but the new positioning could be to change to a nutritional market segment and position the beans as a high-protein food. This may be too major a change!

The marketing innovation can also come from changes in the marketing technology, e.g. changing the communications, the retailers or other organisations in the market channel, the pricing methods. These changes can cause innovation in the methods of marketing. In evaluating the marketing innovation, sales and profits predictions are often used, but it is also important to study the consumers' reactions to the innovation, difficulties in accessing the market and the company's capability in entering the market.

2.2.4   Unification – the combined innovation strategy

In developing the company's innovation strategy, aspects of innovation in the product, technology and marketing strategies are combined with the innovation possibilities. The company has to decide which is the 'lead' innovation and then choose the other strategies to complement it. One company may decide that the critical innovation is to change from providing food for people to providing health to people. This will need to be combined with a major raw material and processing technology change from general food technology to pharmaceutical technology. The knowledge, safety and ethical standards in selecting raw materials and in controlling processes will need to be higher. The company may need to acquire or merge with a pharmaceutical company to gain the knowledge and the facilities. The marketing strategy will also change – communication through the medical profession instead of TV advertising around general viewing programmes; selling through specialist health boutiques in the supermarket and through pharmacies; a different brand.

Another company may aim to stay as an energy food supplier for children, teenagers and young adults, and to increase their market size have decided to enter the international market. This will lead to a standard processing line (or kitchen facilities if they are in the food service industry) easily adapted to different national infrastructures; strong raw material and ingredients specifications; understanding of cultural needs as a basis of the advertising and public relations; building of an international brand.

This interrelationship of the various strategies is mapped out in Fig. 2.1, so that an interlocking overall business strategy can be built up.

2.3   Building up the innovation strategy

There are now a number of innovations that have passed the initial screening against the company's aims and then had their relationships with the product, marketing and technology strategies assessed. These need to be brought together into the final innovation strategy as shown in Fig. 2.8. Whatever direction its innovation strategy may take, a company needs the knowledge and techniques to create, design and develop the innovation, as well as the resources and implementation skills to bring it to fulfilment. There are many innovation strategies, and they can be combined in various ways, but they need to be analysed on their predicted outcomes before they are accepted into the company. So there is a need in building an innovation strategy firstly to study the total system and the company's situation in it and then to predict the changes that may occur in the system, and to state the optimal situation for the company in the future. In developing innovation strategies, a food company will consider its

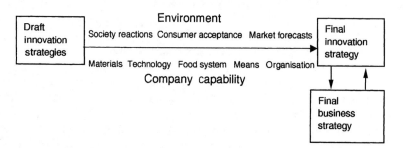

Fig. 2.8   Building the final innovation strategy.

raw materials, technology, markets, targeted consumers and their wants and needs, but need also to set out clearly:

• the company's place in the food system;
• the company's means of achieving the innovation aims;
• the company's organisation and resources for innovation.

2.3.1 The company's place in the food system

The two main food channels are the fresh product channel and the processed product channel. The fresh products channel has increased a great deal in importance in the last few years and it is predicted to grow further because of improved distribution technology and consumers' evident wish for fresh products. The processed products have been the mainstay of the food industry because of their enhanced storage life and the amount of variation that can be achieved in the products. The next decision is to decide on the stage of the food system: production, ingredient processing, manufacturing, distribution or retail. There is increased innovation in the production sector with many new types and varieties of fruits and vegetables; farming of an increasing variety of fish; organic farming; new types of animals. This is as well as genetic engineering, which up to now has concentrated on farming methods, such as resistance to herbicides and higher yields, rather than potential for product innovation. Innovations from the ingredient processors have increased markedly and this is an increasingly powerful part of the food industry. Food manufacturing has mainly concentrated on incremental changes, with some new innovations such as UHT processing and extrusion. The retail sector is a continuous area for innovation, both inside the supermarkets with own label products, organic products and boutique stalls, and outside with the increase in food stores associated with petrol stations and the rising growth of takeaways and restaurants. Some possible innovation strategies for the various stages in the food industry are shown in Box 2.2 (Earle 1997).

Vertical integration has been an important innovation strategy in the past, for example in the chicken industry, and in large multinationals which have combined ingredients processing with food manufacturing. Recently in some companies there has been a breaking of the integration with selling off the ingredients processing section by large food companies and of contracting farmers instead of owning farms in the production, processing and marketing integration. Retailers increasingly have a high degree of integration, although not always ownership, with production, processing and manufacturing, and are more strongly involved in innovation in the food system. Food manufacturers are increasingly directed in innovation by the food ingredients' processors and the retailers. It is interesting to speculate how the food manufacturers will develop innovation strategies in the future; it would appear that today's strong influence on their innovations of retailers and ingredients suppliers may make them redundant in innovation or spur them into new directions.

Box 2.2   Some possible innovation strategies for the various stages in the food system

Food service – Fast foods is an area which will develop further in the global scene, with international foods from fusion of meals and snacks from different countries.
In some countries, particularly the USA, the fast food companies could develop more fresh take-home meals or part meals.

Retailers – New developments will start in the USA to cope with the changing consumer needs. There could be innovation in types of stores, the present supermarkets' domination changing into different types of stores – fresh food markets, convenience stores, take-home meals outlets. Increasing use of the Internet in retailing.

Manufacturers – In seeking 'total food technology', the manufacturer could seek innovations in the retail sector, developing new retail outlets which they could own or be in joint ventures with other manufacturers or retailers. Two possible innovations are:
* marketing a specialised group of nutritionally designed products through nutrition boutiques;
* cooperation with fast food outlets to develop a new combined manufacturing and retail system to provide fresh meals or part meals for taking home.

Ingredients' processors – This is an innovative sector at the present time and one can only see them increasing their industrial marketing to have cooperative programmes with their customers, both food manufacturers and food service, and increasingly involving farmers and fishers.

Farmers and fishers – They could increasingly manage a 'fresh' chain from the farm and sea to the 'fresh' supermarket and to the food service outlet. There could be closer relationships with the food processors in developing new food ingredients by animal, plant and fish breeding. Ownership could ensure that new varieties and more sustainable production methods are linked directly to the needs of the consumer.

Source: After Earle, 1997.

2.3.2   The company's means of achieving the innovation aims

In the innovation strategy, the company needs to decide the means for achieving the innovation: grow own technology, acquisitions, mergers or licensing. These are all methods of bringing the innovation to fruition and the choice depends on resources in the company, time available, costs, risks involved and the probability of success.

If the decision is to develop the innovation within the company, there has to be the decision on whether the change has to be incremental or discontinuous. This means that the management decides if the innovation is to grow from the present base or if this is to be a completely new direction – maybe a new plant or a new market or a new product platform. In the industry, is it strengthening its position, changing position or moving out? Is the company organisation staying the same, gradually changing or completely changing? There also needs to be a specification of risk – high, moderate or low risk.

This is really setting the company philosophy for innovation. A large company may say that it has different types of innovation in different parts of the company – some strategic business units may be high risk, discontinuous change, growing their own technology; other strategic business units can be low risk, incremental changes, acquisitions. But usually the company has one philosophy; there may be venture parts of the company that have a different philosophy. The degree of risk in an innovation strategy varies with the company; two different companies may decide to develop the same product for the same market – for one it is high risk and for the other it is low risk (Souder, 1987). Two companies developing frozen bread dough and two developing low-fat beef are compared in Table 2.5.

Table 2.5 Innovation strategies and their risks in different companies

Frozen bread doughs
The innovation is frozen bread dough as a consumer product in supermarkets

- A small baker marketing bread in its local area is looking at an innovation strategy for marketing frozen bread doughs to supermarkets nationally. This is a high-risk venture as both the technology and the market are new to the company.
- A large baking company which is marketing cakes, biscuits and frozen pastry to supermarkets is considering marketing bread doughs to supermarkets. This is a low-risk venture because it has the technology and the market already and it is a new product to expand the range.

Low-fat beef
The innovation is a production method for growing beef cattle to produce low-fat beef

- A group of farmers is setting up a new processing and marketing cooperative to market the beef in the local market as gourmet products for high-class restaurants. This is an innovation with a high capital cost for plant, but a low risk as the farmers are already selling beef in this market, and know it well. The risk is that the market may be too small to carry the capital cost.
- A meat company, with processing facilities and marketing system selling beef to hamburger processors in an overseas market, is seeking to set up a marketing system in the overseas country through meat importers and restaurant distributors. This is a high-risk venture as the company does not know the marketing system for beef to restaurants, and also the consumers' requirements for low-fat beef which is grass-fed. There is no capital cost for equipment, but costs in setting up the marketing system.

2.3.3   The company's organisation and resources for innovation

It is important that the company sets out the innovation strategy clearly for all to read because it is going to be the basis of the criteria for decisions in setting up the product development programme and also during the individual product development projects. The innovation strategy is the basis for all innovation in the company, and the basic strategy leads to the company direction for innovation. For example many large food companies want to keep their position in the industry consolidated with some growth, and this would likely be associated with incremental changes, low risk, gradual company organisation change, and internal technology with acquisitions when necessary to acquire new technology. The small company with a new technology and wishing to grow would combine discontinuous change, with high risk and major company change, and would grow its own technology. It is important to recognise the interrelationship of the innovation strategy with product development. All product innovations of course lead to new products, but most processing innovations and marketing innovations and even some of the organisational innovations lead to new products. So the innovation strategy has to be studied and incorporated when building the product strategy.

The innovation strategy as a basis for product development defines:

- the innovation areas and the types of innovation;
- the overall aims of the innovations;
- the growth in sales revenue/profits expected;
- the aims of the innovations in growing the present markets or diversifying into other markets;
- the resources and the timing available for the innovation strategy and the individual innovation programmes;
- the company organisation for innovation.

In the innovation strategy within the business strategy, there is a need for top management to outline the type of organisation, and the resources of people, finance, time and equipment, which will be provided for the development of the innovation. An innovation strategy without defining the means for carrying it out is apt to be slow in actioning, and some parts of it may never be put into action. The Board of Directors needs to balance the company's ideal innovation strategy and the company's capabilities and resources before finalising the innovation strategy.

2.4   Getting the innovation strategy right

The company in its yearly development of business plans may be adjusting or rolling forward the present innovation strategy or/and developing a new innovation strategy. Developing a completely new innovation strategy could result in forming a new company or a new strategic business unit in the present company if there is a major change requiring new technology and marketing. The rolling forward of the present innovation strategy is an incremental process. A change of innovation strategy results in major company changes and is obviously expensive in resources of people, time and money. Hence the reason for a number of companies not considering new innovation strategies and becoming conservative and stagnant. Whether it is a rolling change or a new innovation strategy, there is a need every year:

- to determine if the focus of the innovation strategy needs to change;
- to study the balance of innovation areas in the strategy;
- to analyse the innovation areas both operationally and strategically;
- to determine the company's capabilities and organisation for the innovation areas.

The innovation strategy for the company is a portfolio of strategies that needs to continue achieving the overall aims of the company in the changing business strategy.

2.4.1   Analysis of mix of innovation projects

In the area of new product innovations, there is a need to analyse the mix of new products to see if the new product relates to the present product mix and to the company's innovation strategy position. Cooper (1998) identified four types of companies in innovation strategy development:

1. Prospectors – the industry innovators.
2. Analysers – the fast followers.
3. Defenders – the holders of secure positions.
4. Reactors – the responders to competitive pressures.

His analysis of their project types is shown in Table 2.6.

The emphasis here is on products but it is indicative for all innovations. There are companies seeking innovations that will change the company; others that wish to innovate in their present area and situation. Souder (1987) called these respectively promotive and restrictive organisations. In the promotive organisation, growth and innovation were the important goals; acquisition and product

Table 2.6   Project types by business strategies

Project type	Prospector	Analyser	Defender	Reactor
New-to-the-world (%)	30	6	7	0
New-to-the-firm (%)	15	16	17	8
Additions to existing product line (%)	22	42	40	48
Improvements to existing products (%)	11	16	11	13
Repositionings (%)	8	8	9	11
Cost reductions (%)	15	17	21	12
Number of firms	30	22	22	4

Source: From *Product Leadership: Creating and Launching Superior New Products* by Robert Cooper. Copyright © 1998 by Robert G. Cooper. Reprinted by permission of Perseus Book Publishers, a member of Perseus Books, LLC.

diversification were cited as means. Growth and innovation were ranked higher as goals than market share maximisation, profit maximisation and company stock (share price) maximisation. In restrictive companies, market share, stock price and profit maximisation were often ranked higher than growth. It is important to see that the mix of proposed innovation projects fits into the company's overall desire to be a prospector, analyser, defender or reactor.

Think break

1. Where would you place your company – prospectors, analysers, defenders or reactors?
2. Does analysis of your new products in the last five years agree with this?

2.4.2   The company's capabilities and organisation

The company organisation is also a necessary part of the analysis for the innovation strategy – is it a centralised, rigid top-down organisation or a fluid organisation with lower-level managers in major decision-making positions over resources and direction? The type of organisation has a major influence in deciding whether innovations are suitable for the company. The knowledge and the resources in the company are also determining factors. If the company does not have the knowledge or the ability to collect and analyse information to create the knowledge, then the innovation strategies are restricted. There needs to be a long-term commitment to technology and technological knowledge to build strongly innovative strategies. Also if there is not sufficient discretionary capital for new ventures, then there is difficulty in funding the more innovative strategies. Souder (1987) summarised some of the qualities of an innovative organisation:

- Willingness to accept change, altered behaviour and disruption.
- Long-term commitment to technology.
- Patience in permitting ideas to gestate, and decisiveness in allocating resources to these ideas having the greatest commercial prospects.
- Willingness to confront uncertainties and accept balanced risks.
- Alertness in sensing environmental threats/opportunities, and promptness in responding to them.
- Openness of internal, cross-departmental communications; diversity of internal talents and cultures; existence of many external contacts and information sources.
- A climate that fosters the natural confrontation and resolution of interdepartmental rivalries and conflicts, and the development of reciprocal role-persons.

This checklist for studying the innovation characteristics of a company has not been bettered over the years, and should be regarded as fundamental to the evaluation of the company for innovation.

2.4.3 Strategic and operational analysis

This is a very important step in ensuring that innovation can be successful in the company. It is the time when the people to be involved in the projects are brought together with the people who have been developing the business strategy and the innovation strategy. It is both a creative and an analytical exercise. The creative abilities of the designers and developers will start giving 'flesh' to the innovation strategy. The outcomes of the various sub-innovation strategies and then the innovation strategy as a whole need to be predicted.

This needs to be an interactive, multifunctional, multidisciplinary activity in the company, so that the various departments and people who are going to be involved in the projects are knowledgeable about the strategies and have been involved in prioritising them. There may be a need for consultants to provide information and for facilitators to conduct the discussion but this has to be an internal, creative activity. The company group as a whole needs to feel that it has been involved in developing the innovation strategies so that they will take them on enthusiastically through the next stages. The Board members need also to be involved in some of the discussions so that they have an understanding of the knowledge and the abilities for innovation in the company. They also indicate where the company is focused for the future and the goals that they see the company has to achieve. Such an interactive building of innovation strategies can be simple in the small company – it probably does it continuously over cups of coffee or pints of beer. But even a small company must do it formally at least once a year. In the larger companies, it may be more difficult but with interactive computer systems the discussions can evolve without too many large meetings.

For major strategy changes, this identification and ranking of the innovation possibilities can take some time. It is essentially a creative process followed by an analytical step, which is reiterated time and time again. Everyone needs to have the same interpretation of the proposed innovation strategy and the necessary outcomes. For ranking, various techniques may help from simple scoring on the ranking factors, to use of the Delphi method. Again there needs to be discussion on differences in the scores, and re-scoring until agreement is reached. Some factors to study in the combined discussions are shown in Fig. 2.9.

It is important to identify innovations that:

• will fail;
• cannot be accommodated in the company;
• will need an effort beyond the resources of the company;
• will take too long to complete or have an indeterminate end point;
• will cause a problem because there is not the necessary integration of design, production, marketing.

It is also important to identify innovations where the technology is uncertain, or where the transfer from basic or strategic research to development needs advanced and difficult technological research.

Predicting the outcomes of the innovation strategy, in particular the prediction of success, may be intuitive and subjective at this stage. But of course there are levels of success that will need to be predicted if decisions are to be made on the projects. Both the resources available and changes in the environment will affect the outcome of the innovation – changing a probable success into an actual failure. So the important success targets for the company are identified, and then the external and internal environments. The individuals in the group make predictions on the outcomes for each success measure. They may be subjective descriptions such as:

Complete dud　Doubtful　Should be OK　Probable success　Out-of-this-world
Or on scales:

Definitely a failure _____ Definitely a success

Alternatively the Q sort method can be used in which a set of cards, one for each individual strategy, is given to each group member. Group members sort them individually into five categories from definitely a success to definitely a failure, or just into two categories – Yes, a success/ No, a failure (Green *et al.*, 1988). It is important after each rating of the company measures to show the participants all the scores, and then to repeat the scoring. If the scores are widely different and consensus is not being reached, then further discussions need to be held. It is important not to do just a success or failure for the overall innovation, but to do it on the individual measures to see where differences are occurring.

The problems in these subjective measures are that some innovations, which could be successful, are dropped and that some failures are carried on. It is better at

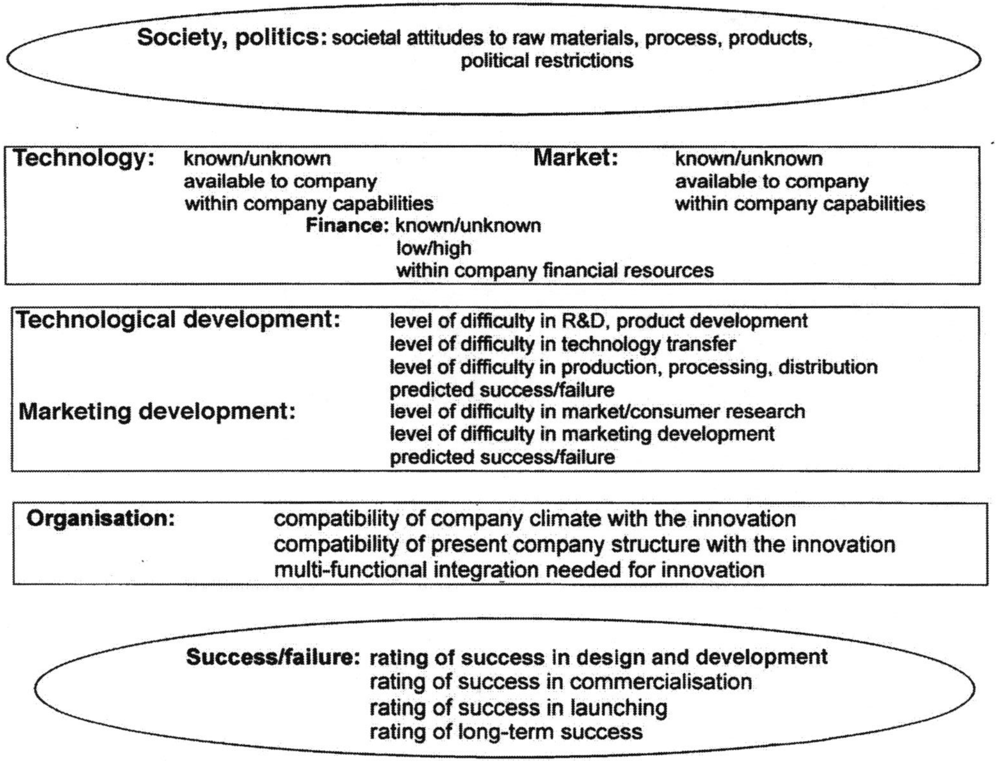

Society, politics: societal attitudes to raw materials, process, products, political restrictions

Technology: known/unknown
available to company
within company capabilities

Market: known/unknown
available to company
within company capabilities

Finance: known/unknown
low/high
within company financial resources

Technological development: level of difficulty in R&D, product development
level of difficulty in technology transfer
level of difficulty in production, processing, distribution
predicted success/failure

Marketing development: level of difficulty in market/consumer research
level of difficulty in marketing development
predicted success/failure

Organisation: compatibility of company climate with the innovation
compatibility of present company structure with the innovation
multi-functional integration needed for innovation

Success/failure: rating of success in design and development
rating of success in commercialisation
rating of success in launching
rating of long-term success

Fig. 2.9 Factors for discussions on innovation strategies.

this preliminary stage to give the project the benefit of the doubt and keep it in for the later stages. Another problem might be that all the projects are mediocre, but one feels that as they are ranked, the top five should be chosen. It is very important to be critical and to recycle rather than go on. It is much cheaper to recycle than to take below-standard strategies on to the next more expensive stages. Maybe you need other people in the group to give more creative and useful ideas!

> **Think break**
>
> Choose two recent products developed by your company, and score them for chances of success:
>
> - as if this prediction had been done at the outset of the development; and
> - after the development was concluded, with the benefit of hindsight.
>
> Do this using three different methods of scoring. Score some of the factors in Fig. 2.9, as well as the overall prediction of success.

2.4.4   Quantitative analysis of most suitable strategies

After ranking, the innovation strategies that could lead to success are identified. Now the predicted outcomes and inputs need to be more detailed. Usually this means more financial analysis and determination of the probabilities of achieving these outcomes. A range of predicted sales revenues and the related costs of development and launching the innovation, need to be determined so that possible outcomes such as break-even times, return on investment (ROI), present values, can be analysed. The sales revenues and profits can be predicted for 3–5 years or the life of the innovation. Some important outcomes and costs, and their relationships to probabilities of success and project timing are shown in Fig. 2.10. Increasing the money spent on the project can reduce the times and may increase the sales revenues and the probability of success, so it is important to make predictions on these inter-relationships.

The pipeline timing for the innovations needs to be predicted to ensure there is a flow of innovations throughout the future years and that innovations are not jumbled for both timing and resources. There is a need to predict the resources needed for the innovations: raw materials, plant, equipment and distribution system; but most important are the human resources. There is a need to predict the skills and knowledge needed for the innovation and to relate this to the skills and knowledge available in the company. If they are not available, how can they be met? It is surprising how little consideration has been taken of this in recent years – much knowledge and skills have been lost permanently and it is difficult to find new skills and knowledge. In the middle of the last century, there was a philosophy to keep the experienced people going a little longer as you bring in the new people, gradually absorbing the old knowledge and skills into the new minds as well as

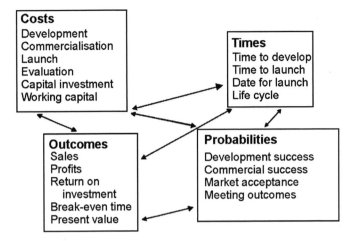

Fig. 2.10 Quantitative analysis of innovation strategies.

acquiring new skills from outside. For example, one salt manufacturing company, in changing from processing in open salt pans and marketing salt bricks and sticky crystalline salt, to triple-effect evaporators and free-flow salt, kept both going side-by-side so that the old operators and their customers were kept as the new customers and operators developed. There was only one problem – marketing sold the new salt to stall holders in the markets in West Africa; they still wanted the old sticky salt because peaks of salt could be formed in the tin cans used to measure salt for sale! This was an important selling point. So research had to start again – on making the running salt into sticky salt!

To sum up, it is important that the innovation portfolio:

- is balanced in levels of innovation, in timing of development,
- achieves the company's objectives,
- is readily acceptable to the markets,
- blends with the societal and political needs and attitudes,

and that:

- resources are available, in particular knowledge and skills,
- company cooperation is organised,
- company personnel and organisation can make the innovations happen.

The innovation portfolio is the basis for the next few years, which can roll onward with yearly tweaking, and with major changes perhaps every 5–8 years. But the major changes need to be developing through the years and not be suddenly introduced. If there is a dramatic change caused by a major advance in technology or a major social upset such as a war or a major entry into the industry, then there does need to be a fast reaction in the innovation portfolio and a dramatic change in the company. Emergency reactions are part of developing innovation strategies and portfolio.

2.4.5 Decisions

Decision making is the key activity in innovation from the business strategy to the evaluation of the results of the practice of the innovation. At this stage it is major decision making of the top management who must:

- accept the innovation strategy into the business strategy;
- provide the resources;
- set up the organisational structure for the innovation; and
- determine the measures against which the innovation has to be judged throughout its development and in the final application.

Top management is given the knowledge to do this, but it must decide what knowledge is needed. Knowledge costs money and usually the depth and width of knowledge are set by the money that top management makes available. It is important that this triangular relationship between knowledge, finance and decision making is understood by both top management and the people providing the information. There can be excess costs, inadequate information and poor decision making!

There are 10, 20, 50, maybe even 100 innovation strategies in a large, multinational company. How can they be compared and the decisions made? The decisions can be made on the financial analysis alone but this is dangerous at this early stage. The top management needs also to be given scoring on the other measures, which have been given as important aims for the company. Management can be presented with separate analysis of the different innovations, but needs to be shown the outcomes and inputs of different mixtures of innovations in possible innovation portfolios. It is the total picture that is necessary and not just the individual innovations. Sometimes the directors on a Board make a decision on one innovation strategy at one meeting and another at the next meeting, and the decisions can be counter-productive. It is the yearly presentation of the long-term innovation portfolio that is necessary for good decision making.

Think break

For the two company products that you chose in the last Think break, imagine that you are preparing for a presentation to the company's Board of Directors, so that they can select the most suitable innovative strategy for further development. Outline each innovative strategy and then using some of the factors in Figs 2.9 and 2.10, analyse the viability of each innovative strategy and the suitability for the company. How would you present the innovation strategies and their analysis to the Board?

2.4.6 Total innovation management

This decision-making process leads to the **total innovation management** for the company – direction, areas, resources and timing (Voss, 1994). The innovation

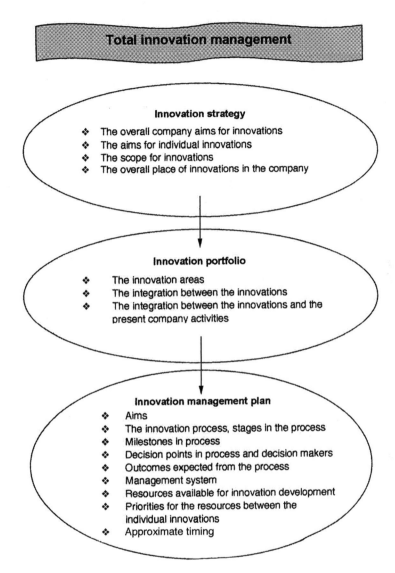

Fig. 2.11 Total innovation management.

strategy sets the direction for innovation, the portfolio specifies the areas and the management plan outlines the process for innovation and the measures for following the innovation as shown in Fig. 2.11. In organising the innovation portfolio, it is necessary to have a careful study of the resources and time available. There is a need to place a priority on the individual innovations and ensure that they are following the company's business aims over time. There is also a need to consider the present activities in the company and to ensure that the innovation portfolio fits in with the use of resources and time. In other words, the innovation portfolio is not a plan on a green field, it is being applied into a present system.

The innovation management plan shows how the company is bringing these innovations to fruition – it sets out the process that will be used, and the methods of controlling the process and the outcomes expected from the process. General milestones need to be clearly spelt out. Again there is a need to show how the innovation management is related to the day-to-day management of the present activities. It is important that company staff recognise how this is to be done at the beginning of the development and not be presented with it late in the development.

Management needs to unite the innovation strategy with the innovation portfolio and the business plan for the present activities to produce the total innovation management plan. Total innovation management includes all the key processes of product development – product design, process development, product commercialisation and product launching; as well as the other innovation areas such as technology change, technology acquisition, marketing change, marketing acquisition, organisational change and organisational acquisition. Each process interacts with each other and the interfaces between them need to be considered in developing the final innovation strategy and management. The aim of total innovation management is to increase the efficiency and the effectiveness of innovation in the company, leading to strong, focused, development of the company. The company stops jumping on bandwagons and buzz words, sometimes diversifying and sometimes returning to core business, sometimes innovative and sometimes conservative. It understands where it is going, how it is going to get there and when it is going to get there.

It is important at this point to analyse the innovations again to see if the decisions to include them need to be changed because of greater possibility of failure, lack of resources or poor timing.

Think break

1. Define innovation summary, innovation strategy, innovation portfolio, innovation management.
2. How could all of these be combined in your company into total innovation management?
3. Do you think this is a useful method for organising innovation in your company?

2.5   Focusing the product development programme

Now that the general areas for innovations have been identified and are securely embedded in the overall business strategy and plan, the product development part of innovation needs to be recognised and developed. The product development may be coming from a major market change or from a new processing development or a new raw material or even a reorganisation of the company into different units or subsidiaries; as well as specific product innovations identified as needed by the company. It is important to recognise that the product development comes from different innovation areas. Also the product development needs to be associated with the present product mix and its predicted future development. This is the start of creative activity in the product area. Given the innovation strategy direction, what can we do in product development?

2.5.1   Relating to the core competencies

It is important that the product development strategy is related to the technology and marketing strategies in the overall innovation strategy. The product development is related to the present core competencies of the company and, even more importantly, the developing core competencies of the company (Katz, 1998). It is also important to identify where the core competencies are in the food system – with the retailers and food service, with the retail or food service manufacturing companies, the ingredient processors, the producers, or the surrounding market research companies, advertisers, university departments, research organisations or consultants. The basic direction in the innovation strategy for product development is to identify how a unique and superior product can be developed to satisfy consumers' known and unknown needs and wants. Some products and their underlying technologies identified by Katz (1998) are shown in Box 2.3.

These examples are mostly large American companies with some European multinationals and Japanese companies, and may not be indicative of the food industry in other countries. But Katz identified some of the key technologies that are the basis of product development in these companies. It is interesting to see for example how rheology in different facets is a common core technology. The core technologies can also be divided into science-based and engineering-based. In some cases the author identified the core competency clearly, in others they were confused – maybe this is typical of companies. Some can identify core competencies, others are less sure. In no place were the marketing and consumer competencies identified – just as important core competencies as is organisational capability in product and processing technologies.

2.5.2   Relating to the product mix

The product portfolio is the collection of products produced by the marine and agricultural farmers and harvesters; manufactured and marketed by the food ingredient processor and the retail foods manufacturer; and for the retailer and food service, the food products marketed. In large companies in the food industry, there are many products in a product mix so that they are usually grouped into product areas, which are further subdivided into product lines. A product line is a group of products that are related, either used for similar purposes or possessing similar characteristics (Schaffner et al., 1998). The

Box 2.3 How major core competencies affect development of hot new products

Products	Core competencies	Companies
Low-fat meat products	Particle size analysis, protein–fat interactions, actual fat reduction in tissues, flavour improvement carbohydrate chemistry	Swift-Eckrich, Kraft Foods, Doskocil Food Service Co., Nestlé, Lean & Free Products, National Starch and Chemical
Fruit and vegetable products	Physical structure, biochemical changes in ripening, flavour chemistry, breeding, biotechnology, enzymes, antioxidants	Kagome Kabushiki Kaisha, Tropicana Products, Ocean Spray Cranberries
Coffee products	Structure and biotechnology of coffee beans, co-spray drying, glass transition technology, particle size management, caffeine effects, compaction	Nestlé, Procter & Gamble, Kraft Foods
Tea	Antioxidants, phytochemistry, flavours, colour development, oxidation and antioxidants, enzymes, cloud emulsions	Lipton, Nestlé, Procter & Gamble, Mitsui Norin Co., Sky Food Co., Coca-Cola
Chocolate	Phytochemistry, cold extrusion, viscosity, low-calorie fats, rheology, flavours	Nestlé, Hershey Foods, FMC Corp., M&M Mars
Dairy products	Texture, flavour, nutrition, foaming, heat denaturation, particle size, protein stabilisation, ultrafiltration, mineral separation, microbiology	Kraft Foods, Schreiber Cheese, P&G, Nestlé, Calpis Food Industry, Danone, GalaGen
Grain products	Rheology, refrigeration, glass transition, retrogradation, nutrition, flavours, extrusion, refrigerated doughs	Nestlé, Kellogg, General Mills, Pillsbury

Source: Based on material from Katz, 1998 by permission of Institute of Food Technologists, Chicago, Ill.

Table 2.7 Characteristics of the product mix

Products	Marketing	Finance
Types of product	Types of market	Sales revenues
Product platforms	Market segments	Profits
Product lines	Consumers	Market potential
Product ages	Industrial customers	
Product images	Food service customers	
Product attributes	Competing products	

product mix is live and evolving. It is currently profitable and as it changes, its profitability needs to continue to achieve the aims of the company. This does not mean that every product in the mix is profitable – there are other aims for products in a mix. They may complement other products, extend a line to give it variety, fill a place in the market, and so on.

The product mix is a mixture of products at different stages in the product life cycle: from new products to products that are at the end of their life cycle and dying. It is this variation of age that gives the mix its evolving character. The product mix also has variations in the sales revenue and the profits: some products are the major revenue earners and some the major profit earners. So the product mix has characteristics shown in Table 2.7. Sometimes products are also grouped according to the types of raw materials and methods of processing and distribution, for example, cereal products and meat products, frozen products and canned products.

2.5.3 Analysis of the product portfolio

In analysing the product mix so as to incorporate the innovation and product strategies for product development, one has to be aware of what changes can do to the product mix in the long term. Rash decisions based only on the innovation strategy -may affect some of the products or even the whole mix, causing imbalance and an overall loss of market potential and profitability. Some important factors to consider are:

- possible changes of product portfolio with time;
- reactive and proactive strategies;
- market change and technology change from the innovation strategy;
- target revenues and profits from business strategy.

The possible changes can be firstly divided into incremental changes and discontinuous changes. What are the products that need some new packaging, an extension of the flavours in the product line, a relaunch as a newer product, a cost reduction, a new image? Do any of these changes relate to an innovation strategy? Is the innovation strategy to keep with present product platforms but add improvements and variety? If the innovation strategy is to move the product portfolio in a new direction – perhaps to a new market – what new product areas could be introduced? So it is a case of balancing the possible product mix changes with the innovation strategy and also with the long-term balance of the product mix.

It is also important to understand from the business strategy, if the company wants to have reactive or proactive strategies. A reactive product strategy deals with problems as they arise. A proactive strategy is planning ahead to take advantage of opportunities.

Reactive product strategy   **Proactive product strategy**
Solving problems   (Market)   **Looking for opportunities**
Me-too products, customer complaints, change   New product line, new product platform,
second on the market, packaging change   superior products, new consumer need

The company may have a mixture of these – most of the resources being for proactive strategies but some resources kept for reactive projects in case unidentified problems arise such as new competing products. Each strategy has its place. The question is not which is right or wrong but which is specific to the overall business strategy. Defensive strategies can be imitating competitors' products, always being second and better – allowing the competitor to be on the market first and then introducing a new product. Another common defence strategy is to respond to consumers' requests – some companies base their products on consumers' complaints.

The proactive strategies may be technology-based, with emphasis on producing technically superior products, or marketing-based, building products to satisfy consumer needs. The innovation strategy is integrated with the technology and the marketing strategies and will identify the possible changes that can be made. Product ideas are developed based on these changes. The product designers need to be involved at this stage, creating ideas for the innovation strategies and gradually developing a library of new products. These new product ideas need to be analysed to see that they satisfy the aims of the innovation strategy. But they must also be compatible with the present and predicted product mix and can fit into marketing and production constraints such as production capabilities and quantities, distribution methods and quantities, product and company images.

Think break

1. Identify some reactive product development strategies that your company used in the last five years. What changes caused these reactions?
2. Identify some proactive product development strategies that your company has used in the last five years? What instigated these product development strategies?

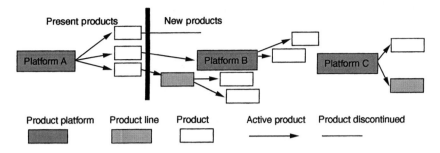

Fig. 2.12   Planning on a product map.

2.5.4   Planning a new product portfolio

After this product-idea generation related to the innovation strategy and a preliminary screening of the new product ideas, a map of the company's product mix for the next few years can be developed. The incremental product changes and the new products are fitted into the product mix over time. The aims of the product mix and the constraints on the product mix are defined, then the actual planning of the product introductions over time developed. This is the blueprint or the map for the future of the product mix and for the product development portfolio (Clark and Wheelwright, 1993) as shown in Fig. 2.12.

At this time the product idea is only a simple description, with identification of the use and some attributes, and may be a relation to competing products. The relationship of the product to the other products in the mix needs to be identified. There is also identification of the target market and the technology area. A range of costs and/or prices may also be identified. The timing of the introduction of the improved products and the new products is also identified.

In studying the proposed product mix, it is useful to divide products into groups according to growth potential, technological capabilities and market position as shown in Fig. 2.13. It is important to analyse the products in this way,

	Weak market position	Strong market position
High growth potential	**High growth potential** **Weak market position** High technological capabilities or Low technological capabilities	**High growth potential** **Strong market position** High technological capabilities or Low technological capabilities
Low growth potential	Low technological capabilities or High technological capabilities **Low growth potential** **Weak market position**	Low technological capabilities or High technological capabilities **Low growth potential** **Strong market position**

Fig. 2.13   New product groupings.

Table 2.8 Categories for new product areas

Type	Description	Level of innovation
New product platform	A completely new technology and/or market	Very high, discontinuous
New direction on present platform	A new product line/product	High, continuous
New-to-the-world single product	A single innovation not related to a platform, new technology	Very high, discontinuous
Revamping a product platform	New focus, add new products, drop old products, another market	High, continuous
Product line relaunch	New packaging, new image, change in product variety	Moderate, continuous
Product relaunch	New packaging, new image, product change	Moderate, continuous
Product line extension	Add new products	Low, incremental
Product improvement	Improve attributes, use, image	Low, incremental
Product cost reduction	Reduce costs of production, marketing	Very low, analysis

comparing the predicted markets with the technological capabilities. If the new product areas seem unsatisfactory or if they present major problems to the company capabilities then they need to be recycled back to the management group that developed the innovation strategy. The remaining product areas provide the basis from which the product development strategy is developed.

2.5.5   Categorising the new product portfolio

The product areas from the innovation and product strategies need to be built up into a new product development portfolio. The new product areas are categorised as shown in Table 2.8. There are many systems of categorising new product ideas, for example new-to-the-world, new product lines, additions to existing product lines, improvements and revisions, repositionings and cost reductions (Cooper, 1998). The categories in the table are useful for the food industry where the product mixes are large and there is continuous change to cope with supermarket wants.

Think break

1.   List last year's product development projects in your company and divide them into the categories in Table 2.8.
2.   Compare this with the previous year's product development projects categorised in the same way.

3. Are they different or is there a typical pattern?
4. Do you think this pattern might change in the future?

2.6 Developing the product development strategy

The first stage in designing the product development strategy is to produce more detailed descriptions of the products, and determine how their development can be organised within the specified resources and any other constraints that may have been identified in the final innovation strategy by top management. The individual projects are identified and their aims, outcomes and constraints. These are developed from the innovation strategy by the product development team and will need to get final agreement from management. The team will have to confirm that the projects are in agreement with the total innovation management programme. It is important that the team predicts the probabilities for success and failure as more knowledge is developed about the project.

2.6.1 Identifying the PD Process, outcomes and activities
To develop this knowledge, the product development team or product development management needs:

- to outline the development needed;
- to determine the outcomes of the different stages of the project;
- to identify the activities needed in each stage of the project;
- to study the present knowledge and resources;
- to identify the knowledge and resources needed;
- to identify problems in design, commercialisation and launching;
- to time the project overall and for different stages.

In outlining the development needed, the team will have the black boxes of the four stages in the product development process – product strategy development, product design and process development, product commercialisation and product launch. From the innovation management plan, it needs to recognise the outcomes needed overall and those needed at different points in the product development process. Then the team can identify the major activities needed in each black box. As will be discussed in the next chapter, it may already have a framework for the product development process for projects at the different levels of innovation, and therefore can relate the project into the particular framework. The team also needs to identify any problem areas in the product development process for each project – any risks of failure in the product or the project.

From this, the team can identify the knowledge and resources needed for each project and relate this to the present knowledge and resources available. Where there are shortfalls, it will need to identify possible sources. In the case of knowledge, if it is not in the company and there is not information outside, the team will need to identify how this knowledge can be created and when it is needed. The team can also start to time the overall project and the stages in the project.

2.6.2   Prediction of success of products

It is very important at this stage to identify what could be major failures. From the top management's identification of the necessary outcomes from the innovation strategy, the requirements of the product mix development and from previous measures used by the company in measuring success in past projects, the team needs to develop a group of measures for those product areas (see Chapter 1 for possible measures of success/failure). They can be quantitative, such as meeting certain sales revenues or profits, product costs, project costs, time for development or time to build sales. They can be qualitative, such as developing a unique or superior product; achieving the quality of execution of the technological activities in development, production and marketing; attractiveness to the market.

For high-level innovation it is important at this time to study the synergy between:

- product and the market;
- technical needs of the project and the company's development, engineering and production resources and skills;
- marketing needs of the project and the company's marketing skills and resources.

The prediction of success at this stage has a wide range of probabilities and is mainly subjective. But it is important that doubtful projects are sent back to the previous decision makers and not carried forward into the later stages. It is important that they are not completely dropped as decisions may be made with insufficient information and sometimes even wrong information.

2.6.3   Types of new product development strategies

Cooper (1998) described the new product development strategy as 'a strategic master plan that guides your business's new product war efforts'. This may be a rather dramatic definition for commercial product development, but it does emphasise four very important points: it is strategic, focusing on particular outcomes; it is an overall master strategy binding product development projects together; it is a guide for the complete product development programme; it is part of the company's business. It is a binding of the product areas into the whole organisation – functional areas, knowledge and skills areas, people. This is why it is important to develop a truly effective product development strategy.

The product development strategy sets out in a master plan, the aim or aims, the projects, the resources and the constraints, so that all involved in new product development are aware of the overall company policy for product development at this time. If the management wants integration of functional areas, more creativity in the company or more efficiency in product development, then the product development strategy can incorporate all of these into the overall aims.

Companies do have different overall product development strategies as shown in Table 2.9. In the food industry, all these strategies can be seen – and companies will say that they are successful for them. Historically there has been a preponderance of the low-budget conservative, which suits a market dominant position. As Cooper (1998) indicated from his studies, this strategy does achieve moderate results; the projects usually have a low failure rate, and the products are profitable – but wonders if the standards of success are high enough. It tends to yield a low percentage of new products in the product mix. It is a 'steady as you go' strategy, which shows no dramatic change.

It is important to consider together the drive from the consumer and the market and the drive from technology change in developing the product strategy (R.L. Earle and Earle, 1999). Balachandra and Friar (1997) suggested that a useful analysis is first to identify the context of the new product – is the

Table 2.9 Some product development strategies

Strategy	Description	Products
Differentiated strategy	Technologically sophisticated Strong market orientation High degree of product fit	Premium priced Unique features and benefits Competitive advantage
Low-budget conservative	Low R&D spending Highly synergistic with present production and marketing	Me-too Undifferentiated Lower price
Technology push	Technology oriented Lacks strong market orientation Lacks market synergy Can be costly	Innovative Technology oriented May not fit consumer needs
Not-in-the-game	Simple, mature technologies Ill-defined market needs	Low technology Me-too Low risk
High-budget diverse	Heavy spending on R&D No direction, focus No synergy New markets New technologies	Innovative products High-risk products May not fit consumer needs

Source: After Cooper, 1998.

Table 2.10   Relative importance of PD factors in different contexts

Contextual variables			Level of importance		
Innovation	Technology	Market	Market factors	Technology factors	Organisation factors
1. Incremental	Low	Existing	Very	Low	Very
2. Incremental	Low	New	Very	Low	Very
3. Incremental	High	Existing	Very	Very	Moderate
4. Incremental	High	New	Moderate	Very	Moderate
5. Radical	Low	Existing	Moderate	Moderate	Moderate
6. Radical	Low	New	Low	Moderate	Moderate
7. Radical	High	Existing	Moderate	Very	Moderate
8. Radical	High	New	Low	Very	Very

Source: After Balachandra and Friar, 1997.

innovation incremental or radical, the market existing or new, the technology level low or high? Using this one can identify the important factors in product development for different mixes of these factors as shown in Table 2.10. These are suggestions by Balachandra and Friar, but it is a useful way to study the product development factors. Ali (1994) also emphasised that in developing a product, it is useful to know for what types of products the company should undertake particular activities. The analysis of environmental and situational factors (firm, project and market characteristics) is a necessary condition for effective planning of new product development.

2.6.4   The overall product development strategy
The product development strategy lies between the new product portfolio and the product development programme as shown in Fig. 2.14. They are interconnected and there is recycling between the three as the final product development management plan develops in the programme. The product development projects are being identified from the product portfolio, and the PD Processes and their management gradually built up. This is a creative as well as a controlled process as the ideas for the products and the projects are being developed.

The aims of the new product development strategy can be specific:

• the structure of the product mix;
• increase the percentage of sales from new products to 30% in five years;
• returns on investment from new products;
• specific products to be launched in each year;
• returns from specific products or sales in a new market.

But they can also be subjective, for example developing the image of a health-providing company, or products of superior quality. They can also be organisational, for example using up the slack production, developing a new distribution system, developing a new subsidiary. As stated previously, there

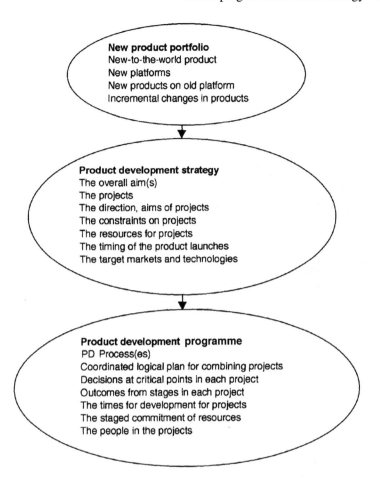

Fig. 2.14   Developing the product development programme.

may be aims for the organisation of the product development, for example being more systematic, separation of incremental and innovative product development, decreased time to market.

The individual projects and their aims, objectives and constraints are also starting to be developed although the detail may come later in developing the product development programme. The aim of the project, the ultimate outcome desired by the top management, is specified. This aim must be definite and not vague, straightforward and not complex. It must mean the same to all people, the management that are confirming it at this time, and the people who are developing the product commercialisation 6–12 months from now. This aim is the guide to the product development planning and also to the decision making during the project. It needs to be agreed in the product development strategy, although it may develop in detail as the product development programme is developed. It needs to state the type of product, the target market, the technology (or method of processing and distribution), the type of plant available, the marketing methods and the size of the market needed. There may be choices in this aim because there is a need still to study several methods of processing and marketing, and even several different markets.

There are also limits or guidelines developing for the total product development programme and the individual projects. They are specified levels of resources, and these need to be allocated to the different projects. The amount of spending on each project is always specified, but there is also a need to recognise the knowledge priorities, in other words people with different types of knowledge. Where are people with specific technological or marketing knowledge needed, where is there a need for creative design, strong financial analysis and consumer research? Where is there a need for different types of managers – senior/junior, leaders/controllers, knowledge leaders/system leaders? The resources of people, finance and equipment are limited and priorities and timing have to be determined. Some projects may have immediate priorities, because they have to be launched quickly or they are major projects that will take some time but need to be started immediately. There is also a need to identify how the new product development is to be achieved – internal R&D, internal product development, licensing, joint venturing or acquisitions (Cooper, 1998).

The timing of the product launches is set, since this is usually critical because of seasonal and other market conditions. The timings of particular stages such as product commercialisation which involve a great deal of resource from the functional departments and also need to be fitted into the present production and marketing, are outlined at this stage as they are the basis for developing the product development programme.

In developing the product development strategies, it is important to specify the type of market – consumer/retail, industrial and food service. In the latter two, the service is as important if not more important than the product and really what is being developed is a product and service strategy. This means that the human factor becomes more significant and the human resource strategy has to be integrated with the product strategy, creating a new product/service that has to include communication and understanding. The service can be an improvement or a new service, just like the products, but consideration in developing the new service has to be given to the consumer and to internal staff participation (Atuahene-Gima, 1996).

Think break

1. Identify a project in your company that includes significant service development as well as product development.
2. What product development strategy can you identify for this project?
3. What are the aims for this project?
4. What are the outcomes identified for the whole project and for each of the four stages?

2.7   Planning the product development programme

The product development programme has to bring the strategy into a new product plan for the next few years. It is the directional and the controlling document for the product development projects (Lord, 1999). From the product development strategy, it can develop a rolling programme, which will be quite specific for timing and results for the next two years, but will be more general for future years. The projects have to be integrated in this programme so that the resources, particularly people and equipment, are being used efficiently. In recent years, there has been more emphasis on the integration of projects as this is where efficiency and improved quality of product development can be achieved. There are problems with the more innovative projects because it is not known how long it will take to create the new knowledge and bring it into the product design and the product commercialisation. But certainly for the incremental changes, this integration of projects can be achieved successfully.

The integration plan needs to take consideration of time, resources and knowledge. In developing the product development programme, it is important to recognise what knowledge is required at the different stages of the project and where this knowledge can be obtained or how it can be created. A great deal of product development knowledge is tacit knowledge in the individual heads and in interactive tacit knowledge in the company. This is a very important consideration in planning the product development programme, especially in large companies. An outsider may be asked to come in because a team is lacking knowledge; the outsider immediately asks why they are not consulting someone in the company who is an expert in this area. In building the programme, there has to be consideration of personnel and in particular their knowledge and skills. How knowledgeable are they in the multidisciplinary skills needed in product development? Seldom if ever, when interviewing people for product development do the company personnel ask how creative they are. They look at their academic record and their experience, but do not ask for proof of their creativity. In other industries, product designers customarily carry their portfolio of new products to interviews to show how creative they are; perhaps food product designers should be asked to do the same.

So the product development programme defines the projects, their integration, their timing, the resources they can use, the people involved in the project. Two other important parts to be included are firstly how decisions are to be made by top management at critical points in the project and how the costs are to be controlled. Critical points are always the points between the main stages of the PD Process. But in large projects there may be intervening critical points. For example a critical point is at the end of product development and process design. Because of the expense of scale-up, there may also be a critical point after the laboratory or small-scale trials to find the optimum product and process. At each critical point, decisions have to be made by senior management on whether the project is to continue as planned, slowed down or dropped. The outcomes of the previous development must provide the information for the managers to make these decisions. For example, at the end of the product concept development, the outcomes include product design specifications, some product mock-ups and a product report. The product report includes the technical feasibility, marketing suitability, consumer acceptance, future predicted project costs, prediction of sales revenue and profits, risks, probability of product success and probability of project success. The details of the product report vary with the company and the project. For the incremental change, management may want only to have the product design specifications, consumer acceptance, product costs, predicted price. It is very important that what is needed as outcomes at the critical points are clearly identified in the product development programme and agreed by the top management who are to make the decisions.

It is necessary to define any constraints on the project, from either company or environmental needs, particularly societal and political constraints. The company or the society may restrict the raw materials used, the political regulations may define some of the properties of the products or the processing method. There may also be cost constraints caused by the price range in the market, and the company's pricing structure. Constraints need to be identified as they outline the ball-park for the product, and the product development.

Finally the product development programme needs to define the measures of success for the individual projects and the programme. The standards for the final evaluation of the success of the products in the market also need to be set, so that the sales revenue, profits and time for sales to grow are all set long before the product is launched. The standard for the project organisation also needs to be set – what is the range of timing, what is the quality of the work expected in the project, what multifunctional integration is expected, what cost over-run can be tolerated?

As described here, developing the product development programme is a complex and difficult task. What has been said is more directly appropriate to large enterprises with multiple projects and large resources, but the principles are just the same, and just as significant, in much smaller companies. The same considerations apply. It is important to cover them comprehensively and carefully so that the possible failures and problems are identified before the major effort and money is spent, rather than in the middle of projects where cost can grow astronomically.

Think break

1. The product development programme includes a number of projects that have to run in parallel, and some that run in series. What problems do you see in planning the product programme to cope with all the projects?
2. How can you set up a system to control the programme, so that it runs efficiently with project stages and the overall projects completed at the right time?
3. How can the knowledge resources, that is the people with the correct knowledge and skills, be encouraged to work creatively so the quality of the project is optimum?

2.8 References

ALI, A. (1994) Pioneering versus incremental innovation: review and research propositions. *Journal of Product Innovation Management*, 11, 46–61.

ATUAHENE-GIMA, K. (1996) Differential potency of factors affecting innovation performance in manufacturing and service firms in Australia. *Journal of Product Innovation Management*, 13, 35–52.

BALACHANDRA, R. & FRIAR, J.H. (1997) Factors for success in R & D projects and new product innovation: a contextual framework. *IEEE Transactions on Engineering Management*, 44, 276–287.

CAMPBELL, H. (1999) *Knowledge Creation in New Zealand Manufacturing.* Masterate Thesis, Massey University, Palmerston North, New Zealand.

CLARK, K.B. & WHEELWRIGHT, S.C. (1993) *Managing New Product and Process Development* (New York: Free Press).

COOPER, R.G. (1998) *Product Leadership – Creating and Launching Superior New Products* (Reading: Perseus).

EARLE, M.D. (1997) Innovation in the food industry. *Trends in Food Science and Technology*, 8, 166–175.

EARLE, M.D. & EARLE, R.L. (1997) Food industry research and development, in *Perspectives on Food Industry/Government Linkages*, Wallace, L.T. and Schroder, W.R. (Eds) (Norwell: Kluwer Academic).

EARLE, M.D. & EARLE, R.L. (1999) *Creating New Foods – The Product Developer's Guide* (London: Chadwick House Group).

EARLE, M.D. & EARLE, R.L. (2000) *Building the Future on New Products* (Leatherhead: Leatherhead Food RA Publishing).

EARLE, R.L. & EARLE, M.D. (1999) Innovation in the food industry. *Food Technology in New Zealand*, 34(6) 11–12, 22.

FRATER, P., STUART, G., ROSE, D. & ANDREWS, G. (1995) *The New Zealand Innovation Environment* (Wellington: Business and Economic Research Ltd).

GREEN, P.E., TULL, D.S. & ALBAUM, G. (1988) *Research for Marketing Decisions*, 5th Edn (Englewood Cliffs, NJ: Prentice-Hall).

HEDGES, A. (1969) Innovation in food marketing & research. *Food Processing and Marketing*, Feb., 64–66.

KATZ, F. (1998) Major core competencies affect development of hot new products. *Food Technology*, 52(12), 46–52.

KUCZMARSKI, T.D. (1996) *Innovation–Leadership Strategies for the Competitive Edge* (Chicago: NTC).

LORD, J.B. (1999) Product policy and goals, in *Developing New Food Products for a Changing Market Place*, Brody, A.L. and Lord, J.B. (Eds) (Lancaster, PA: Technomic).

PINE, B.J. & GILMORE, J.H. (1998) Welcome to the experience economy. *Harvard Business Review*, July–Aug., 97–105.

RIZVI, S.H., SINGH, R.K., HOTCHKISS, J.H., HELDMAN, D.R. & LEUNG, H.K. (1993) Research needs in food engineering, processing and packaging. *Food*

Technology, 47(3), 26S–35S.

ROGERS, E.M. (1962) *The Diffusion of Innovation* (London: Free Press of Glencoe – Collier-Macmillan).

SCHAFFNER, D.J., SCHRODER, W.R. & EARLE, M.D. (1998) *Food Marketing – An International Perspective* (Boston: McGraw-Hill).

SOUDER, W.E. (1987) *Managing New Product Innovations* (Lexington: Lexington Books).

THRODHAL, M.C. (1982) National and multinational opportunities for the encouragement of innovation. *Food Technology*, 36(1), 101–102.

VOSS, C.A. (1994) Significant issues for the future of product innovation. *Journal of Product Innovation Management*, 11, 460–463.

3

The product development process

The PD Process coordinates the specific research activities such as product design, process development, engineering plant design, marketing strategy and design with the aim of producing an integrated approach to the development of new products. The overall aim is to create a product that an individual consumer or a food manufacturing company or a food service organisation will buy. The two parts of product development – the knowledge of the consumer's needs/wants and the knowledge of modern scientific discoveries and technological developments – are both equally important. The PD Process combines and applies the natural sciences with the social sciences to systematically produce innovation in industry.

The PD Process is a system of research for the individual product development project and the product development programme. It varies in detail from project to project but overall retains the same structure of four main stages, subdivided further into 7–9 stages in some product development models (Cooper, 1996; Earle, 1997). The four stages are product strategy, product design and process development, product commercialisation, product launch and evaluation. Between the four stages, there are critical evaluations and top management decisions on the project and the products, called stage gates (Cooper, 1990) or critical points (Earle, 1971). **Critical points** are an essential part of the PD Process. For the critical decisions to be made, certain knowledge has to be generated in the research – the **outcomes** from the various stages. To build this knowledge, specific research is needed – the **activities** of the various stages. The project teams choose different procedures for these activities – the **techniques** used in the activities. There are important interrelationships in the four main stages between:

Critical decisions ↔ Outcomes ↔ Activities ↔ Techniques

This PD Process can be called the Critical PD Process because it is based around critical decisions, and because there is critical analysis of the activities/ techniques and the outcomes throughout the project.

3.1   Product strategy

Stage 1: product strategy, starts with the finalising of the product development strategy and product development programme. Then the aims of the individual product development projects can be set. The project starts with the generation of new product ideas and the outlining of the product design strategy, and ends with the product concept and product design specifications. There is real dichotomy in the decisions and activities; there is on the one hand, the need for freedom to be creative, and on the other, the need to set boundaries in the product design strategy. Before top management can make the critical decision to fund the further stages of the project, or to stop it, or to return it to the team for more knowledge, there are three critical decisions:

1. Is the product concept a unique product satisfying the needs and wants of the target consumer/customer?
2. Will the product concept and the project deliver the financial and other aims set in the business and product development strategies?
3. Does the product concept harmonise with the company's business and environment?

Top management, to make these decisions, needs knowledge on the processing, production, distribution and marketing technologies for the product. Knowledge will be incomplete at this time. The financial predictions (sales revenue, gross profits or margins, the probabilities for success, the returns on investments or break-even times), and future costs and time for the project are very approximate. There will be other specific requirements for each project, such as enhancing health (Ericson, 1997), environmental effects, food regulations and trade barriers. But of course the most important knowledge is the description of the product idea in the product concept and the product design specifications. The project team has to build up this knowledge throughout the stage, and the type of knowledge identified will determine the critical activities that have to be completed in the product development project (Earle and Earle, 1999). The knowledge is built up in substages and decisions are made at the end of each stage usually by product development management, but sometimes by top management if the project is a major innovation and costly.

The substages in Stage 1: product strategy for the individual project are:

- defining the project;
- developing the product concept;
- identification of processes, distribution and marketing;
- development of product design specifications;

- planning of the project;
- predictions of project costs and financial outcomes.

This is total technology research incorporating product, processing and market research with consumer and society studies. At this early stage, the knowledge may be generalised, and the aim is to make it greater in breadth and depth through the later stages of the project. This stage sets the direction for the product development project, and has been identified in much research as most important to the final success of the project.

3.1.1   Defining the project

The aim, outcomes and the constraints have been identified in the product development programme and presented to the product development team or manager for the project. But there is usually a need for further desk research by the team to determine the accuracy of the aim, outcomes and constraints and also to 'flesh them out' to give a more detailed project definition that can drive and control the project (Rosenau, 2000). This is also the time to select a suitable PD Process for the project and to set out an outline project plan.

There are four aspects of the initial research to define the project by developing more detailed aims: product ideas, consumers, technology and market as shown in Fig. 3.1. The research includes all aspects of the PD Process. At the same time the team is developing new product ideas, and relating them to the market possibility, to the technology possibility and to the product possibility. What are the products? Can they be made? Can they be sold? Who wants them? What do they need? These are the types of questions being discussed by the team and it is an important time for team interaction. This is only 'desk research' – using information in the company, outside records, published textbooks and papers, which are easily available. There is a maximum use of tacit knowledge within the group and within the company. Information technology has improved the storage and use of knowledge in product development, in particular the use of product models with a framework of raw materials, ingredients, packaging and production methods (Jonsdottir *et al.*, 1998).

Think break

In a project, the aim was changed from:

Export a nutritional product to Thailand with a market size of $5 million.

to:

Export a protein product, minimum 20% protein, to the Thai middle class, urban market, marketed through gyms and supermarkets; processed in the spray drying plant or the UHT plant and distributed at ambient temperatures. It must have sales greater than $4 million.

1. Study the aims and discuss how the first aim has been improved in the second aim as a focus for the project.
2. How would you improve the second aim to make it clearer for all people in the project?

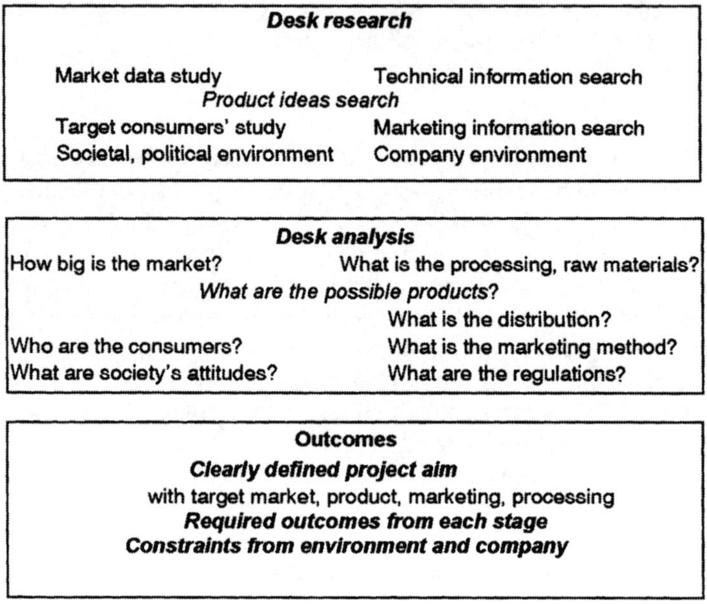
Desk research
Market data study
Technical information search
Product ideas search
Target consumers' study
Marketing information search
Societal, political environment
Company environment
Desk analysis
How big is the market?
What is the processing, raw materials?
What are the possible products?
What is the distribution?
Who are the consumers?
What is the marketing method?
What are society's attitudes?
What are the regulations?
Outcomes
Clearly defined project aim
with target market, product, marketing, processing
Required outcomes from each stage
Constraints from environment and company

Fig. 3.1 Defining the project: activities, outcomes and constraints.

The second aim allows two different methods of processing. In other aims there may be two target markets, or two methods of marketing, as it is not clear at that time just which is the direction to go. Aims can be adjusted during the project but there must be agreed reasons for doing this.

The **outcomes** for the different stages of the PD Process are developed from the aim, the company's PD Process for this type of product, and the decisions that the top management has indicated for different times in the project. In particular the decisions identified are used to determine the outcomes as shown in Fig. 3.2. There are both product and project decisions to be made, the product decisions and outcomes are ovals in Fig. 3.2. The general decisions are similar for many projects but there will also be specific decisions for each project. Therefore other outcomes will be needed. It is important to recognise the decisions that have to be made, and by whom, and to then select the knowledge needed in the outcomes to make these decisions. Outcomes are sometimes called objectives; they are the knowledge goals that have to be reached at the end of the different stages of the PD Process. In some projects, especially large projects,

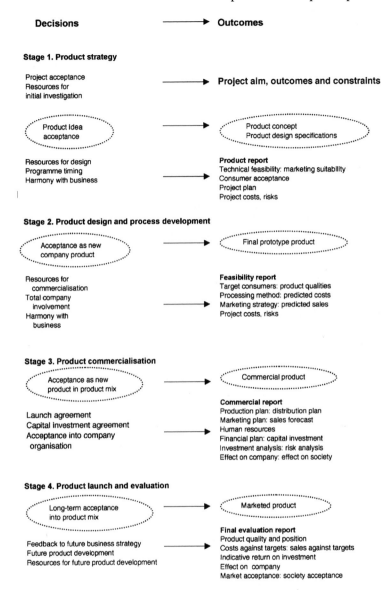

Decisions ──────→ **Outcomes**

Stage 1. Product strategy

Project acceptance
Resources for ──────→ **Project aim, outcomes and constraints**
initial investigation

Product Idea   Product concept
acceptance   Product design specifications

Resources for design   **Product report**
Programme timing   Technical feasibility: marketing suitability
Harmony with business   Consumer acceptance
                          Project plan
                          Project costs, risks

Stage 2. Product design and process development

Acceptance as new   Final prototype product
company product

Resources for   **Feasibility report**
  commercialisation   Target consumers: product qualities
Total company   Processing method: predicted costs
  involvement   Marketing strategy: predicted sales
Harmony with   Project costs, risks
  business

Stage 3. Product commercialisation

Acceptance as new   Commercial product
product in product mix

                          Commercial report
Launch agreement   Production plan: distribution plan
Capital investment agreement   Marketing plan: sales forecast
Acceptance into company   Human resources
  organisation   Financial plan: capital investment
                          Investment analysis: risk analysis
                          Effect on company: effect on society

Stage 4. Product launch and evaluation

Long-term acceptance   Marketed product
into product mix

                          Final evaluation report
Feedback to future business strategy   Product quality and position
Future product development   Costs against targets: sales against targets
Resources for future product development   Indicative return on investment
                          Effect on company
                          Market acceptance: society acceptance

Fig. 3.2   Identifying the outcomes necessary for the decisions (After Earle amd Earle, 1999, by permission of Chadwick House Group Ltd).

critical decisions may be made more often in the project; again these decisions have to be recognised and the required outcomes defined.

It is important to select the outcomes by balancing the need for knowledge against the resources and time needed for the activities to give the outcomes. With the recent emphasis on faster but quality product development, more attention is being paid to selection of outcomes. The choice of outcomes and therefore of project activities depends on the risk of failure the company is prepared to take. Teams often seek extensive knowledge so that they are surer of the whole picture, but this can be expensive and take too long and even sometimes result in failure. There is history of some companies seeking too much information in the test markets, and being overtaken by other companies. Outcomes that are fundamental to the project and whose completion is necessary for the project are always included. Identifying possible outcomes at the beginning of the project and selecting the critical outcomes for the company and its environment, which are within the money and time the company is willing to provide, ensure a project that is efficient and effective.

The **constraints** are any factors defining the area of the project. Some of these, such as financial resources and time for launching, will have been specified in the product development programme. At this time it is important to identify constraints on the product, processing and marketing, and also the constraints placed by the company and by the social and political environment. For example, the constraints from the food regulations and from society's attitudes to production, processing, food additives and safety need to be identified before product design starts. There are sometimes constraints caused by the availability of people and equipment. A checklist for studying constraints is shown in Table 3.1.

The constraints need to be recognised but they must not be too tight as this could stifle the creativity in product design and process development. For example, specifying the protein level as exactly 20% for a perceived consumer need and not a requirement of the regulations could restrict the other product characteristics. But a protein range of 20–30% could satisfy the consumer but allow more freedom in design. It is important to criticise the constraints – are they all needed, are they too tight? Sometimes a company constraint may stifle the project, and it is important to revisit it with management to see if it can be changed.

The aim(s), outcomes and constraints direct and control the project. They are used as factors in screening and evaluating the product ideas and product concepts, and then in evaluating the different prototype products. They are the

Table 3.1   Project constraints: a checklist for product development projects

Product	Processing	Marketing	Financial	Company	Environment
Eating quality	Equipment	Channels	Fixed capital	Strategy	Local government
Composition	Capacity	Distribution	Working capital	Structure	National government
Nutrition	Raw materials	Prices	Investment	Expertise	Industry agreements
Packaging	Wastes	Promotion	Project finance	Location	Farmers' agreements
Shelf life	Energy	Competitors	Cash flows	Management	Economic status
Use	Water	Size	Profits	Innovation	Business cycle
Safety	Personnel	Product mix	Returns	Size	Social restrictions

Source: From Earle and Earle, 1999, by permission of Chadwick House Group Ltd.

basis for identifying the activities and choosing suitable techniques and for the project plan, which directs and controls the process.

The **outline project plan** is based on the PD Process selected for the project and the outcomes identified. The PD Process varies according to the type of product – industrial, consumer and food service, and also whether the product is incremental or a major innovation. The activities are selected to give the outcomes previously identified. Choice of activities is not only determined by the knowledge needed in the related outcome, but also by the resources and time available. The **description of the activity** defines the outcome needed, the time frame to be met and the resources that can be used. The outline plan is set up so that everyone in the project can identify their place in the project and what they are aiming to achieve. They can start to select the techniques for their section of the project, particularly for the early stages.

Think break

In Chapter 3, we are going to do the initial stage of a PD project, either a project from your company or using the Case Study in Section 7.4. Obtain from the management of your company the general aim, constraints and resources for this project. In this Think Break, search for more information and develop the final aim(s), outcomes and constraints for management's approval.

1. What is the market type – consumer/retail, consumer/food service, business to business/industrial, business to business/food service? Identify the target market, its possible size, needs and competing products. Use Fig. 3.1 as a guide, try to find information to answer the market/consumer questions.
2. Identify the type of product development in the project – me-too, improvement, product line extension, innovation on the same product platform, a new platform; and also the type of market. Then select/design the PD Process.
3. Using Fig. 3.2, identify the possible decisions to be made and then discuss them with management. Select the final decisions.
4. Determine what knowledge is needed to make these decisions and then select the outcomes that are needed for the decisions at the various stages of the project.
5. What are the principal constraints already identified for this project – economic, physical, political, social? Now use the checklist in Table 3.1 to discover any other constraints that might be important. Rank the constraints from critical to not important and select the final constraints for the project.

3.1.2 Developing the product concept

The food industry has seldom used the word design except as related to packaging and to advertising. The development of the product has usually been called 'product development' and had connotations of laboratory formulation and sensory panel. But today, there may be real benefits in adopting food product design and in associating food product design with other areas of design.

The product is an amalgam expected by the consumer of the hard values or the basic qualities and the soft values or the differentiating qualities such as aesthetic appearance and environmental friendliness. Product design, or the product creation process, is therefore an amalgamation of the disciplines of consumer and market research, technology and engineering research with design practice as shown in Fig. 3.3. Product design is an essential part of the product creation process in equal cooperation with engineers, marketers and consumer researchers (Blaich and Blaich, 1993). All come together in the technology of the product.

- Consumer researchers build the consumer/product relationship throughout the PD Process.
- The market researchers analyse markets and design the marketing and distribution methods in the market strategy.
- The food engineer and technologist research the product and the process together in co-engineering and design the production and physical distribution methods.
- The food product designer researches the social and cultural backgrounds and designs the holistic product.

It is important that these are all integrated from the beginning of the PD Process. As the product concept and the product design specifications are built up, all aspects are brought together; then as the project progresses, the people involved understand what is needed in the design of product, production and marketing to satisfy the consumers' needs, wants and behaviour.

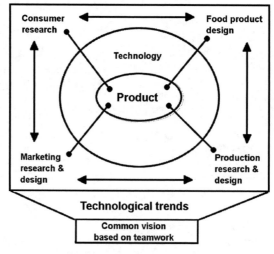

Fig. 3.3 Integrating the main disciplines in product creation (Source: After Blaich and Blaich, 1993).

The product design process is subject to a set of requirements (product design specification), including basic and desirable product functions, performance, aesthetics and cost (Dasgupta, 1996). This is common to many industrial areas but is now only becoming accepted in the food industry. Firstly, the consumers and product designers, very often with marketers, come together to develop a product concept, and then the technologists and engineers are brought in to develop the product design specification. Some of you may be thinking that this is sounding rather complicated, but actually you do it all the time but may be doing it without clear directions. Designers and the consumers have difficulty in working in the abstract and there is some design taking place either in drawings, computer descriptions or 'mock-up' products. Creativity starts here. It is a useless exercise for marketing to work alone with consumers to develop a product concept and then hand it to the food designer/technologist and say make this! That seldom leads to unique products. There needs to be cooperation among marketing, consumer and the product designers (or food technologists/ product developers as they are often called in the food industry).

The areas in building the product concept for design are (Ulrich and Eppinger, 1995):

- identifying consumer needs;
- establishing target product brief;
- analysis of competitive products;
- concept generation;
- concept selection.

The project team works between these areas. Firstly they study consumers, trying to build their needs into more specific terms in the product brief, and at the same time studying the competing products. Then they go back to the consumers with more defined product types to generate specific product concepts. Finally they work the product concepts into more specific and detailed product descriptions and go back to the consumers to find their reactions.

The product concept progresses through the product development project from the original idea to the final product specifications controlling production and the final product proposition that is the basis for the marketing. It is refined and expanded in two different ways because of the different end uses – in a technical, quantitative description and in a consumer-based, in-depth, description as shown in Fig. 3.4.

The outcomes needed in the first stage are the design product concept and the design product specifications. These start from a name or a simple description in the product development programme, and firstly the team generates ideas for the product and then with consumers builds simple product idea concepts. After evaluation these are reduced to one or two product ideas, and research with consumers and the market gradually builds up the product concept for design. This is then integrated with the processing and marketing technologies, and the product concept is built up by product concept engineering into metric descriptions in the product design specification. The design product concept is

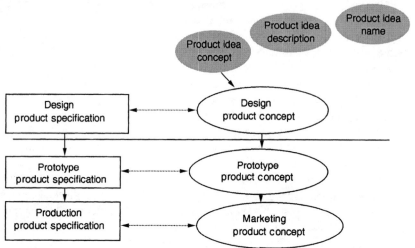

Fig. 3.4   Product concepts and product specifications in the product development project.

the consumer's description of the product and includes the product characteristics, benefits and position in the market as identified by the consumer. The
design product specification is developed from the product concept with
reference to the technical aspects of the product, processing and distribution. It is
the precise definition of what the product has to do, it is metric and has a value
(Ulrich and Eppinger, 1995).

A product has several layers and these are being built up gradually during the
product development. There is the company's basic functional product, the total
company product (with packaging, aesthetics, brand, price and advertising) and
the consumer's product (which relates it to the competitors, the environment, the
media, the society, as well as its communication and use) as shown in Fig. 3.5.
There is a continuing interaction between these three layers of the food product,
and therefore between the four groups of people – consumers, product designer,
technical and marketing – during the development of the product concept and
the product design specifications.

To research the products, there is a need to identify the following:

- **Product morphology**, the breakdown of a product into the specific
characteristics (or attributes) that identify it to consumers or/and business
customers. Determined by analysis of the product family and the individual
product (Schaffner *et al*, 1998).
- **Product characteristics** (or attributes), the features identifying the product
to the company, the market and the consumer. Identified by consumers and
designers in the creation of the product concept.
- **Product benefits**, the product characteristics important to the consumer.
Identified in the consumer/product designer discussion groups. The product
benefits are in four main areas – basic product benefits, package benefits, use

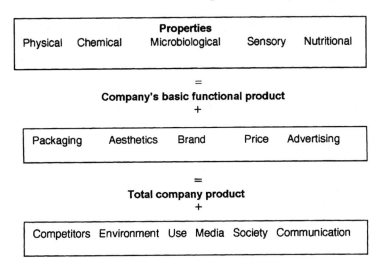

Fig. 3.5 The total food product (Source: From Schaffner, Schroder and Earle, *Food Marketing: An International Perspective*, © 1998, by permission of the McGraw-Hill Companies).

benefits and psychological benefits – and these need to be integrated into the final consumer preference.

• **Product profile**, the group of product characteristics which is the unique identification of the product – it is the product's DNA or fingerprint.

In developing a product strategy to introduce meat pies to Malaysia by a New Zealand company, the activity was to identify the product benefits required by Chinese and Malaysian consumers; three techniques were used – focus group, consumer survey, and multidimensional scaling (MDS) as shown in Table 3.2.

Table 3.2 Product benefits for meat pies in Malaysia

Multidimensional scaling*	Focus group[†]	Consumer survey*
Taste (sweet–savoury)	Convenience	Taste
Product type (bread–non-bread)	Freshness	Cleanliness
Origin of product (local–foreign)	Smell	Freshness
	Local flavour	Healthy (good for you)
	Healthiness	Convenient to obtain

* Chinese and Malay women in Malaysia.
[†] Malaysian students in New Zealand.

Source: After Lai, 1987.

The consumers compared the pie, particularly in the MDS, against the sweet and savoury baked/fried snacks already eaten in Malaysia. The MDS identified the main characteristics, and the focus group and the consumer survey identified general product benefits. To design the products more information was needed, and a sensory ideal product profile was identified by a small group of the consumers tasting the preliminary experimental products. The consumers' sensory characteristics were 7 for the pie top, 5 for the pie bottom and 14 for the pie filling. The five scales for the pastry bottom with the consumers' ideal scores are shown in Fig. 3.6.

The scales with their ideal points were included in the product design specification. The product profile needed to be analysed in two ways – what do the consumers mean by an ideal score of 5 for the pastry thickness? Can a physical measurement mimic this sensory characteristic? It is easy for thickness. Texture can also be measured in a physical instrument, but it may be necessary to train a panel to judge 'flour smell' and 'oiliness' unless a chemical test can be found for them.

The important product benefits may include the type of raw materials and processing, as in organic foods and environmentally friendly foods, as well as the recognised consumer concerns of nutrition, safety, eating qualities and the psychological benefits such as prestige and fun (Earle and Earle, 2000). There has been a concentration on sensory benefits as shown by the rapid development of sensory science but this needs to be made much wider to include all benefits. An example of a product concept strongly based on the psychological needs is described in Box 3.1, a product concept for pet foods.

The concept of the package often follows the more traditional path of industrial design, developing a product architecture that defines the major subsystems of the package such as the inner, outer, closure, seals. Product architecture can also be useful in building up products such as complete meals with various meat, vegetables and noodles in some type of display pack.

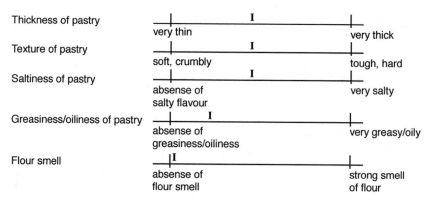

I is the ideal score

Fig. 3.6   A product profile for pastry (Source: After Lai, 1997).

Box 3.1   Four-legged trends

How many times have you seen a cat prey on a chicken or, heaven forbid, a turkey? But then cat food with blackbird or field mouse doesn't sound very discerning to a petfood shopper. Let there be no doubt that Britain's petfood shoppers are discerning and willing to show the colour of their money to satisfy their pet's taste buds.

The dominant trend over recent years has been the humanisation of petfood. Supermarkets stock an awesome display, encompassing not just a plethora of brands but also variants. A cosseted cat can start the day with a bowl of muesli and a splash of cat milk, enjoy chicken in jelly for lunch and perhaps have some tuna for supper. Meanwhile, the family dog can enjoy beef chunks at noon and 'a complete dry meal' to round off the day. In essence, the petfood sector has expanded to cater for owners' perceptions of what their pet requires. This is echoed in pack design.

In design terms, humanisation manifests itself by mimicking the same brand-building cues used for human brands. For example, Trix dog snacks bear a striking visual similarity to the human treat Minstrels, or equally they could be mistaken for a beef flavour packet of crisps. Similarly the packaging of Whiskas cat milk seems to draw inspiration from Carnation long-life milk. There is also a move towards injecting 'appetite appeal' into petfood packaging, with stylised displays of the product depicted on-pack. The use of expensive illustrations and top food photographers confirms this move. It has reached the point where the only difference between human and petfood packaging is the animal images on-pack.

The way forward for packaging design in the petfood sector is to aim for the right balance between traditional petfood brand values and those of the tinned food destined for human consumption. Yes, appetite appeal is a very important sales tool in this arena, but the trade-off shouldn't be a loss of whimsy and humour.

Source: From Petrie, 1995 by permission of *Marketing Week*, published by Centaur Communications (London).

Think break

For the project identified in the previous Think Break:

1. Generate five ideas for new products within the area of the aims.
2. Consider these product ideas against the aims and constraints for the product and choose the three most suitable products.

3. With some consumers build simple product idea concepts for these product ideas.
4. Expand the product idea concepts with knowledge of the processing, marketing and the technical characteristics of the product. Select the two most promising product idea concepts with the consumers.
5. Finally with the consumer group, build product concepts for design of the two remaining products.

3.1.3 Product design specification

Building the product design specification from the product concept includes both research and design. Market research provides more details about the target market characteristics and size, the methods of marketing that might be used and the position of the product as compared with the competitors. The market study is progressing into consumer and retailer surveys in consumer marketing and customer surveys in industrial marketing. The technical research involves the searching of the scientific and technological literature, including patents, as a preliminary investigation into the possible products, processing and physical distribution. The designer is starting to create the products and often needs to make models so that ideas on the product characteristics can develop. The modelling can be on paper or computer, and some preliminary laboratory research makes the products on a small scale. Of course in incremental development, the basic product is already known and both the marketing and the technical research, and product model building are much less and indeed may not be done at all.

The product design specification has for a long time played an important part in design in other industries and now is considered the area that has a major effect on quickening development and ensuring product success. The use of computer techniques such as CAD (computer aided design), CAID (computer aided industrial design), CAM (computer aided modelling), especially with more modern versions, has given the opportunity to design on the computer and to present the product ideas on the computer to other project members and even to consumers. The computer designs can be transferred into engineering design and linked to small-scale production units producing the experimental prototype for the consumer to discuss. Some of the newer tools in product design are shown in Table 3.3.

All of these new developments are changing industrial design and making it quicker. These techniques can be used for food packaging and for a structural food such as a loaf of bread and snacks. Snacks have already been designed using earlier CAD versions. Word descriptions of food product characteristics have been used in computer techniques such as conjoint analysis for a number of years to build and evaluate food product concepts (Moore *et al.*, 1999). The question is how far can the food industry use computer design techniques in building up product concepts and product design specifications? Certainly the

Table 3.3   Tools at the cutting edge of product design

3D solid modelling software
Describes both the exterior and interior of the product in three dimensions

Virtual-reality design tools
Aid interaction of the computer models in a manner that resembles real life using
stereoscopic eyewear which tracks with the computer

Rapid-prototyping
Tests new design concepts with models using plastic materials such as polyamide/epoxy
resins

Collaborative design tools
Use an internal Net or the Internet so that people can design together.

Source: After Schmitz, 2000.

personal computer is being used actively in the product concept stage – Internet
for desk research, software for interpretation of statistical market research,
computer-based literature searches and databases (Hegenbart, 1997). Newer
developments are the use of detailed product models of present and past
products to use as an information base to design new products (Jonsdottir *et al.*,
1998).

What are specifications for product design? The product concept states
clearly the needs and wants of the consumer or customers, but it does not
provide specific guidance for design of the product in technical terms. It is often
subjective and leaves room for different interpretations. Product concept
engineering interprets the consumers' product characteristics into measurable
terms, metrics, which can be tested in the product prototypes to see if the design
is meeting the specification. An individual specification consists of a metric and
a value, for example protein content between 20 and 30%; or thickness between
0.1 and 0.2 cm, or an ideal target value with an acceptable range, for example,
strength of onion flavour, 7, range 6.5 to 7.5 on a linear flavour scale. Metrics
and their values should be:

- critical to the consumer;
- consumer-acceptable ideal value and range of values;
- practical and capable of being achieved.

The product design specification is a set of individual specifications. Too
many metrics should not be included, as this will limit the area in which the
designer works and cause problems with too much testing. Only the metrics
recognised as important by the consumer, or needed for the consumer such as
safety, or for food regulations, are usually included, but sometimes there may be
specifications dictated by the process or the distribution. Also it is important to
choose metrics that are achievable, for example it may not be possible to choose
vitamin C as a metric because heat processing conditions needed to ensure a

critical metric safety (microbiological) value, will destroy it. And metrics must be practical, for example there may be no measure for spicy hotness in a food so the acceptance of different levels in the new product have to be tested with consumers during design.

Choosing metrics and their values is simple if it is an incremental product or a copy of a competing product in the market. The metrics are already identified and the values can be chosen by competitive or company product benchmarking (Ulrich and Eppinger, 1995). With the radical innovation, there is not sufficient previous knowledge and there will be a need to continue the metric identification into later stages of design. As prototypes are developed and tested both technically and by the consumer, the metrics for the consumer-identified product characteristics are built. The design specification evolves to the product prototype specification at the end of the design process, so it does change, but care must be taken that critical metrics are neither dropped nor changed in value without consumer acceptance of the change. Factors sometimes causing changes in metrics are costs, availability or variability of raw materials and processes, new competing products, contradictions between product characteristics, difficulties in design. Nothing is black and white: usually various forms of grey have to be accepted.

Think break

1. Evaluate the two product concepts remaining after your work in the last Think Break, for marketing and production suitability. Make a checklist of all the important factors to consider in marketing and production and score the two product concepts.
2. Calculate a prediction of the possible sales volumes, prices and sales revenue for the two product concepts.
3. Do an evaluative comparison of the two product concepts and select the best product concept.
4. For the remaining product concept, write down the product benefits identified by the consumers and the other critical product characteristics you have so far identified. Suggest a metric for each product characteristic – this can be a physical, chemical, nutritional, sensory or microbiological metric.
5. What are the product characteristics for which you have not identified a metric? Can you create an empirical metric for them?
6. What are the raw material, processing and distribution requirements that need to be included in the product design specification?

3.1.4   Product feasibility and project plan

From the detailed knowledge, a more quantitative comparison can be made of the ideas for the new product. The consumer study gives in the product concept a comprehensive description of the product characteristics wanted by the target consumers who are more clearly identified. The market research gives an indication of the probable sales of the product, the position of the product in the market, the possible prices, promotion and market channels. The technical study describes the possible products, processes and the probable costs and time for development and production. By a qualitative evaluation of the suitability of the product concepts and a quantitative estimation of the profits and costs ratio, and by predictions of the probabilities of successful development and launching, the most suitable product concepts for development can be selected.

The various activities needed for the project are firstly developed in the outcomes and then in the building of the product design specification. They are all brought together and integrated in the **operational plan** for directing and controlling the project. For the plan:

- list all the major activities;
- place them in a logical sequence, noting activities that run in sequence, in parallel, and those that need to be integrated (**project logic flow plan**);
- time each activity from start to finish (**project scheduling plan**);
- identify the money, resource needs, personnel, for each activity (**project resource plan**);
- identify activities that are critical for time and resources (**critical path network**).

Review the network so that it meets the required launch date and is within the resources designated for the project (**project operational plan**).

3.2   Product design and process development

The themes for Stage 2: product design and process development, are integration, creativity, systematic planning and monitoring. Food product development is process-intensive, the characteristics of the product are highly constrained by the processing. Therefore the process and the product are developed together. This tight integration of process development and product design, called concurrent or simultaneous engineering, is becoming more important because of the time and cost constraints on getting the product to the market (Fox, 1993; Stoy, 1996). Jonsdottir *et al.* (1998), reviewing concurrent engineering in seafood companies, defined the overall goal of concurrent engineering as quality, cost, schedule, product user requirements and reduction of the time the product takes to reach the market. They emphasised the information technology applications in product models, in particular the knowledge of the product's functional and structural characteristics, and the development of a system model that secures the integration and reuse of knowledge in the different stages of the product development process. The concurrent design also integrates with marketing and production (Hollingsworth, 1995) as shown in Fig. 3.7. Often in incremental development, the production

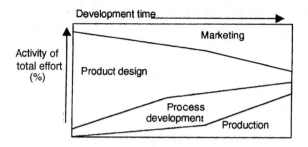

Fig. 3.7   Integration in product development.

plant is already in place, and the product has to be designed for that plant and the process can only be varied between narrow limits.

The company's identity or company's image is the sum of product design, communications design and environment design (Blaich and Blaich, 1993). Communication design directly supports the product in the marketplace with branding, packaging, advertising and promotion; therefore it needs to be closely integrated with the product design. Environment design is a concept that is not always considered, but it does influence the product and communications design, and the final acceptance of the new product. If a company wants to communicate the appropriate perception about its products, it must concern itself with the entire milieu surrounding the products, both inside and outside the company. If the company image diffused to the employees and the customers is quality, the new product is also seen as quality; if it is fresh and innovative, the product will be recognised as excitingly new. The company and distribution environments give the company and its new products an 'image' to the customers. Therefore product design needs to be integrated with communication and environment design throughout the design process.

3.2.1   Stages in product design and process development

The stages of the product design and process development are shown in Fig. 3.8; the activities are in the boxes, the outcomes in the ovals.

At the beginning of Stage 2, product design is the major part of the work, with process development considered in the design of the product. As the project progresses and the area for the product is more clearly defined, the study of the variables in the process becomes important so as to achieve the optimum product. The variables include both input and output variables.

- **Input variables:** raw materials (type, quality, quantity) and processing (types of processing, processing conditions).
- **Output variables:** product qualities and product yields.

The two main areas for research are formulation and processing; the first studying the type and quantities of raw materials and the second studying the

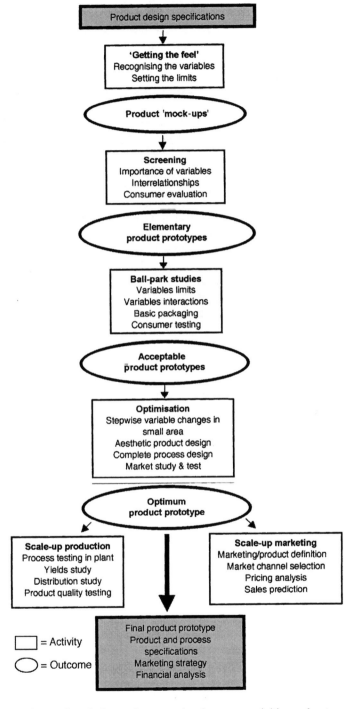

Fig. 3.8   Product design and process development: activities and outcomes.

effects of changing the processing conditions, but it is important that these are not studied separately as they are strongly interactive. The design is a continuous study of the relationships between the input variables and the product qualities, so that the final product is the optimum product under the conditions of the process. The prototype products are tested under the individual specifications set for the product design, so that product testing needs to be organised along with the product design and the processing experiments (Earle and Earle, 1999). Regular consumer testing of the product prototypes is necessary to confirm that the product has the characteristics identified in the product concept and not characteristics that are undesirable to the consumers.

Seldom does product design and process development occur in a straight line as in Fig. 3.8. There is back cycling because the prototype product is not completely acceptable to the consumer or the costs are not within the cost limits, or the chosen equipment cannot produce the product at the right yield or quality. It is important in each of these steps that there are technical, consumer and cost testings (Earle and Earle, 2000). The costs develop in stages from an identification of the parts of the company's cost system for this type of product and the limits for the various costs. Usually at the early stages, the raw material costs, their limits on the formulation, and the general costs of manufacturing are identified. The product, packaging and processing costs can be determined during the experimentation for the optimum product. After yield results during the production scale-up have been obtained and costs of marketing predicted, the total costs can be assessed.

3.2.2   Important factors in product design and process development
In food product design, there are some important points to consider:

Raw materials and ingredients
In many industries, there is increasing recognition of the place of suppliers in product development. In the past, the manufacturing company studied the effects of different raw materials and ingredients in the development of the product, and then produced specifications for the raw material/ingredient. Today, there is an increasing emphasis on working with suppliers in product development, and this is prevalent in the food industry (Hood *et al.*, 1995). The ingredient supplier is introduced to the initial problem in the product design specifications and then cooperates in developing the solution. This is sometimes called the 'black box approach' and it is claimed to reduce the time for the project (Karlsson *et al.*, 1998). Certainly the ingredient processor can be developing the process for the ingredient at the same time as the manufacturer is developing the consumer product. There needs to be a good relationship between the supplier and the manufacturer for this codevelopment to be successful. The food ingredient suppliers have actually gone further than this and developed the ingredient, the manufacturing process and the consumer product and handed this to the manufacturer. The reason for this may be the greater knowledge of product development in the food ingredient companies.

Quantitative techniques to integrate product and processing
In the past 20 years, there has been an increasing use of experimental designs
and statistical analysis in food design and process development (Hu, 1999).
There is software available that indicates suitable designs for the experimentation and analyses the results. Techniques such as linear programming
have been used in animal feeds and petfood formulation for many years but
have been slow to be used in human foods. Some of the problems in using
quantitative techniques have been the variety of critical product characteristics,
the poor definition of some characteristics and non-linear relationships
between processing variables and product qualities. Food product design is
complex but with increasing knowledge of the reactions in processing and new
software, quantitative techniques will be increasingly the norm, but this will
need increasing level of knowledge of the product designers and process
developers. Hegenbart (1997) noted in product formulation, the use of
spreadsheets to calculate formula costs, electronic information sources for
ingredient supplier details, and company database of in-house ingredients; and
in product testing the use of software for prediction of microbial growth in
food and for sensory testing.

Aesthetic skills in product design
In the design of food, there has been extensive use of sensory science in
developing a sensory product acceptable to the consumer. The industrial
designers have not been greatly involved in the design of the appearance,
colour, shape, but there has been interest in recent years (Pearlman, 1998;
Capatti, 2000). Extended design is most immediately applicable to haute
cuisine, but enters also into such items as extruded shapes and packaging. The
package design is often by industrial designers and therefore relates to the
artistic environment of the time. Airline meals (Kabat, 1998) and restaurant
meals are influenced by aesthetic design and we have seen this with
development of art nouveau, post-modern and other influences in meal
presentation. Today, many food products are completely artificial, in that they
are made from processed ingredients, and their design can be varied according
to aesthetic environment. This is the area where aesthetic design can be a strong
part of design – the question is how to encourage the industrial designer into
food design or for the food designer to adopt some of the practices of industrial
designers.

Values of the product characteristics
It is easy to spend a great deal of time designing a product characteristic that is
of no importance to the consumer. Technical characteristics are often beloved by
engineers in design but are of little consequence to the consumer. They may of
course be an integral part of the product and therefore need some concentration
in design. Value analysis or value engineering relates the cost of a product
characteristic to its importance; and then selects the characteristics with the
greatest value. There is a need to recognise the main aim of the product, for

example long life, and then to identify the characteristics of the product that relate to this, such as low water activity and controlled atmosphere, and then the cost of achieving them. There will be other characteristics, such as convenience, sweet fruity flavour, which also need to be fulfilled and other characteristics of less critical importance. The cost of these characteristics in the design can be determined to see if the cost is too high for the product characteristic, in other words above the value to the consumer. The highest valued characteristics are then the major part of the design.

Ergonomics
A neglected area in some food design, particularly in packaging, is ergonomics, the relationship of the physical product to the person (Ulrich and Eppinger, 1995). An example of poor ergonomics is an aerosol can for depositing a dairy cream on a cake or a dessert, that is mostly used by women and children, but cannot be held and used in one hand by them. Food is opened from a package, used in cooking, served and eaten; so design needs to take into consideration the physical aspects of the product and their relationships to humans using and eating it in all these steps.

Semi-production plant facilities
The stumbling block in technology transfer is the movement of the product from the laboratory to the full-scale plant. This is caused by various factors such as lack of processing knowledge of the food designer, the change in the processing conditions as equipment is scaled up, the difference in process control in the experimental and production plants, the transportation by pumps and lines in the production plant. Some products made and poured from a bucket or a jacketed pan will collapse when pumped around a factory. Many of these problems can be studied in a semi-production plant, without incurring excessive costs in materials and processing. When new products are based on incremental product changes, a semi-production plant can be used for a number of years and so the capital costs are paid back.

Internal and external capabilities
In the past, the aim was to have and build up the necessary expertise inside the company; then in the last ten years there was a popular movement to contract expertise from outside the company. On the one hand there is a need to have the activities of strategic importance inside the company so that the direction of the project is maintained. But on the other hand, there is a need to accept opportunities when they appear and if expertise is not available internally, to go out and buy it. Usually it is agreed that it is best to have an internal product development process championed, directed and understood by people inside the company, and to buy expertise from outside as needed. In other words have the company define the decisions, outcomes and activities in the PD Process, but contract out some of the tasks used in the activities.

Review and control of design process

The design process delivers the optimum product in the predicted time and costs – too idealistic? Yes, the design process is creative and working in the unknown, so it is difficult to be specific about product quality, time and costs. But there is a need to follow the product by regular testing – by the design group in the beginning and by consumers as the prototypes become more refined – to see that it is delivering the product. There also needs to be a time and resource plan which can be reviewed at different times in the design process by peer review to see if the project is effective and efficient (Fox, 1993). Problems will be encountered and there needs to be a recognised method of problem solving available to solve the problem quickly before the project collapses.

3.2.3   Conclusions to product design and process development

It is important that there is a clear end to this stage, and also the knowledge available to make the decision to go on or stop the project before the more expensive next two stages. This may not be the time to commercialise or the time to launch, so the project has to be shelved; or it has to be admitted that the product did not fulfil the expectations and the project must stop. Five important outcomes are:

- clearly defined final product prototype with consumer acceptance;
- product specifications including processing method, physical distribution;
- market strategy including distribution, promotion, pricing;
- prediction of investment needed and financial outcomes;
- probability of achieving project completion and financial outcomes.

Think break

1. For the product design specifications you prepared in the last Think break, identify the stages in designing the product prototypes and developing the process.
2. Create the basic product options by doodling on paper or computer or on the bench, evaluate them and select the most suitable basic product.
3. Identify the raw materials and processing variables related to the specified product qualities, and outline an experimental programme to identify the ranges of variables where the optimum product could lie.
4. Design an acceptable aesthetic product using the basic product, including appearance, shape, colour, sensory attributes and relating the product to the present culture of the target consumers.
5. Identify the packaging needs for the product, including protection and use, and also the needs for promotion of the product.
6. Combine all the knowledge you have so far created, and develop the final design for total product and package.

3.3   Product commercialisation

Stage 3: product commercialisation, is full scale-up of both production and marketing. These two developments need to be integrated throughout product commercialisation. Also design continues for the product, the production and the marketing, leading into the operational production and marketing. There is a need for integration, between the design and the operations. Product commercialisation ends with full integration of the product, production and marketplace. So the important factor in commercialisation is integration. Other factors to consider are the costs and the time. The costs really start to increase at this stage – maybe a plant has to be designed, built and commissioned; or fast-food outlets designed and built, or new distribution facilities built, all having a high capital cost. The risk of high financial losses increases as shown in Fig. 3.9.

There are four important stages in product commercialisation:

1. Setting up the commercialisation.
2. Design of marketing, production and distribution.
3. Testing of marketing, production and distribution.
4. Final integration of marketing, production and finance.

3.3.1   Setting up the commercialisation

The first activities in the product commercialisation are to agree on the aim, the resources and the final definition of the product and consumer relationship by developing an integrated project plan, and finalising the market and the product as shown in Fig. 3.10. All the people who are to be involved in the commercialisation need to be in the discussion, together with the product designers, so that there is technology integration between the design and the commercialisation. The

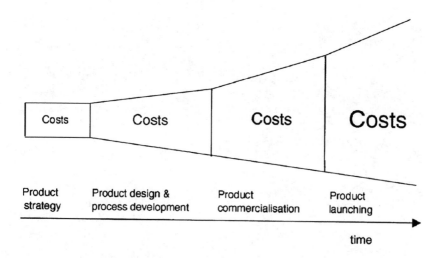

Product   Product design &   Product   Product
strategy   process development   commercialisation   launching

time

Fig. 3.9 Increasing costs in the product development process.

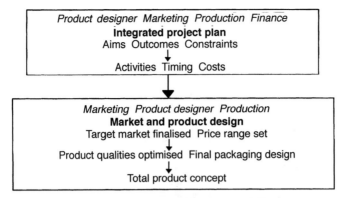

Fig. 3.10   Setting up product commercialisation.

business strategy is revisited at this stage to ensure that the product is still in harmony with the business. The aims and outcomes are becoming more specific because of the increasing knowledge created in the product design. It is very important that the aims, constraints and outcomes for the product commercialisation are considered in a combined discussion so that the different groups are not going in diverse directions and working towards different outcomes. For a drastic example, marketing and production may be aiming for different sales volumes, or marketing may be working to a price outcome not related to production's cost outcome. From the joint agreement on aims and outcomes comes joint identification of the necessary activities and then integration of the activities in the project plan. New constraints may have appeared because of competitive actions or changes in raw material availability, or changes in the finance for capital investment and some of them may have become critical. It is important to revisit and re-identify the critical constraints. Finally the timing and the costs for the various activities are identified so that the combined plan for commercialisation can be as efficient and effective as possible.

The other consideration in setting up the product commercialisation is to finalise the product and relate it to the target market. The total product concept needs to be built up from the market and product design (Earle and Earle, 2000), defining the core product, the total company product, the consumer's product concept and the society's product concept. There may be a need for some further product design to optimise the total product concept.

3.3.2   Commercial design

There are four types of design in product commercialisation – marketing, product qualities, physical distribution and production plant, as shown in Fig. 3.11. This is a time for many creative activities and they can career off into different directions. Nothing is worse than product qualities at variance with the marketing image of the product; for example product designers designing high-vitamin dog food and marketing building an image of a high-protein food. It is

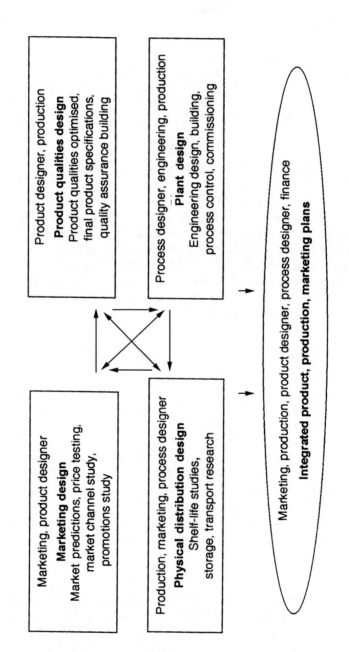

Fig. 3.11 Design in product commercialisation.

too late when the advertising designs come out and the designers say that that product is not what we designed! There needs to be close integration during the design and a final integration in the operational plans.

These are the general areas of activities but the choice of specific activities and techniques depends on (Earle and Earle, 2000):

- the type of product (incremental, innovation);
- the type of marketing (consumer, industrial, food service);
- the amount of learning needed by the company, the distributors, the consumers (high learning, low learning);
- scale of entry (local, national, international);
- the time (long, short) and timing (wide range, crucial).

For example, the time for the launch could be crucial because of competitive activity or the season, but the design has taken longer than expected, so the product commercialisation has to be rushed and the risk taken to drop some of the important activities. Companies often drop test marketing and business analysis when rushing to launch. But in all projects, time is expensive during commercialisation – an extra two weeks may make the costs shoot well over budget. So it needs to be well controlled.

Creating knowledge is another important aspect of the marketing and production design – as in all other designs. This is a major area of industrial research. For the incremental new product, the company has a great deal of past production and marketing knowledge and it is a case of fine-tuning the knowledge to include perhaps some production improvement and some new competitive marketing activities. But for the product innovation, it is a learning experience for company staff, distributors and consumers. The path of diffusion of the new product is identified, through all functional groups and top management in the company, the sales staff, the storage and transport operators, the retailers, the buyers, the users and the final consumers of the food. The learning experiences of all participants need to be incorporated in the activities in the final plan. Costs, revenues and profits are now assuming major importance and need to be followed carefully in the designs so the final financial plan is acceptable to the top management and the launch agreed.

3.3.3 Testing

The final product testing includes many aspects of the product:

- technical product qualities – core product qualities, packaged product qualities, agreement with regulations, services with the product;
- consumers' product concept – acceptance, competitive difference, uniqueness, aesthetic worth, brand attitude, product worth;
- marketing's product – product image, product position, promoted product, product price, retailers' product image;
- company's product – market share, sales revenue/profits, product effectiveness in business strategy, product problems, company fit;

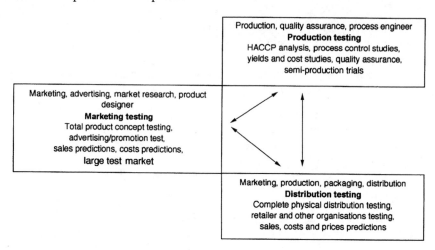

Fig. 3.12   Testing in product commercialisation.

- society's product – reliability, truthfulness of claims, protection from defects, value for money, social responsibility, environmental responsibility.

Combined with the testing of the product, there is production, distribution and marketing testing as shown in Fig. 3.12. Consumers test at least the total company product, with the packaging, advertising and public relations material; if time and cost allows the total product and the total marketing is studied in a test market. This can be an individual market or it can be the first phase in a roll-on marketing programme. Other important testing in the food industry is distribution testing which tests both the changes in product during transport and storage, and also the reactions of the retailers to the product. All food deteriorates with time – some in a few days, some in a year, and the effects of the temperatures, humidity, atmospheres and time during the transport and storage before sale has to be predicted. This is related to the label of the food with 'best by' dates.

An important aspect of testing today, which will increase with the trends into nutriceuticals, is ethical product testing. Ethical testing is related to a particular society; and the type and degree of testing depends on the ethics of the society. Basically people want to trust the food industry: firstly not to harm them and in fact to improve their health, and secondly not to use fraud and deceit when providing them with food. It is not ethical for the food industry to claim a lack of knowledge when being criticised by the society for unethical behaviour. When launching a new food product onto the market, the company must have extensive and detailed knowledge of the product's benefits and defects, of the raw materials and ingredients, and of the truth of the advertising claims. Always, the company knowledge must be more than the general knowledge in the society and in particular the consumers' knowledge, and the company must be willing to provide their knowledge. The company must not deceive any one about either

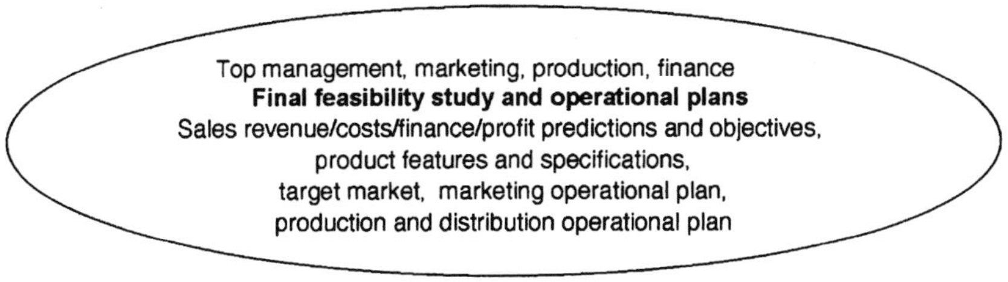

Top management, marketing, production, finance
Final feasibility study and operational plans
Sales revenue/costs/finance/profit predictions and objectives,
product features and specifications,
target market, marketing operational plan,
production and distribution operational plan

Fig. 3.13   Integration of product commercialisation.

the benefits or the defects or the problems associated with the product (Legge, 1999).

3.3.4 Final integration

The final step in product commercialisation is to bring together the knowledge from the design and the testing and to decide if the product is feasible; if it is, how it should be launched on the market. Integration is vital at this stage so that the launch can be efficient and effective (Andreasen and Hein, 1987). Obviously good decision making by top management is also vital, but management can only make decisions with the knowledge provided. The integrated knowledge is shown in Fig. 3.13.

The strategic orientation and the organisational capability are detailed at this stage. It is useful to develop a method of problem solving which can be introduced to everyone before the launch. Problems nearly always do occur in a launch and it is necessary to have a method of solving them to reduce both the chances of failure and the time taken for problem solving.

Think break

1. For your product designed in the last Think break, identify the aims, constraints and outcomes for the product commercialisation.
2. According to your expertise and knowledge, design the production, distribution or marketing. Ask some colleagues with different expertise to design the areas outside your knowledge.
3. Integrate the three design areas to give the total product/production/distribution/ marketing of the product commercialisation.
4. Evaluate the integrated design for its effectiveness in achieving the project aims and for obeying the constraints on the project.

3.4   Product launch and evaluation

'Effective product launch is a key driver of top performance, and launch is often the single costliest step in new product development. Despite its importance, costs and risks, product launch has been relatively under-researched in the product literature' (Di Benedetto, 1999). How true this is. Much of the research has emphasised the 'fuzzy' front-end activities and there is little on the critical back-end activities; in fact many PD Process models show seven or nine steps but only one for product launch!

There are three important parts of the launch – strategy, activities and demand outcomes (Guiltinan, 1999). The demand outcomes sought from the launch of the new product set the basis for strategy and the activities, and of course in the actual launch the strategy and the activities determine the sales outcome! This interrelationship between strategy, activities and demand outcomes is the major basis for planning the launch. The other important factor to consider is the evaluation and control of the launch; no matter how extensive the predictions for a launch, the unexpected always happens and there is a need for an evaluation and control plan.

3.4.1   Demand outcomes from the launch

The general demand outcomes include trial and repurchase, customer migration, innovation adoption and diffusion. The choice of demand outcome depends on the relationship between the consumer and the new product. Trial and repurchase, if the product is acceptable, is usually the buyer behaviour with incremental food products where the risk of purchase and eating is perceived as small. Buyers recognise the product as related to other products, the price is small and there is no great loss to the consumer unless there is a problem with food safety. Customer migration, the movement of competitors' customers to the new product, is the desired demand outcome when the product represents a significant improvement or change. The new product has a greater value for the consumer than the competitor's product and the ability to replace the existing product. Some of the situations for selecting particular demand outcomes are shown in Table 3.4.

Innovation adoption and diffusion are chosen where the product is new to the market and the consumer. This follows the traditional innovation curve with the

Table 3.4   Demand outcomes for product launch

Demand outcome	Product development project
Trial and repurchase	New product in existing market
	Line addition in existing market
	Emphasis on selective demand
Customer migration	Product improvement
	Emphasis on replacement demand
Innovation adoption and diffusion	New-to-the-world product
	Emphasis on primary demand, adoption and diffusion

Source: After Guiltinan, 1999.

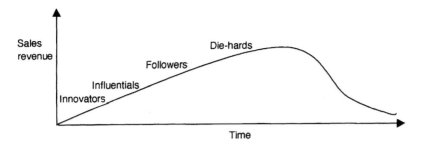

Fig. 3.14   The product diffusion cycle.

innovators, influentials, followers and die-hards as shown in Fig. 3.14. The diffusion curve can vary a great deal: the initial sales may be very slow and then there is a sharp rise, or there may be a fast initial rise and then a plateau. The shape of the curve is related to the product/consumer relationship but the launch tactics can affect it. The consumers may take some time to recognise and want the product so there is a slow uptake; or the product may fulfil an important need of the consumer, so they buy it immediately and sales increase rapidly. However, after the innovators have bought the product, the influentials and followers may take some time to buy and there is a plateau in the sales curve. Promotion and advertising can make consumers aware of the product more quickly than by word of mouth, and so they will buy earlier and the rate of sales growth will increase. Another marketing method to quicken the sales rate is to give consumers samples to taste in the supermarket; this gives them the opportunity to try the product at no cost, and if they accept the product they are encouraged to buy it. In launching, it is very important to understand the consumer/product reaction and how launch strategies and tactics affect it.

3.4.2   Launch strategies

The launch strategy can be described as the marketing, production and distribution decisions to introduce the product to the market and to start to generate sales. The launch strategies include the targeting strategy, the timing strategy and the product's innovation level. The perceived innovation level depends on the target market and also the competing products in the market. The target market can be a mass market or a niche market; the choice often governed by the size and resources of the company. New products may be aimed at a market segment, which is likely to be attracted to the new product, and then may be expanded to other market segments. Mass customisation in which the product is modified for specific groups of consumers is also another possibility. So there must be a strategy for reaching the target market segments.

Another launching strategy is to lead or to follow the competitors. This is an important timing strategy. With an innovation, the costs of being the pioneer can be high and if sales growth is slow then it takes some time to recover these costs.

But of course if there is a reasonably fast sales growth, then the product can win the major share of the market for a long time. With the incremental product, it is important to do this continuously with the succeeding products, so that the market share is either held or grows. To increase the market share, people of course make a product change so the new product must have their desired benefits and also needs marketing tactics to encourage them to make the change.

Hultink and Robben (1999) grouped the launch strategy decisions as:

- strategic product/market decisions – relative product innovativeness, targeting, introduction objectives and product newness;
- timing-related decisions – timing of market entry, speed to market.

These decisions have to fit into the company environment, its capabilities and resources, and the working environment of technology, market and competitors, as well as the surrounding societal environment. Successful launches were found to be related to perceived superior skills in marketing research, sales, distribution, promotion, R&D and engineering (Di Benedetto, 1999). Having cross-functional teams making key marketing and manufacturing decisions, and getting logistics involved in early planning, were strategic activities that were strongly related to successful launches.

3.4.3   Launch activities

There are two important decision areas for activities:

1. Marketing-mix decisions – relative distribution and promotion expenditures, relative breadth of product assortment, distribution channels used, marketing communications channels used, branding and pricing.
2. Production and distribution decisions – raw materials quality and quantity, production outputs, product quality, inventories, logistical times and quantities for delivery.

There are many activities in these two areas as shown in Table 3.5 and the problem is to choose the activities and integrate them. Very often there is emphasis on marketing tactics in the launch (Guiltinan, 1999) but the production and the logistic tactics can often make or break a launch. The launch strategy integrates the launch; and the launch tactics need to integrate the production, distribution and marketing activities so that they are focused on the same desired demand outcomes. Common pricing tactics are market skimming with a high price, and market penetration with a lower price. In choosing one of these, there needs to be consideration not only by marketing of the demand but also by production and distribution on the capability of producing the volumes and the costs of inventory in storage. Good launch management with control of activities and in particular of their timing and costs is the basis for a successful launch. The timing of the launch is all-important and sometimes activities have to be shortened to achieve this timing; this may cause problems for staff but it usually results in a successful launch unless everything falls apart!

Table 3.5 Activities in product launch and evaluation

Marketing organisation	Production organisation
Organising for the launch	
Finalise promotion	Design, build, commission plant
Media advertising contracted	Quality assurance finalised
In-store material prepared	Raw materials contracts
Sales presentation to staff	Physical distribution contracts
First introduction to retailers	Production finalised
	Market channel/physical distribution organisation
Product launch	
Launch targets finalised	
Complete selling to retailer	Produce the product
In-store material distributed	Distribute the product
Merchandising in supermarket	Check product quality in supermarket
Release advertising	Check product safety
Release product for sale	
Product launch evaluation	
Merchandising	Improving production efficiency
Advertising	Reducing product quality variation
Sales recording	Checking product in distribution
Buyers' surveys	Checking product in retailers
Competition study	Improving distribution efficiency
Marketing costing	Production and distribution costing
Financial analysis of costs, revenues	
Analysis of production, distribution, marketing	
Comparison of actual results with targets	
Adoption of product into the company	
New phase of advertising	Standardising production
New phase of in-store promotion	Total quality management in place
Pricing revamping	Raw material procurement revised
Sales recording	Output increased
Future costing	Costs reviewed
Sales analysis	Logistics optimised
Buyers' studies	Retailers' handling optimised
Future developments of product, production, marketing	
Financial analysis of investment, costs, revenues and profits	
Future returns on investment predicted	

Source: After Earle and Earle, 1999.

A very important aspect of launching is logistics; the aim is to have sufficient product on the shelves but not for too long. In the past, there was the practice of filling up with product the pipeline from the factory to the retailers' shelves, according to the predicted sales demand. This meant having a large inventory, which was costly, and in the case of introducing product lines did not allow for different rates of uptake of the individual products. A lean launch strategy based on logistics and supply chain collaboration can greatly reduce the costs and the risks of the product launch (Bowersox *et al.*, 1999). The lean launch strategy is based on response-based logistics, a flexible and responsive system with agile supply and manufacturing which can react quickly to real-time information from point-of-sale data transmitted via electronic data interchange (EDI) and Internet communications. The aim is to plan for lean inventory and to focus on in-stock position to support product successes, reduce stock of product failures and manage stock for niche markets. This gives better management of start-up costs, as sales will more rapidly balance costs and make a profit more quickly.

3.4.4   Evaluating and controlling the product launch

Evaluating and controlling the product launch is critical to success. The launch involves people and functions from all parts of the company, and the organisation of these people and their actions is complex. A well-planned organisational structure plans the activities, and can also quickly respond to problems caused by product quality, competitors' reactions and non-predicted consumer behaviour. Changes to activities or their timing can be made during the launch to counteract any problems arising.

Companies tend to build a 'launch' structure and use this for successive launches. For example in the food industry, it used to be a big TV campaign, backed by in-store promotions and simultaneous wide distribution in super-markets. With changes occurring in the food system, there is a need to be more adaptable so that the organisation system is permanent but the activities and techniques are selected for each project. In other words, the adaptability of the earlier stages of the PD Process is transferred to the launch. It is the most expensive stage and therefore requires the greatest knowledge from past experience. It also creates a great deal of knowledge, which should be captured for future launches.

Targets will have been set for the launch: short-term targets of sales volumes, sales revenue and market share, and long-term targets of a certain profit and return on investment and a time to recover the development and launch costs. Quantitative recording and analysis systems are set up to continuously analyse the sales and to improve the sales predictions. As the launch proceeds, the evaluation will become more definitive as more accurate data accumulate, and more realistic predictions of future cash flows can be made. The data necessary for the evaluation include production costs, prices, unit sales, sales revenues, marketing costs, company costs and finance costs. This is not just a recording system, it is also the basis for action during the launch.

One of the most difficult decisions is to change/not change the activities and the timing. If one reacts to every out-of-target result, then the whole system may get out of control; if a decision is delayed, the opportunity for success may be lost. It is important to follow trends and make decisions on these trends, not on spot data. The raw materials and direct processing costs are continuously

Table 3.6   Costs, finance and market monitoring during launch

Costs data

Finances	*Analysis*
Raw material costs	Cost trends
Production costs	Costs breakdown
Distribution costs	Production efficiency
Advertising and promotion costs	Distribution efficiency
Product losses costs	Marketing efficiency
Wastes costs	Additional operational costs
Company costs	

Financial data

Finances	*Revenues*	*Analysis*
Total costs of launching	Total sales revenues	Gross profit/loss
Cost of financing		Profit margin
Capital investment		Pay-back time project
Working capital		Pay-back time launch
Financial condition		Return on investment
of the company		Additional capital investment
		Additional working capital

Market data

Sales, marketing	*Analysis*
Sales total volume	Market share overall
Prices, range, specials	Market share in individual retailers
Sales individual retailers	Per capita sales rate
Buyers' purchasing patterns	Purchase/repurchase pattern
Competitors' sales	Ratio of sales against competitive products
	Predicted future sales

checked to see if they are improving and are within or better than target. The distribution costs, delivery times and product losses during distribution (which also are an important cost) need to be recorded regularly. There needs to be systematic monitoring during the launch in costs, finance and market as shown in Table 3.6.

Following sales is only one of the outcomes that need to be monitored and controlled during the launch. It is necessary to check how the product is performing in distribution, storage and retail outlets – is the quality correct, is the product becoming unsafe, are there many product rejects in the system? The retailers' and the consumers' attitudes to the product are monitored – how has the retailer placed and promoted the product? How much are the consumers buying and rebuying? What do consumers like/dislike about the product? The answers to these questions are crucial to the future of the product and need to be found in retailer and consumer surveys during the launch. Some important factors to follow are summarised in Table 3.7.

Table 3.7   Monitoring of production, distribution and marketing

Production	Distribution
Raw material quality	Delivery times
Raw material availability	Delivery quantities
Process variations	Product losses
Yields	Quality of product on delivery
Waste – processing material,	Quality of product on sale
product, packaging	Inventory in company stores
Quality of product	Inventory in customers' stores
Equipment breakdowns	Breakdowns in delivery
Response of staff	

Marketing

Retailers	*Consumers*	*Prices*
Reaction to delivery times	Consumer awareness	Price range
Product returns	Consumer reaction	Specials prices
Shelf space	Consumer buying	Price/demand relationship
Promotion	initial	
Prices	re-buy	
Orders	Consumer segments	
	Relationship to other products	

Advertising and promotion

Coverage	Effectiveness	Communication
Impact	Precision	Focus
Reinforcement	Relevance	Emphasis
Retention	Acceptance	

Think break

Using the knowledge from your product commercialisation in the last Think break:

1. Outline the demand outcomes wanted from the launch of your product.
2. Develop for your product launching
   (a) relative product innovation level strategy,
   (b) targeting strategy,
   (c) timing strategy.
3. Outline the critical monitoring points in the launch and describe the information you would collect at these points.
4. Discuss how you would analyse this information and use it to make decisions on the launch.

3.5   Service in product development

The previous sections discussed new product development in general, but sometimes new services have to be developed as well as the physical product. In these cases, a product and a related service are developed together. In developing new industrial products, the new product benefits and the service to the customer need to be developed in tandem to give the optimum integrated product. In food service, the service component is of major importance; for example in a high-class restaurant, the service of individual attention, and in a take-away chain, the service of fast convenience, is all-important. The basic four-stage PD Process is the same for new product and service development, but the activities and the organisation can be different. An integrated process including product and service needs to be used where product and service are developed together.

3.5.1 Services

Services are intangible as compared with tangible products. They are intangible experiences that are produced and delivered simultaneously. An important feature of the service is that it is adaptable and can change with different customers; this is just about the opposite of the tangible product, which stays the same for all customers. Service customers want individualised experiences that yield strategic benefits for them as individuals. The customer and the company employee are parts of the service; it is their interaction that is the service. Therefore service quality is highly variable, but it is very important, as the customer reaction is immediate (Terrell and Middlebrooks, 1996).

The customer's concept of the service includes the company's service system as well as the company skills and knowledge. From the customers' point of view, important features of a service are (Walton, 1992):

- treatment of the customer;
- speed and convenience of service;
- price of the service;
- variety of services;
- quality of the tangibles that accompany the service;
- unique skills that constitute the service offering.

From the company's point of view, a service has three parts: the service itself, the augmented service (service firm's reputation, quality of the interaction with the firm's system and staff) and the marketing support (Storey and Easingwood, 1998). The total service is shown in Fig. 3.15. In developing a new service, the three layers of the product have to be integrated so that the total optimum product is achieved.

3.5.2 New service development

With growth in the service industries in the 1980s and 1990s and the increasing need for new 'products', there grew an interest in the method of developing new services. In the 1980s, new service development started with the basic model of Booz, Allen and Hamilton (1982). A typical process was to develop a business

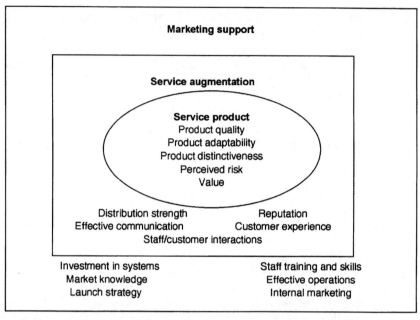

Fig. 3.15   Augmented service product (Source: After Storey and Easingwood, 1998).

strategy, develop a new service strategy, idea generation, concept development and evaluation, business analysis, service development and evaluation, market testing and commercialisation (Bowers, 1989). As service development grew, specific service development processes were produced (Johne and Storey, 1998). Scheuing and Johnson in 1989 proposed four stages – direction, design, testing and introduction, and identified a 15-step sequence of activities.

- Direction (or service strategy): formulation of new service objectives and strategy, idea generation, idea screening.
- Design: concept development, concept testing, business analysis, project authorisation, service design and testing, process and system design and testing, marketing programme design and testing, personnel training.
- Testing: service testing and pilot run, test marketing.
- Introduction: full-scale launch, post-launch review.

This model emphasises the intricate interplay between the design and testing functions during the design of a new service. The involvement of the operations personnel and the users is an important feature in the design stage. Customer participation is an essential part of new service development. Employee participation is also necessary, as the front-line employees are delivering the service. They are psychologically and physically close to the customer, and can identify customer needs and problems. If employees and customers are involved in the development, they will also behave knowledgeably and willingly in the delivery of the service. Therefore the design process in service development

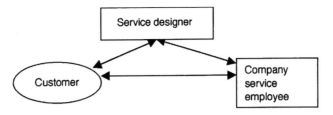

Fig. 3.16   The design triangle in service development.

needs to include the designer, the customer and the employee as shown in Fig. 3.16. The designer, customer and the service employee are usually not individuals, but three separate groups – the customer's group, the development group in the company and the company marketing organisation.

The question is how well can new service development fit into a service development process? Edvardsson *et al.* (1995) concluded from their studies that innovation of new services is an extremely complex process when it comes to planning and control. Their four stages were idea phase, project formation phase, design phase and implementation phase; they suggested it was not a sequential process but an interacting process. The stages overlapped and could not be clearly identified. Edgett (1994) found that the launch plan must be part of the development process. It needs to be well planned and coordinated, with communication materials and marketing targeted correctly and backed with sufficient resources. New service development can be a planned process following the four stages of the PD Process, but there is significant iteration in and between stages because the strong involvement of customers and service employees does not allow a rigid sequential structure. Two important stages are the service strategy and the service design.

Service strategy is the direction for development based on the business strategy. Some of the service innovation strategies include positioning, process, new service, employee/customer relationship and communication (Stinson, 1996). Examples in the food industry are:

- commodity meat suppliers repositioning to ingredient meat suppliers with products tailored to high class hotels and restaurants;
- restaurant changing from table service to self-serve;
- frozen food company introducing a new home delivery of frozen meals, nutritionally balanced for different age groups;
- starch ingredient supplier developing a buyer contact group with recognised product development skills;
- soy products company opening an on-line data base so that their clients can formulate new products.

Service design includes service concept, service system, service process (Edvardsson and Olsson, 1996). The service concept is a description of the customer's needs and how they are satisfied in the form and the content of the

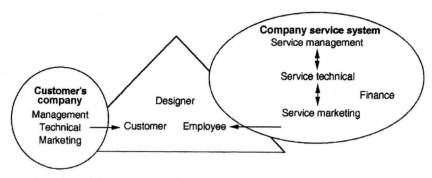

Fig. 3.17 Company service system and customer group within the design triangle.

service. As in product development, the customers have needs, wants and expectations. Expectation is critical in service – as can be seen from the attitude of any diner in an expensive restaurant, or a child in McDonald's. Expectation is based on the customers' needs and wants but it is also influenced by the company's image or reputation on the market, the customers' previous experience of the service company, the service company's marketing. The customer perceptions of the company and the service, especially as compared with competitors, have to be taken into consideration in the design of the service.

The service system in the company is mainly the people in the company, not just the front-line staff but the whole chain of customer relations in the company. The customer is relating to the technical resources and administrative routines and procedures in the company, as well as the marketing personnel. There is also a relation with finance as they are setting limits on prices, financial contract and investment. The whole company system is a part of the service design as shown in Fig. 3.17. The service system includes the resources available for the service development and operation. The finance section is involved not only in allowing the resources but also in setting the financial arrangements with the customers. Marketing has a key role in building up the part of the customer in the service, in particular to inform, educate and give them the skills to take part in the service. The technical people are involved in designing the hardware and software supporting the service.

The service and the customer outcome are generated in the service process. The customer is present in the process and affects the result (Edvardsson and Olsson, 1996). The nature of customer contact is a factor in the design – is it mail, telephone, face-to-face; long-term or short-term relationship; casual or a contract? There are individualised customer experiences in which the company may wish to be involved or keep at a distance. The behaviour of the customer must be taken into account as the service process is built up. In designing the service process, a framework of activities is built up from the customer introduction to the service to the customer outcome of the service. The service process consists of a clear description of the various activities needed to generate the service – service company staff, the customers, the physical/technical environment and the organisational structure. The service process depends on the resources – people, knowledge, skills – in the company and how they are organised. The customers also have knowledge, skills and procedures that need to be taken into account in the design of the actual process for delivering the service. The service process designed is a framework, but it will vary with every customer; every customer makes it an individual customer process.

In developing from the service concept to the new service, there is constant interaction between the service concept, the system and the process; and testing of various combinations with the customers and the employees. This gradually expands with increasing numbers of customers into pilot testing, test marketing and the final launch.

3.5.3   Industrial food products and services

There are two different groups in industrial business-to-business relationships – the industrial buyer (food processor or manufacturer) who employs raw materials and food ingredients in manufacturing a food product, and the industrial supplier of raw materials or ingredients (farmers, primary processors, ingredients processors). There is a great variety of buyers and suppliers, and also a wide variety of products. The product development varies from a branded coffee for one-person coffee bars, which is similar to consumer product development, to the highly specific ingredient for one large multinational food manufacturer. But there are some general factors to consider in developing new industrial products (Schaffner *et al.*, 1998).

The types of products
These could be raw materials from farm and sea, specialised commodities, bulk industrial products, partially processed materials, processed products, processed speciality products.

The industrial food-product characteristics
The industrial product can be divided into the tangible product, the uses of the product and the services that are marketed with the product. Some important features of industrial food products are shown in Fig. 3.18.

There is a tangible product that has specific composition, microbiological levels, physical properties and sensory properties, and there is the customer's product which includes the qualities directly related to the buyer – their uses and also the quality of the derived product made from the raw materials, usually the consumer product. In the customer's product there are also special features, quality and specifications, packaging and branding. Services included implicitly or explicitly with the industrial product can be reliability, safety, availability and replacement, technical information and help, delivery and credit. The service product can also include some or all of the features in the service augmentation and marketing support shown in Fig. 3.15. Products are not just a physical entity

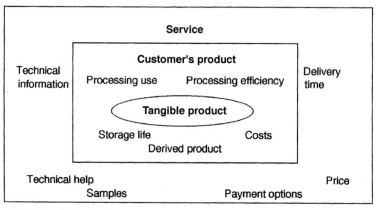

Fig. 3.18   The industrial food product (Source: From Schaffner, Schroder and Earle, *Food Marketing: An International Perspective*, © 1998, by permission of the McGraw-Hill Companies).

but an array of economic, technical and personal relationships between buyer and seller.

Industrial buyers

Industrial buyers can be grouped together as market segments. The buying company can be buying directly for their own use or for reselling to the users. The users can be segmented as shown in Table 3.8.

Think break

A large oils and fats ingredient company plans to develop a new pastry margarine product and is trying to identify a target market and new products. Possible target markets are pie manufacturers, frozen pastry manufacturers, croissant manufacturers, biscuit manufacturers, cake shops, small retail bakers, supermarket bakeries and hotel patisseries.

1.   Choose some suitable segmentation factors from Table 3.8 and assign the target markets into the segments. Select what you think are the two most suitable segments for a new pastry margarine product.
2.   Identify possible new products, both incremental and innovations, for each segment.
3.   Evaluate these new products and select the two most promising ideas.
4.   Identify the most important 'customers' to include in the design process for these two new product ideas.
5.   What tangible product qualities would they need?
6.   What services would they need?
7.   Sketch the complete product concepts for the two products.

Table 3.8 Methods for segmenting industrial buyers

Stage in the food chain: primary processor, secondary processor, caterer, retailer.
Type of processing: for example baking, freezing, dry mixing, sterilisation.
End consumer products: for example snack foods, takeaways, breakfast foods.
Size: number of employees, amount of capital, turnover per year, production volume.
Technical knowledge and skills: high technology, average technology, craft.
Usage rate: large, medium, low; regular, variable.
Type of purchaser: new, old, repeat, contract, casual.
Organisational structure: private or multinational company, farmers' cooperative.

The needs and wants of the buyers

All buyers are interested in firstly the ease of using the ingredient in the process and secondly the cost and quality of the final products. Although the buying action is logically based on these needs, there are still some psychological reasons for buying. Basic needs and wants of the industrial buyer are shown in Table 3.9. Actual needs and wants do vary with the different people in the buyer's company. For example:

- Production personnel – delivery time, reliability in supply, constant quality, ease in processing.
- Product development personnel – ease and shorter time for development, final product qualities.
- Quality assurance personnel – raw material specifications, ISO standards, narrow range of quality variation.
- Purchasing personnel – reliability of supply, price, size of delivery, regular deliveries.

In looking at these needs, one can see that there is an emphasis on service as well as the product, and this reinforces the need to develop the service with the

Table 3.9 Needs and wants of the industrial buyer

Availability	**Use**
Ease of delivery	Convenience in processing
Ease of storage	Uniform, stable, processing
Ease of ordering	Technical simplicity in processing
Reduced risks	**Costs**
Safety	Costs, discounts
Financial losses	Value
Product failure	Payment method
Staff failure	Payment time
Equipment failure	
Knowledge	**Outcome**
Technical information	Production of uniform, acceptable products
Formulations	Satisfactory sales and profits
New and improved consumer products	Competitive advantage
Help in processing	Few equipment problems
Information on derived products	Efficient staff use
Marketing help with derived product	

---

product. In developing industrial products, there is a need to identify the important people in the buying company as regards this type of ingredient and to find from them their needs and wants in the new ingredient, and decide how their needs and wants relate to the buying company's critical needs. In other words, the product development team in the supplier's company needs to understand the buying company's overall needs in product and services, and also the needs and wants of some individuals.

The PD Process

The PD Process is therefore a combination of product and service development. In the past, these have been done in sequence, completing the product development process, and then starting the service development. This leads to an increased time for development and also sometimes to a lack of harmony between the product and the service. In Fig. 3.19, there is an attempt to combine the product and service development processes to give an integrated product and service. The integrated product/service development process is particularly useful when new products are being introduced with a new service process and a new service system. The service system may already be in place and a new product and a new service process will be developed. This still means integration of the two development systems.

De Brentani (1995) has suggested three successful scenarios for industrial service development:

- Customised expert service: expert capabilities and resources providing customers with customised and high-quality service.
- Planned pioneering venture: pioneering new service ventures aimed at attractive, high-volume markets.
- Improved service experience: enhanced speed, good service quality and reliability.

Think break

The sales office of a large flour miller has just received a bread-baking mix from the production department. Recently there was a marked increase in the number of small hot-bread shops and the salespeople think that these small bakers might be a good market for this product. A salesperson knows a small baker and takes a bag of the mix to him. The baker promises to try it and in a day or two the salesperson has a telephone call from the baker to say that the product was a failure – there were difficulties in processing and the final loaf was small and hard.

1. How might the salesperson have handled this better?
2. How did the company go wrong in its industrial product development process?
3. Suggest a product/service integrated product development process for this company to ensure more successful new industrial products.

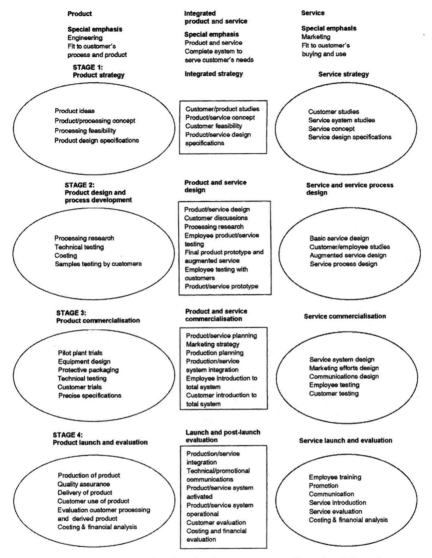

Fig. 3.19   The product/service development process for industrial products.

3.5.4   Food service development

In food service, there are three participants – the food manufacturer/food processor, the food service organisation and the customers – and two product development processes – the food manufacture and the food service. The two product development processes may be working in parallel or in sequence. The supplier's product development usually follows the standard sequence of the industrial product development process; in food service, product development is a major part of menu planning. The food service product involves the dishes offered and the service delivery of those dishes in the dining/eating environment. Both the food service operator and the customer want service as well as the product. Therefore food service development is a complex interweaving of product and service, through two development projects. There may even be another commercial customer between the food service and the customer; for example, in flight catering there are food manufacturers, flight caterers, the airlines and then the passengers. In developing new in-flight meals, the airlines regard new meals as service development, the catering services as service and product development, the food manufacturers as product development with some service development. Some new developments for in-flight meals are shown in Box 3.2 to illustrate the variety of development taking place.

Food service development is usually based on menu planning, which has five major aims: creative, nutritional, marketing, economic and logistical (Roberts, 1997). This food design is strongly aesthetic, but there is also a price direction and a serving need. Today, there is an increasing inclusion of nutrition into the design aims. The basis for the new development is the design of dishes, which are combined to give the new menu or in the case of institutions a whole meal structure (Ngarmsak, 1983; Roberts, 1997). In some instances such as takeaways, there is only one dish to be designed, although this has to be related to the overall takeaway product mix. Usually there is an existing menu, which can be improved by adding new dishes, or which can be used as part of a new menu. Development of new dishes is the basis for the menu change as shown in Fig. 3.20.

The supplier can give the new ingredient to the menu planner and let them take this through their PD Process of ideas, idea screening, recipe formulation, trial dishes, trial dish evaluation, standard recipe, menu design, menu trials,

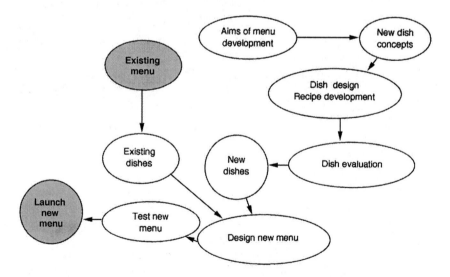

Fig. 3.20   Menu planning for new menu (Source: After Roberts, 1997).

Box 3.2   Developing new products and services for in-flight catering

Product: Meal components

Delta Daily Foods are a medium-sized food manufacturing company based in the Netherlands producing food items for both flight catering and supermarkets. They have developed a system for freezing individual vegetable and meat products, moulded in specific shapes that can be assembled by hand and even robot machinery into main dishes for in-flight trays.

Service: British Airways 'well-being in the air' concept

Based on extensive research into food trends, nutrition, macrobiotic diets and the oriental art of well-being, this takes the form of an advice pack for passengers on how to prepare for air travel, what to eat and do on board, and ideas concerning exercises and relaxation. Meals are designed to blend with this concept.

Process: Materials-handling system

SAS Service Partners and British Airways central production unit at London's Heathrow airport applied technology developed in the motor industry to flight catering equipment handling. This entails moving food trolleys on hooks suspended from a moving beltway from the unloading dock, through the wash-up area and into storage.

Software: Electronic reproduction of meals

Abela/Gate Gourmet developed a software package that produces electronic reproduction of images for catering. This system stores detailed recipes and dish specifications, along with full-colour digital menu pictures originally captured on video camera. The coded, kitchen-proof, keyboard enables chefs to access any menu or dish and enter the number of meals required. The system then computes the amount of each ingredient required and provides full specification and digital image on a colour printer.

Source: After Jones, 1995.

menu launch, or they can work with the menu planner in developing the meal ingredient or meal part. This combined product development occurs in stages as shown in Fig. 3.21. These are the overall activities in the two interacting PD Processes, but of course there are variations caused by the different situations.

The menu planner in the PD Process can be the product development manager for a large chain, the owner of a restaurant or the senior chef managing

Supplier PD Process	Interface	Food service PD Process
Stage 1: Product and menu strategy		
Exploratory market research		
	Needs of chefs/menu planners	
Product ideas		Menu ideas
Preliminary ideas		
Product concepts		Ideas for dishes
	Product concept agreement	
Development of design specifications		
Stage 2: Product and dish development		
Product design		Dish ideas screening
Samples for testing		Recipes formulation
	Product characteristics agreement	
Product improvement		
Process development		Trial dishes
Costing		Trial dishes evaluation
		Costing
	Development agreement	
Stage 3: Product commercialisation and menu planning		
Product technical testing		Final standard recipe
Production trials		Menu design
Quality assurance		Menu consumer trials
Product specification		Menu improvement
Production and delivery planning	*Product specifications agreement*	Final menu planning
		Risk analysis
Costing and pricing		Costing and pricing
	Contract negotiations	
Stage 4: Product and menu launching		
Production		Storage method
Quality assurance		Cooking methods
Delivery		Serving methods
Production analysis		Menu introduction
Customers' reactions	*Product quantities agreement*	Consumers' reactions
Evaluation and costing		Evaluation and costing
	Contract concluded	

Fig. 3.21   Developing meal components and menus.

a hotel, restaurant or institutional kitchen. The wants and the abilities of the menu planner are important in planning the activities in the PD Process. Two other important groups are the consumers and the providers of information to the menu planners, such as other suppliers and their professional associates. The management of the food service company strongly influence the overall product and service, particularly as regards price and choice of supplier. There are two important relationships: supplier/food service and food service/consumer. In designing an ingredient, the supplier has to bring these two relationships together, preferably by conducting research with both consumers and food service outlets; or if this is not possible, by obtaining consumer information from the food service company. Product development activities in food service are also influenced by:

- menus – menu analysis, menu planning, menu changes (type, timing – periodical, continuous);
- food service company – outlet type, meal periods, size, development capability, skills and knowledge, needs, wants;
- supplier company – type, size, development capability.

The products from the supplier to the food service include basic ingredients, meal components, partially prepared–not cooked meals, and pre-cooked, complete meals. The benefits of new products that the menu planner/chef usually identifies are in the areas of ease of use, safety, prestige of product and reliability. Two fundamental needs are value and risk; increased value of the dish or decreased costs is wanted, but risks of failure and indeed food poisoning are always present.

The benefits identified by chefs at the product concept and product development stage for two meat products in the hotel and restaurant market in Melbourne, Australia, are shown in Table 3.10 (Roberts, 1997). The study compared a meat product with little processing (thin beef slices) with a meat product with moderate processing (fricadelle, an alternative to the beefburger). The sliced beef was a basic ingredient, and the fricadelle was already prepared and only needed grilling. The chefs were looking for ingredients that would save time but also could be used for different dishes. In this situation with chefs selecting the products, beef slices were favoured over the fricadelle. It was interesting to see that there was a change in attitude between the written product concept and the actual prototype. The quality of the beef slices increased, but that of the fricadelle went down.

The risks were also studied in these two products. The important risks identified by the chefs were increased staff costs, food safety risk, too high use of one piece of equipment, increased storage capacity required, high financial losses, chef skills vulnerability, poor peer recognition and failure of the product in the marketplace.

Table 3.10   Product benefits identified by menu planners for two meat products

| Product benefits | Percentage of respondents scoring highly | | | |
| | Tender beef in thin slices | | Fricadelle | |
	Product concept	Product prototype	Product concept	Product prototype
Save time	94	81	61	70
Versatility	68	68	42	40
Value for money	65	58	55	27
Quality	26	55	23	23
Need	55	32	39	17

Source: From Roberts, 1997.

The stages in the adoption process used by menu planners are product awareness and interest, product concept, prototype trial and product adoption (including post-purchase evaluation). For awareness and interest, direct word-of-mouth communication between developer and adopter is important. Concept evaluation is a vital stage in the new product development process for satisfactory development of product specifications. The decision to try the new product is often based on cost, quality, need to save time and risk involved. The quality is often related to the consumers' needs as well as the chef's needs, so consumer testing is necessary. It may be organised by the supplier so that the food service has evidence on how the product is accepted by the consumers, but of course the chefs will also trial it themselves, probably in a blackboard menu. Product adoption by the food service company may not be systematic, but a case of trying it in the kitchen and giving opinions on the dish's acceptability and the cooking benefits and problems.

3.6   Where is the product development process going?

The project development process has settled into a well-proven stage pattern with critical reviews at each stage (the Critical PD Process). These reviews give opportunities for careful examination of progress, which if passed leads to the next stage and if not to abandonment or recycling as appropriate. Generally a four-stage PD Process is clear and sufficient, though in some projects substages may be necessary, especially for major innovations taking time for development. The importance of the activities and their sequence within the stages are determined by the level of innovation, the resources available to the company, the timing of the project, the company's risk level, and the knowledge and skills in the company.

One important factor is the degree of novelty, ranging from product improvement to a major product innovation. There can be a standard PD Process with a reduced number of activities for each project where there is:

- more or less continuous modification of an existing product line with fairly minor changes to produce variations on products;
- processes, equipment and markets are substantially unaltered; and
- no major shift in structure and organisation of the company's product development.

If the company has a data recording and storage system for product development, there may be sufficient knowledge of consumers, markets, products and production to reduce the research in the first stage, and also in the product commercialisation. Even the product launch can be a standard procedure. For these incremental changes in products, there can be a standard PD Process which is steadily improved after the analysis and evaluation at the end of each project. The efficiency and the effectiveness of the product development process can be improved over time.

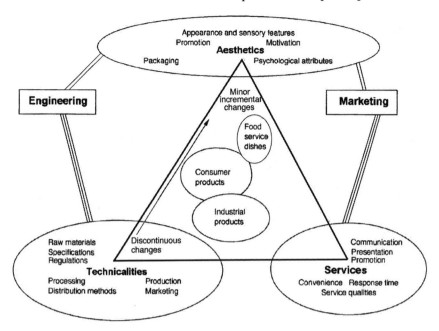

Key:
Strength of relationships: —— weak; ═══ moderate; ▅▅▅ strong

Fig. 3.22   Indicative influences affecting activities choice in the PD Process.

For major discontinuous changes, there is a need for more exploratory activities in the first stage of the PD Process, and also because of the large costs involved there will be more project and business analysis throughout the project. The decisions are major because of the resources needed, and therefore a great deal of knowledge is required which usually has to be created in the project. A consistent, logical process is needed, but it cannot be highly structured because of the unknown nature of the project. The process is usually more exploratory and less customer-driven than the typical incremental product development process. It concentrates in the first stage on recognising the application of developing technologies in new products for the company, so there is an early design of product prototypes before opportunity analysis, assessment of market attractiveness, market research and financial analysis (Veryzer, 1998). Technological research is necessary to identify what is possible, before the consumer can study product ideas and develop product concepts. In the later stages, there is important design of the plant, production and the market strategy, which again needs design ideas, evaluation and then application.

Figure 3.22 indicates how new product placement can often be related to the three major components – aesthetics, technicalities and service. Placement yields more useful lines of emphasis in choosing, planning and executing activities in the product development process than just putting products into categories of industrial, consumer and food services, because of the wide variations in products in these categories. Placement aids the selection of activities and ensures a product development process that runs smoothly and with a better probability of success.

A very important influence is the market. On the one hand for a consumer market, the influences of aesthetic factors, which persuade customers into trying a new product, can be much more significant than technical considerations. Hopefully their acceptance leads towards brand acceptance and therefore consistent support with little further effort needed on choice. On the other hand, industrial products move to a much more stringent, technical scrutiny. Careful definition of specifications, examination of pricing and longer-term contractual detail lead to dominance by technical considerations with little or no emotional overtone, and with a close eye on service and on convenience-of-fit to further processing or manufacture. Food service industries again emphasise technical detailing of ingredients, but in developing dishes and meals quite often aesthetics are a major thought; consistency, reliability and service are critical factors.

3.7   References

ANDREASEN, M.M. & HEIN, L. (1987) *Integrated Product Development* (Berlin: Springer).

BETZ, F. (1998) *Managing Technological Innovation* (New York: Wiley).

BLAICH, R. & BLAICH, J. (1993) *Product Design and Corporate Strategy* (New York: McGraw-Hill).

BOOZ, ALLEN & HAMILTON (1982) *New Products Management for the 1980's* (New York: Booz, Allen and Hamilton).

BOWERS, M.R. (1989) Developing new services: improving the process makes it better. *Journal of Services Marketing*, 3(1), 15–20.

BOWERSOX, D.J., STANK, T.P. & DAUGHERTY, P.J. (1999) Lean launch: managing product innovation risk through response-based logistics. *Journal of Product Innovation Management*, 16(6), 557–568.

CAPATTI, A. (2000) Food design. *Domus*, 823(Feb.), 68–78.

COOPER, R.G. (1990) Stage-gate system: a new tool for managing new products. *Business Horizons*, 33, 44-54.

COOPER, R.G. (1996) Overhauling the new product process. *Industrial Marketing Management*, 25, 465–482.

DASGUPTA, S. (1996) *Technology and Creativity* (New York: Oxford University Press).

DE BRENTANI, U. (1995) New industrial service development: scenarios for success and failure. *Journal of Business Research*, 32, 93–103.

DI BENEDETTO, C.A. (1999) Identifying the key success factors in new product launch. *Journal of Product Innovation Management*, 16(6), 530–544.

EARLE, M.D. (1971) *The Science of Product Development and its Application in the Food Industry* (Palmerston North: Massey University).

EARLE, M.D. (1997) Changes in the food product development process. *Trends in Food Science and Technology*, 8, 19–24.

EARLE, M.D. & EARLE, R.L. (1999) *Creating New Foods: The Product Developer's Guide* (London: Chadwick House Group).

EARLE, M.D. & EARLE, R.L. (2000) *Building the Future on New Products* (Leatherhead: Leatherhead Food R.A. Publishing).

EDGETT, S. (1994) The trials of successful new service development. *Journal of Services Marketing*, 8(3), 40–49.

EDVARDSSON, B. & OLSSON, J. (1996) Key concepts for new service development. *Service Industries Journal*, 16(2), 140–164.

EDVARDSSON, B., HAGLUND, L. & MATSON, J. (1995) Analysis, planning, improvisation, and control in the development of new services. *International Journal of Service Industry Management*, 6(2), 24–35.

ERICSON, O.P. (1997) Survey offers full plate of food product ideas. *Food Product Design*, 7(9), 16, 19, 20.

FOX, J. (1993) *Quality Through Design: the Key to Successful Product Development* (London: McGraw-Hill).

GUILTINAN, J.P. (1999) Launch strategy, launch tactics and demand outcomes. *Product Innovation Management*, 16(6), 509–529.

HEGENBART, S.L. (1997) Computers: yesterday's novelty, today's development partner. *Food Product Design*, 7(4), 81, 82, 84, 85.

HOLLINGSWORTH, P. (1995) The slimmed-down shape of new product marketing. *Food Technology*, 49(3), 68, 70, 72.

HOOD, L.L., LUNDY, R.J. & JONSON, D.L. (1995) New product development: North American ingredient suppliers' role. *British Food Journal*, 97(3), 12–17.

HU, R. (1999) *Food Product Design: A Computer-Aided Statistical Approach* (Lancaster, PA: Technomic).

HULTINK, E.J. & ROBBEN, H.S.J. (1999) Launch strategy and new product performance: an empirical examination in the Netherlands. *Journal of Product Innovation Management* 16(6), 545–556.

JOHNE, A. & STOREY, C. (1998) New service development: a review of the literature and annotated bibliography. *European Journal of Marketing*, 32, (3/4), 184–251.

JONES, P. (1995) Developing new products and services in flight catering. *International Journal of Contemporary Hospitality Management*, 7(2,3), 24–28.

JONSDOTTIR, S., VESTERAGER, J. & BORRESON, T. (1998) Concurrent engineering and product models in seafood companies. *Trends in Food Science and Technology*, 9(10), 362–367.

KABAT, J. (1998) High flying food. *International Design*, 45(6), 64–65.

KARLSSON, C., NELLORE, R. & DAVIES-COOPER, R. (1998) Black box engineering: redefining the role of product specifications. *Journal of Product Innovation Management*, 15(6), 534–549.

LAI, PAI WAN (1987) *Development of a Bakery Snack for Export from New Zealand to Malaysia*. Thesis, Massey University, Palmerston North, New Zealand.

LEGGE, J. (1999) *Product Management: Sharpening the Competitive Edge* (South Yarra: Macmillan Education).

MOORE, W.L., LOUVIERE, J.L. & VERMA, R. (1999) Using conjoint analysis to help design product platforms. *Journal of Product Innovation Management*, 16(1), 27–39.

NGARMSAK, T. (1983) *A System of Meal Planning for Self-improvement of the Diet of Villagers in Northeastern Thailand*. PhD Thesis. Massey University, Palmerston North, New Zealand.

PEARLMAN, C. (1998) Food for thought. *International Design*, 45(6), 47.

PETRIE, R. (1995) Four-legged trends. *Marketing Week*, 18(13), 42–43.

ROBERTS, L.M. (1997) *A New Beef Product Adoption Model for Hotels and Motels in Greater Melbourne*. PhD Thesis, Massey University, Palmerston North, New Zealand.

ROSENAU, M.D. (2000) *Successful Product Development: Speeding from Opportunity to Profit* (New York: Wiley).

SCHAFFNER, D.J., SCHRODER, W.R. & EARLE, M.D. (1998) *Food Marketing: An International Perspective* (New York: McGraw-Hill).

SCHEUING, E.E. & JOHNSON, E.M. (1989) A proposed model for new services development. *Journal of Services Marketing*, 3(2), 25–34.

SCHMITZ, B. (2000) Tools of innovation. *Industry Week*, 249(10), 57–66.

STINSON, W.S. (1996) Consumer packaged goods (branded food goods), in *PDMA Handbook of New Product Development*, Rosenau, M.D. (Ed.) (New York: Wiley).

STOREY, C. & EASINGWOOD, C.J. (1998) The augmented service offering: a conceptualization and study of its impact on new service success. *Journal of Product Innovation Management*, 15(4), 335–351.

STOY, R. (1996) Assembled product development, in *PDMA Handbook of New Product Development*, Rosenau, M.D. (Ed.) (New York: Wiley).

TERRELL, C.A. & MIDDLEBROOKS, A.G. (1996) Service development, in *PDMA Handbook of New Product Development*, Rosenau, M.D. (Ed.) (New York: Wiley).

ULRICH, K.T. & EPPINGER, S.D. (1995) *Product Design and Development* (New York: McGraw-Hill).

VERYZER, R.W. (1998) Discontinuous innovation and the new product development process. *Journal of Product Innovation Management*, 15(4), 304–321.

WALTON, T. (1992) Where's the design in service design? *Design Management Journal*, 3(1), 6–8.

4

The knowledge base for product development

The ability of a company to build a knowledge core and continuously create new knowledge is critical to the success of product development. There are four areas where knowledge is needed for product development:

- the different cultures of the world, their needs, wants and attitudes, and how they can assimilate and absorb new products;
- basic knowledge and skills of present raw material production and food processing;
- high technological knowledge and problem-solving skills to develop new technologies;
- product development systems and organisation.

Basically this is applying the total technology concept to food product development – society, company environment, company resources, knowledge, organisation, techniques and the practice of product development. Management selects and integrates the knowledge in the company, and provides the conditions for knowledge to be created. There has to be a communications system in the company so that knowledge spreads and grows throughout the company. Knowledge is dynamic, causing change. It is important to recognise that knowledge is not just information and databases, but it is part of the active development in the company in organising the present system and activities, and also in developing new systems and activities. Information can be the basis for revealing and creating knowledge, but knowledge is in people – in their heads, in their problem-solving skills. It is in their understanding of the interaction between technology and society and also of the specific interactions of the consumer and the product, the worker and the processing plant, the salesperson and the retail outlet, the cook and the kitchen, and so on.

Knowledge causes change; information is the basis of change. Today, there is increasing emphasis of this being a 'knowledge society', as if knowledge is something new. Knowledge has been around for a long time; there are periods when it increases and sometimes, as in the Dark Ages, when it seemed to lose ground. What is different at the beginning of the new millennium is that communication between people has been made much easier; and communication does increase knowledge if the information is absorbed and used in the minds of people. But what does this increasing interchange mean to the food industry?

4.1   Technology, knowledge and the food system

Technology takes knowledge and creates products, processes and services for the use of people. At the heart of technology lies the ability to recognise a human need or desire (actual or potential) and then to devise a means – an invention or a new design – to satisfy it economically. Having done so, the model or prototype has to be scaled up and adapted to become a marketable item. The process of turning the full-scale product into something that satisfies market requirements of safety, cost/profit effectiveness and customer acceptance is a difficult one (Cardwell, 1994). A company not only has to have a store of knowledge but it has to create knowledge during the development of the product, process and service. It also has to connect different types of knowledge during the development – technological, commercial and organisational. After the development, it has not only transformed the knowledge into practical applications but it has increased its own store of knowledge by the knowledge it has created.

Two types of knowledge are recognised – disembodied (before and during development) and embodied (after development). The disembodied knowledge goes eventually to the embodied product in product development:

Disembodied knowledge → Disembodied innovative activities → Embodied product

That is:

Tacit knowledge in people's heads + Explicit (codified) knowledge in records → Knowledge creation in PD Process → New product

There are four important areas of disembodied and embodied knowledge: technology, technological change, innovative activities and technological indicators that are important for product development (Evangelista, 1999), as shown in Table 4.1. A company has a stock of technological knowledge, and then generates more knowledge during its innovative activities to produce productive assets, including products, plants and marketing systems.

In product development, as in all engineering and design, there is a major use of the knowledge that is in people's heads from their education and more importantly from their experience – called either tacit (as used in this book) or embedded knowledge. There is also use of recorded knowledge in reports,

Table 4.1 Concepts of technology

Disembodied

Disembodied technology: stock of technological knowledge both embodied in people and expressed in a codified form.

Disembodied technological change: process of advancing technological knowledge.

Disembodied innovative activities: activities carried out at the firm level to generate or develop new technological knowledge.

Disembodied technological indicators: R&D expenditures and personnel, design and engineering activities, patent and licence counts, technology flows measured by the technological balance of payments and bibliometric data.

Embodied

Embodied technology: stock of technological productive assets consisting of machinery, equipment, plant and operating systems (both tangible and intangible).

Embodied technological change: accumulation of new technical assets (machinery, equipment, plant and operating systems).

Embodied innovative activities: innovative activities consisting of the use or adoption of new productive assets with enhanced technical and technological performances compared with those used before.

Embodied technological and innovative indicators: investment in new machinery, equipment and plant incorporating new (or not yet used) technologies; indicators measuring the adoption and diffusion of embodied technologies.

Source: From Evangelista, 1999, by permission of Rinaldo Evangelista and Edward Elgar Publishing Ltd.

textbooks and journals, called either explicit (as used in this book) or codified knowledge.

4.1.1 Knowledge in the food system

In a study of the Italian industry, Evangelista (1999) placed the food and drink industries in the investment intensive sector. The other sectors were:

- R&D/investment intensive;
- R&D (research and development) and D&E (design and engineering) innovators;
- technology users.

In his investment intensive sector, investment activities play an important role, while research, development, design and engineering play marginal roles. Process innovations are very common and innovation performance is linked to investment in technologically new machinery and equipment. Other processing industries, chemicals and sugar in the investment intensive sector and pharmaceuticals in the R&D/investment intensive sector had higher research, development, design and engineering activities. Pharmaceuticals had high R&D and D&E expenditures accompanied by medium or high levels of investment in machinery, innovation being clearly oriented towards the introduction of product innovations. Comparing companies in Europe in Table 4.2, this greater emphasis

Table 4.2   Product and process innovations in European companies

	Percentage of firms introducing		
	Product innovation	Process innovation	Product and process innovation
Mechanical machinery	92.8	69.8	62.6
Chemicals	91.6	75.5	67.1
Food, drink & tobacco	70.3	93.6	63.9

Source: From Evangelista, 1999, by permission of Rinaldo Evangelista and Edward Elgar Publishing Ltd.

on process innovation in the food industry was clearly shown (Evangelista, 1999).

One recognises that food manufacturing is essentially a supplier-dominated industry with ingredients from the chemical industry and large food ingredients processors and equipment from mechanical/electrical manufacturers. Knowledge is bought in by food manufacturers from the suppliers, there is often less creation of knowledge than in the supplier industries (Hood *et al.*, 1995). This knowledge generation and transfer is emphasised at the food congresses where a large number of suppliers not only exhibit their products and equipment but also give or sponsor many of the papers at the meeting. An interesting recent example demonstrating the limitations of product development when relying heavily on outside sources of technology was shown by Martinez and Burns (1999) when studying the Spanish food and drink industry. They found product technology was predominantly in-house generated, process technology combined internal development with external acquisition mainly from equipment suppliers. Purchase of equipment emerged as the main source of external technology acquisition as opposed to information gathering procedures. This reliance on externally generated technological developments had brought about low levels of technological independence in general and process technology in particular. The importance of in-house technological capabilities in product and process innovation, indicates the problems in product development a company and indeed an industry faces if it relies largely on external sources as opposed to internal developments. Is it time for food manufacturing to include more R&D and D&E in product development so as to produce a more sophisticated technological content in consumer food products? The food manufacturing industry is probably never going to be a high technological industry but there is a need for a different balance between R&D, D&E and capital investment in plant as these are joint determinants of the performance of companies. Wallace and Schroder (1997) made the following statement which the management of food industry development might ponder:

Research and development in the food industry is a well-recognised case of market failure with its private costs and benefits differing from its social ones. The end result is an under-investment in R&D by private firms and attempts to justify government supporting it. The question is how to solve this dilemma. In the meantime, increasing masses of scientific and technical information and analysis are being super-imposed on a world wide background of rapid legal, political and social change.

Organisations can be grouped as functional, processed-based and societal knowledge-based. This means that a company can be based on functional departments such as marketing, production; or it can be an integrated technological entity; or lastly it can be a technological entity integrated into society. Is the food industry, which has been mainly functional, moving towards an integrated technological organisation with management based on societal knowledge? If so, the knowledge needed in the industry will have to increase exponentially.

4.1.2   Creation and movement of knowledge in the food system

The passing of knowledge between suppliers and food manufacturers emphasises that one cannot think of a part of the food industry by itself. In knowledge creation, each part of the food system is affecting knowledge in another part. In primary production, knowledge creation has been very much government-financed and often government-led. In early years, farming and fishing were essential for the production of food for the population, and were often the occupations of many individuals and families. Governments therefore felt that R&D in food production was their social responsibility. Today scientists in private and publicly managed agencies do significant basic and applied research. Governments are still funding agricultural research from government revenues and often organise agricultural research. For example in the United States, the US Department of Agriculture is still a major player in agricultural research and State governments are also involved. Internationally, there are also United Nations organisations and other world governmental agencies funding and organising agricultural research. The roles of the different public agencies and private firms are intertwined in complex ways (Alston *et al.*, 1997). Surprisingly, research for the fishing industry has never been so extensively government funded, and one might think that the over-fishing and lowering of fish stocks has been due to lack of knowledge as much as human greed.

Distribution research has also been an area of government research for many years because of the need to store and transport food to urban areas, and internationally. So knowledge increase in the food system is still dependent on governmental funding and support, except for the food ingredient processing and consumer product manufacturing which have been among the low spenders on R&D related to sales among the industries based on process engineering. This may be due to its only recent emergence as a science-based industry, the

Government - funded research

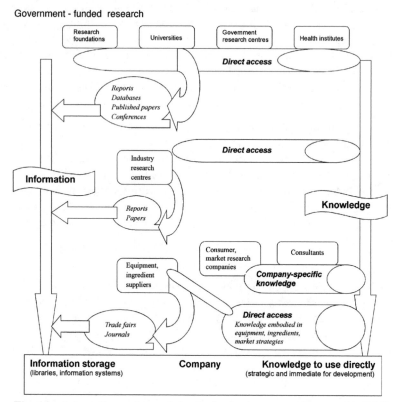

Fig. 4.1   Information and knowledge in the food company from outside R&D.

marketing domination in many food companies, the difficulty of controlling intellectual property in the food industry, and the small margins on which the food industry works (Earle and Earle, 1997). Much of the knowledge in the food manufacturing company has been created in incremental product development, which unfortunately has often not been recorded so it is not an explicit knowledge base for future product development. Much of the private knowledge in the food industry is in the large multinational companies, and tends not to go into the public arena even for the teaching of students in food science/food technology/food engineering.

Knowledge for product development in the company can be acquired from outside R&D. It is important to identify the direct access to knowledge and also the indirect access through information as shown in Fig. 4.1. Many government agencies provide information in reports, databases and published papers, which can be developed into useful knowledge by the company. This information can be stored in libraries or other information storage facilities and on the Internet. But the company can also work directly with government research agencies, consultants, ingredients/equipment suppliers, and consumer research companies, to develop specific knowledge for the company.

It is important to recognise the science and technology information tracks so that they can be tapped into as problems arise in product development. Research in industry is focused primarily on advancing technology to fulfil changing consumers' needs, whereas in universities and in many research institutes it is focused primarily on advancing either science or generic technology (Betz, 1998). In the science track, the knowledge is published in peer-reviewed journals and is eventually summarised in textbooks and taught to students, although with modern funding in universities a significant amount of the knowledge is not published but is transferred directly and exclusively to the sponsors of the research. From an understanding of the current state of scientific knowledge, researchers in engineering and technology advance the knowledge in their disciplines by research on the basics of the technologies. This basic technological knowledge is published and taught to the next generation of engineers and technologists, and transmitted to their counterparts in industry in conferences and journals. In the early years of a new technology, a company works mostly with knowledge discovered during the industrial development. Gradually technological knowledge sources are built up and these can be used in later development projects. A combined knowledge of the food system, and in particular the company's segment of it, is built up over the years by the company's R&D and its experience in marketing, production, distribution and engineering. This is the basis for future product development.

The company also looks for knowledge from its competitors, by studying their actions and products in the marketplace and their production, raw materials and processing. Most industries work from a similar technological base; 80% of the knowledge is known by everyone, maybe even more. In product development it is the extra 10–20% knowledge that makes the competitive edge, but the company also needs to have the capability to use fully the basic knowledge.

The company is creating knowledge along the whole system from the initial R&D to the final outcomes of the product in the market (Quinn, 1992) as shown in Fig. 4.2. Knowledge is being created and then extended in the next stage

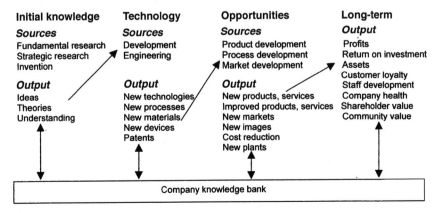

Fig. 4.2 Movement of knowledge through the company (Source: After Quinn, 1992).

where more knowledge is created. Even at the final stages where the new development has become a commercial reality, there is still knowledge being created about the product, production and marketing. Although there is a clear movement of knowledge from one stage to the next, there also needs to be interconnecting communications of knowledge between all stages so that the new knowledge is shared and the total company knowledge grows. There is also a need to evaluate the use and creation of knowledge in product development; usually the embodied knowledge, particularly the product and its success in the marketplace, is used as the indicator of the knowledge achieved in the project.

Companies have difficulty in relating the knowledge created by fundamental research to the company's final profit. But fundamental research can be evaluated on the new knowledge and understanding that is passed on to product development. Just ask the product developers what it would cost them in the long term if the fundamental or the strategic research disappeared; or if strategic research improved its performance what extra value would that give to development! In the food industry in the past 20 years, R&D has tended to be either dropped or reduced – one wonders how the company valued this asset, and how much it cost them to buy in this knowledge in the following years, and how many opportunities were lost. It is important that each knowledge-creating area is evaluated regularly to find which area is performing in creating knowledge that leads successfully to the long-term goals.

Invention is difficult to place in the knowledge flow because it is based on observation of what is happening maybe in a technology or in the community, unlike science, which is trying to discover new knowledge. Invention is not necessarily limited by the extent of scientific knowledge; inventors rely on their accumulated practical knowledge and their own intuition (Cardwell, 1994). Invention requires some conceptual or imaginative creativity. To make an imagined transformation physically real, it also has to have the necessary technology, knowledge and skills. So it is an idea that has come to its time – the idea may have been imagined a long time before but cannot be made real unless the various factors are present in people's knowledge and skills (Mitcham, 1994). The concept of invention is the opposite of the incremental change. As well as taking place in an individual's mind over a short period, it can develop in a group through time together, but not substantially through systematic design. It is intuitive or even accidental events that lead to invention. The food industry has in the last 60 years been looking for the magical invention of a major new method of food preservation, but it has not come. There have been many improvements in drying, freezing, chilling and heat sterilisation, but there has not been the invention of a completely new method. Atmospheric control has been the one new preservation method that has gradually grown as packaging technology and inert gas production have improved. Although scientists have been studying it for over 60 years, the scientific knowledge has grown very slowly, but it is now expanding in combination with chilling for long-term storage and transport of vegetables, fruit and meat. Other methods, such as irradiation and the use of gases such as methyl bromide, have been used in food preservation, but they are rather blunt instruments that certainly did not fit with the societal environment.

Think break

Consider your company and its sources of knowledge for product development:

1. Identify a new product that has come from an invention inside the company. What knowledge did the company need to bring this invention to a commercial product?
2. Choose a product that is being developed at the present time. Identify the tacit knowledge that was used in the first stage of the product development process, and the people who supplied this tacit knowledge.
3. Choose a product that has been launched. Identify the explicit (codified) knowledge that was used in the final stages of this product commercialisation and launching.
4. Describe how in your company the knowledge created in the product development project is saved as tacit and explicit knowledge for use in future projects. Discuss how the saving of this knowledge might be improved in the future.

4.2   Knowledge management or knowledge navigation?

Technological capabilities in product development consist of the resources needed to generate the technological opportunity and manage the technical change, including skills, knowledge and experience, and the institutional structures and linkages. Technological knowledge is usually the most important. A large part of technological knowledge in product development has a tacit nature, being incorporated in people skills, competencies and organisations. Tacit knowledge is often not codified and is largely company- and indeed often area-specific, and may be difficult to transfer to explicit knowledge. Learning is often the central method for passing tacit knowledge and building it in the product development team.

There is also an ever-increasing bank of explicit knowledge used in food product development, from consumer changes to advancing technology, and it is difficult to find all the appropriate knowledge for a specific project. It is not sufficient just to have storage systems for information; there need to be clear paths to find and assess total knowledge in different areas of the company and indeed outside the company. Knowledge navigation is a better description than knowledge management; knowledge navigation includes the strategic directions for knowledge as well as the knowledge systems. One of the key roles of top management is to create a culture and environment that is conducive to knowledge capture and knowledge sharing. Management leads the company into strategic directions for knowledge.

4.2.1   Strategic directions for knowledge

It is management's role to ensure that there are the technological knowledge and capabilities to fulfil the company's overall innovation strategy and to implement product development strategies for the company. It is important to understand where they have been, where they are at present and where they are going. In the 1990s, there was a spurning of historical knowledge, which ended many times in 'reinventing the wheel'. Today, there is recognition that there is a need to store a reasonable percentage of this knowledge in a codified form for the future, because of the much greater turnover of staff and the loss of tacit knowledge. Total quality management introduced much more recording of production and product quality information. Now improved information systems make it much easier to store and retrieve knowledge of formulations and processing differences. Product formulation is an area where there have been attempts to develop recording systems which can be used in later product development. For example using case-based systems, the records of previous formulations – both successful and unsuccessful – are used as a knowledge source with the product properties and their specifications, which can be retrieved to find a possible formulation for a new product (Rowe, 2000). Over the years, this becomes a valuable source of company explicit knowledge, which can lead also to fundamental knowledge in the specific area of the company. This is taking the tacit knowledge learned by experience and building it into generally available explicit knowledge.

Management needs to ensure that there is the needed knowledge in the company for their product development plans to be carried out. But there is always the question of how much money should be spent on knowledge in the company both in people's minds and in recorded knowledge – how much on people and how much on an information technology system? Then how should this be split between the different stages of the product development project? If one looks at the stages of the product development project and the expenditure of money and man-days (Cooper and Kleinschmidt, 1988) in Fig. 4.3, then we could say that the knowledge created is related to the man-days expended in gaining it. Figure 4.3 shows how the expenditure increases as the project goes to commercialisation, but the proportion of man-days spent was greatest in the product design and testing stage. There is a large capital expenditure in the later stages of the PD Process at the latest stages, but it is interesting that in Cooper and Kleinschmidt's study there was not a related increase in knowledge as epitomised in the time spent by people in the project. Management needs to study the pattern of knowledge creation by people in product development and decide if it is optimal. Management has also to ensure there is sufficient communication in the company to make full use of the present knowledge in the company.

There is always a need to identify the knowledge needed in the future, both short term and in the long term; there may be a need to create new knowledge in the present product development project. In the incremental innovation strategy, this is building a bank of knowledge for future projects. But when innovation is more discontinuous, maybe to a new product platform or a new processing technology, then there is need for a new knowledge base. This can be a difficult

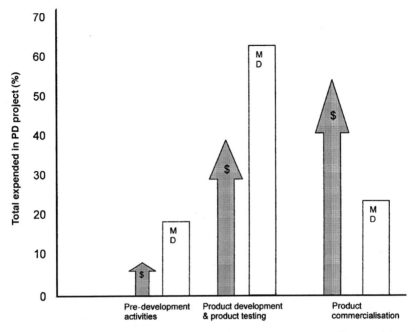

Fig. 4.3   Man-days (MD) and mean expenditures spent in PD Process (Source: Adapted from Cooper and Kleinschmidt, 1988).

and indeed impossible task if management has not been planning ahead. It may be impossible from the present knowledge level in the company, and therefore food companies often fall back on capital investment, buying equipment and knowledge from the equipment manufacturing company.

The management needs a strategic knowledge policy for the company that identifies the knowledge areas and also the dissemination of the knowledge within the company. The communications policy must ensure that the knowledge is not embedded in departments, but can be made available and integrated by the product development team. This again emphasises the need for product development to be integrated throughout the functional areas in the company. The basis for all knowledge is people, and management has to see that people with the necessary knowledge, skills and capabilities are in the product development team, and that they are able to create knowledge in the project as it is needed and communicate this knowledge for future projects. Is management transmitting this to the human resources group and is it prepared to employ people with the necessary skills, and reward them for their skills? Tissen (1999) suggested that these creative, innovative people need to be thought of as highly as soccer players, with high transfer fees and high salaries. Management also needs to provide the information system that selects, collects, integrates and analyses information and also has an interface with the product developers that leads to efficient recovery of the specific information. This is far beyond the information system in many food companies at the present time, but companies should be aiming for it. It is a significant factor that can make product development more effective and efficient.

Another factor that management needs to consider is the direction for the company's knowledge. This grows from the base of present knowledge, which may lead to skewed directions in building up the future knowledge. For example the company can be directed by:

- craftsmen and rely on tacit craft knowledge, knowledge which is based on doing and remembering;
- accountants and rely on financial knowledge;
- engineers and rely on scientifically analysed practical knowledge;
- marketers and rely on social/personal interactions in a marketing situation;
- scientists and rely on scientific logic and method.

There are always several forms of knowledge in the company, but the dominant knowledge gives the direction to the company, and the other knowledge follows it

---

Box 4.1   Formation of a product strategy

1. Develop sophisticated scenarios for the competitive environment of today and the future.
2. For each of the scenarios, describe the ideal successful companies within the scenario and their attributes, in particular the advantaged and base knowledge incorporated in their products/services and throughout their value chains (advantaged knowledge – knowledge that does or can provide competitive advantage; base knowledge – knowledge internal to a business that may provide short-term advantage, e.g. best practices). Also decide on the depth of each knowledge that is needed.
3. Determine who are the current and potential future knowledge leaders in developing and applying the advantaged and base knowledge elements identified. For your company, identify the specific internal individuals who possess the knowledge. Outside the company, it is best to specify institutions/companies and even individuals who possess the knowledge.
4. Decide where the ideal company should source its knowledge, both internal and external to the company. Decide on the depth of a particular knowledge that should be inside the company and the source for the extra knowledge needed.
5. Choose the ideal company to model your company upon, and develop the business strategy and routes to reach that ideal. Plan how you are going to acquire and maintain people with the necessary knowledge.
6. Establish the effects on shareholder value of the particular area of advantaged knowledge.

Source: From Clarke, 1998, by permission of Research Technology Management.

and often is at a much lower level. This is seldom realised by the directors who sit on Boards of companies and give the knowledge direction that is followed by the executive directors and then the rest of the company. It is important that a wider knowledge direction is recognised and set for the company. The company needs to develop a knowledge strategy as shown in Box 4.1.

Think break

Select four food companies, two manufacturing consumer foods and two processing food ingredients for food manufacturers, with which you are familiar.

1. Decide what are the overall directions for the companies, and then decide what are the major and minor knowledge areas in the company.
2. What are the companies' most important innovations during the last five years?
3. How do these innovations relate to their knowledge areas?
4. Now looking at your own company, identify the different knowledge areas and discuss firstly the importance of each and then the use of them. Scale the importance and use on the following scales:

Not important _____ Very important
Seldom used _____ Always used

5. Identify the future innovation directions of the company and decide what knowledge will be needed for these future innovations.

4.2.2   Knowledge systems

There is a need to select a system for knowledge, but what is it to be? The first general concept is a combination of the traditional and the new; but the short answer is that the Western ideology of knowledge may prevent this. One of the knowledge bases for processing technology is science; however, Western science as well as not appreciating technology even finds it hard to tolerate technology that it can neither comprehend nor appropriate (Marglin, 1996). This has presented problems in food knowledge because it has led to definitions of food science and food technology as being different, with one thought of as superior knowledge to the other. This is quite basic to Western thought, with ideas of *episteme* and *techne*:

*   *episteme* is knowledge based on logical deduction from self-evident first principles;
*   *techne* reveals itself only through practice, its theory being implicit and usually available to practitioners.

Techne is embodied as well as embedded in a local social, cultural and historical context (Apfell-Marglin, 1996). *Techne* knowledge is geared to creation and discovery rather than to verification; it recognises a variety of avenues to knowledge; the test of knowledge is practical efficacy. This knowledge split between *episteme* and *techne* was epitomised in food industry knowledge by the craftsman and the food scientist.

But now, knowledge and action are increasingly based on a combination, a synthesis between *episteme* and *techne*. In the food industry it will be the ability to synthesise a method of product development that combines logical thought with action in building knowledge, so that greater knowledge develops and therefore more advanced products. As Apfell-Marglin (1996) noted 'a particular system has its own theory of knowledge, its rules for acquiring and sharing knowledge, its own distinctive ways for changing the content of what counts as knowledge; and finally its own rules of governance, both among insiders and between insiders and outsiders.' Food industry management can do this by making strong access links into the universities and the research centres, and at the same time providing an atmosphere and organisation to create new knowledge. This again needs the adoption of total technology as a basis for company management and in particular innovation management. The dominance of one function has led to a lack of true development in the food industry.

- The domination of the financial knowledge system led to cost cutting, staff redundancy and mergers, which in the end decreased the total knowledge in the company and the industry.
- The domination of the marketing knowledge system led to deterioration in technical ability and plant.
- The domination of the production knowledge system led to deterioration in the consumer/product relationship and loss of competitive strengths.

For successful product innovation, there is a need today for a knowledge system which integrates and does not allow domination; which accepts and uses the logical thought and principles of science but actively creates knowledge by venturing into unknown futures. Product development is a process that is built on this type of knowledge system as shown in Fig. 4.4.

The knowledge capabilities in product development are related to all the functions in the company, R&D, intellectual property, engineering, purchasing, quality assurance, rapid testing of the product, distribution system, personnel, environmental relations, and so on. Everything and everyone need to be included as shown in Box 4.2. This is an example of both collecting the company's information and of using it to develop new knowledge, new products and new restaurants.

The knowledge system relies mainly on three human factors: cognitive understandings, learned skills and deeply held beliefs of individuals (Quinn, 1992). Quinn chose the term cognitive understanding instead of knowledge to emphasise that what is needed is a perceptive and understanding knowledge. The PD team needs the know-how for an activity, and also needs the skills to perform the activity. But if there is a lack of self-belief, will or motivation to succeed, then the activity may be completed at a lower level and in a longer time. The company has to have the know-how to solve the product development problems, the skills to use

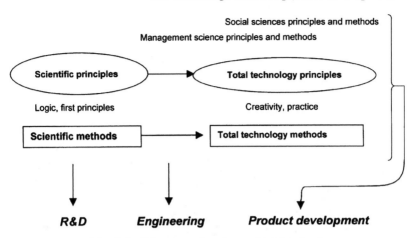

Fig. 4.4   Science, engineering and total technology.

this know-how and develop the commercial product, and the belief in the product that motivates them to lead the project to product success. Bringing the three together in people leads to outstanding product designers, process engineers, marketers, production staff and financial experts. Bringing the three together in the company Board and in management leads to an outstanding company. The knowledge system for product development depends less on providing capital and physical resources (although they are still needed) than on finding and educating people to develop the knowledge, skills and attitudes they need for product development in general and for specific tasks in product development.

The knowledge system also needs to share knowledge, and to provide structures such as teams to encourage this sharing. Knowledge grows when shared, some people would say exponentially. The company knowledge base increases with time, the next project starting from a higher knowledge base than the previous project. Sharing is an excellent way to create knowledge; people with different knowledge and skills, talking, interacting and working together rub ideas off each other so that original ideas form. People with specialised knowledge need to be educated to share their knowledge with other people, so that they can increase their own knowledge as well as blend in with other specialist knowledge in the company. One of the great hurdles to knowledge growth is knowledge snobbery, one type of people thinking they are superior to others. In a commercial company, which has to deliver successful technology – products, markets and production – it can be a complete disaster. The aim for success is interwoven, forward-looking knowledge. It must be realised that there is a certain limit to the amount of knowledge that people can carry; some can work only in one area, others may manage two areas, the outstanding people three or four. Information overload can swamp people. But everyone can integrate knowledge, if it is in a basic form without speciality details and jargon.

It is important to identify what are the key knowledge areas to have in the company and concentrate on them. Knowledge can be bought from outside to

---

Box 4.2   Integration in food service development

General Mills Restaurant Group (GMR) approaches technology in the broadest possible terms.

At the strategic level

It uses its databases with conceptual mapping techniques to define precise unserved needs in the away-from-home eating market. GMR's technologies can determine not just whether people want Italian food, but whether they want Italian fast food, dinner-house, mid-priced or up-market; combinations of foods, prices and values.

- Concept development with inside and external chefs and restaurateurs.
- Concept evaluation with models from databases to select type and situation of outlet.
- Optimum sites and architectural designs using 'other' technologies.
- Optimise site development and construction using PERT (performance evaluation and review technique) and other operations research tools.

At the operations level

By mixing and matching very detailed performance data from its own operations and laboratory analyses, GMR can select the best individual pieces and combinations of kitchen equipment to use in light of investment considerations, performance characteristics, operating costs, repair needs, flexibility for different menus, systems fit with other pieces of equipment.

- Facility layout using own experience and data.
- Equipment design using its laboratory data and equipment manufacturers.
- Raw material sources and availability identified from databases and satellite earth-sensing systems.
- Raw material preparation and handling with suppliers, for maximum market value and minimum cost.
- Restaurants functioning with integrated electronic point-of-sale and operations management system directly connected to headquarter's computers. Satisfaction tracking surveys with customers.
- Monitoring and analysis of quality, sales and operations.

Source: After Quinn, 1992.

---

fill the gaps; either by employing new people with the knowledge, or contracting out to consultants and other companies. The choice depends on the long-term future plans of the company. If the problem is not likely to be met again, at least far into the future, then it is the time to bring in the consultant; if it is going to be a major area for the company, then it is more efficient to bring the knowledge into the company. At all times, there must be sufficient integrated base knowledge inside the company to understand the knowledge needs and to make the decisions on where the necessary knowledge can be found.

Think break

1. Cognitive understanding (knowledge), learned skills and deeply held beliefs of individuals are identified as important for success in product development. For each of the four stages of the product development process, identify what you think are the most important cognitive understandings, learned skills and deeply held beliefs.
2. Knowledge sharing is important for growing knowledge in a company. Identify areas in product development where knowledge is shared in your company and areas where it is not shared. How could you extend knowledge sharing in your company?

4.3   Necessary knowledge for product development

To change the product idea concept into a new product, knowledge of the raw materials, processing, product qualities, consumer/product reactions, marketing and the general environment is needed, as shown in Fig. 4.5. These knowledge areas are all interacting. For example, processing knowledge affects the knowledge of raw materials; if low temperature drying were chosen, the microbiological quality of the raw materials must be carefully controlled to ensure safety in the product. So it is not a case of seeking knowledge specifically in one area, but interacting this with knowledge in another area. The consumer may wish to have the liquid in a bottle, but only cartons can be used in the processing line, so one has to discover how the consumer reacts to a carton and how they would accept cartons. The descriptions of

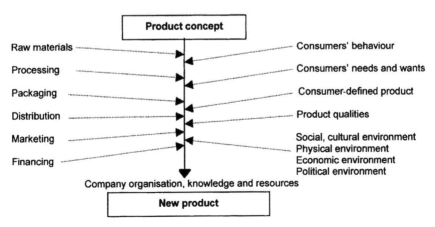

Fig. 4.5   Knowledge for conversion of product concept to new product.

the product by the consumer have to be changed into quantitative terms in the product qualities; so there is need for knowledge of the consumer's needs and wants on product attributes and also the methods of measuring these attributes. Central to the activities are the two important areas of the consumer and the technology, but there also needs to be knowledge of the environment. Knowledge of the consumer is detailed in Chapter 5.

4.3.1   Technological knowledge

The general knowledge areas important in technology are (Gawith, 1999):

* knowledge of science, mathematics, social sciences;
* knowledge of techniques, testing, modelling, interviewing, manipulating tools, materials and data;
* knowledge of procedures and processes;
* knowledge of generic concepts and ways of thinking.

In product development they can be grouped under products, raw materials, processing, packaging, distribution and marketing as shown in Table 4.3. This shows the wide variety of knowledge that is needed in bringing the product from the concept to the actual product.

For example, consider the development of a protein food. Consumers want a high-protein food, but what does that mean in percentage of protein? They want a red colour but what is that measured on a colorimeter? They want a crisp texture but what is that measured on a texture meter? If the protein content is to be 15%, then it is necessary to know the protein content of the raw materials; if the colour has to be a certain red, then the red pigment in the raw materials needs to be identified and measured. There may also be a need for a certain protein; in breadmaking, there is a minimum amount of wheat gluten to give the structure of the leavened bread; in sausage making, only a certain amount of offal can be used because of its poor water-holding capacity. So the type of protein, the quantity and sometimes the amino acid composition need to be specified in the product and the raw materials. Different processing conditions will denature the protein to different extents; limits are set on the processing variables so that the product has the desired nutritional properties. Browning, the combination of amino acids and simple carbohydrates, decreases the value of the protein so the packaging needs to stop absorption of water and also there need to be limits set on the storage conditions of temperature and humidity. If the product has achieved a certain nutritional protein value, then this knowledge is supplied to the consumer in promotion and public relations. So finally the consumer receives the product, but needs to know how to handle it so that the final food eaten has the protein nutritional effect that the consumers desired.

This example gives some idea of the knowledge from many disciplines, which has to be integrated in product development. If there are many specialists from different areas, the problem is how to combine their knowledge throughout the project. If there are not many people in the company, the problem is how to fill the gaps in the knowledge.

Table 4.3 Types of technological knowledge in product development

Product qualities
Properties: appearance, size, shape, sensory; nutritional, compositional
Use: safety, ergonomics, preparation and serving, eating
Product limits: legal, price

Raw materials
Properties: type, production method, chemical composition, traces of pesticides and herbicides, toxicity, nutritional composition, sensory and physical properties, microbiological counts
Price: price range, relationship of price to quality
Raw material limits: caused by processing needs, product structure needs, other product properties, quantity available; minimum and maximum needed in the product, effect of processing on the raw material, legal limits on use

Processing
Unit operations: heating, pasteurisation, sterilisation, freezing, chilling, drying, mixing, tumbling, pumping, conveying, packing
Unit processes: gelatinisation, hydrolysis, browning, denaturation, oxidation, death of microorganisms, growth of microorganisms, vitamin destruction
Processing variables: temperature, water activity, atmosphere, time
Costs: raw materials, processing, factory, distribution, marketing and administration
Processing limits: temperature range, rate of increase/decrease in temperature, viscosity range, mixing speed range, basic equipment design

Packaging
Packaging materials: film, cardboard, metal, glass
Packaging type: bottle, carton, pottle, can, sachet
Packaging method: hand, continuous, automatic, aseptic
Packaging limits: shelf life, protection

Distribution
Transport: roads, rail, sea, air
Transport conditions: time, temperature, humidity, vibration, handling, costs
Storage: ambient, chilled, frozen, atmosphere controlled
Storage conditions: time, temperature, humidity, atmosphere, handling, costs
Distribution limits: shelf life; protection from contamination, breakage, deterioration; available transport and storage; timing of transport; costs

Marketing
Market channel: product flow through market channel, people and organisations (retail outlets, wholesalers, agents, ingredient suppliers)
Market channel requirements: size, weight, availability, price, display and information
Promotion: media advertising, public relations, in-store promotions, free samples, competitions
Promotion needs: create awareness, encourage to buy, education, creating a product image
Pricing: customers' product value, costs, price range, price discounts, competitive pricing
Marketing limits: channel availability, channel controls, competitive actions, promotion availability and costs, customer needs and attitudes, legal controls on marketing.

Think break

1. For the product design and process development of peanut butter slices in a vacuum pack, similar to cheese slices, decide on the knowledge needed for product qualities, raw materials, processing and packaging. How much of this knowledge is already available, and what would need to be created in the product development project?
2. Three different varieties of mangoes have been selected for export to Japan from Thailand (see Section 7.2). What knowledge would be needed for packaging, distributing and marketing in Japan? How would you conduct the development to either find or create this knowledge?
3. Packaging is an area where there has to be close collaboration of food manufacturer and packaging company. It is recognised that for the food manufacturer (Belcher, 1999):
    (a) purchasing wants easily exchangeable packaging that can be quickly delivered by several companies,
    (b) marketing wants a package that is a communication vehicle and that can convey an image,
    (c) legal department wants a package that definitely protects the product from contamination, and is deemed a safe product for the consumer so as not to incur any liability on the company,
    (d) the food technologist is interested in the package for what it can do to protect the product quality, safety and can enhance the new technologies that are being employed.

Discuss how the packaging company can deliver this wide variety of knowledge to the manufacturer and create new knowledge with the manufacturer.

4.3.2   Knowledge of the environment

Knowledge of the social, cultural, physical, economic and political environments is very important in product development, but is a massive task even in the largest companies who can have staff devoting their time to it. Fortunately many of the social and cultural changes are slow moving so there is time to predict where they are going. But of course political changes can be quite fast, especially where revolutions occur. Where does information come from – people, media, magazines, reports, journals and the Internet? How are we to convert it to useful knowledge in product development?

Information in many of these environmental changes does spread around the world quickly because of modern communication, but often it is in the 'communication bites' beloved of the media, which do not tell the whole story and in some cases distort the information. The large companies have staff placed in close proximity to parliaments and politicians so that they are not only actively looking for information but are indeed influencing the trends. But this is often a long way from the person managing product development on a daily basis, so one can find two opposing directions in the same company. For example, the product developer with close customer and retailer relationships in Europe may be designing natural, organic products, while the political lobbying staff may be trying to influence politicians to accept genetic engineering.

There are many economic reports and physical climatic reports around the world; it is not difficult to find information of possible increases in economic status of peoples, which will lead to a different food choice, and on effect of climate change on food choice – less hot soup and more ice cream. Economic change is now occurring in China and it is not difficult for the food companies to predict the food changes and the possible products that can be developed for this market. There are many predictions of climate changes, which may not be precisely reliable on timing, but still give the direction of change.

So how is this information on environmental changes developed into knowledge in product development? Many of the overall changes are incorporated in developing the business strategy, innovation strategy and therefore the product development strategy. So these overall changes are incorporated into the product development planning. But there are specific effects on the product development project and these must be carefully noted in planning the important decisions at each stage, so that the team can find the knowledge to meet these decisions. There need to be people in product development teams who are outward-looking and aware of what is happening in the environment and have the ability to bring this into the product development programme and projects.

4.3.3   Sources of knowledge
In product development, there is a continuous development of knowledge:

past knowledge → present knowledge → future knowledge

There is recognition of the past knowledge which needs to be kept either in a company's memory or in its databases – it is unrewarding to keep 're-inventing the wheel'. One must not cling to the old knowledge as sacrosanct, but as a building base for the present and the future. At the beginning of the product development project, one is judging the knowledge available at present and assessing this against the knowledge required throughout the project.

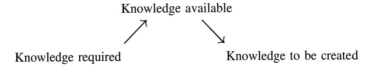

The **knowledge already inside the company** is the first type to consider. Basically this is inside the heads of individuals, but together it becomes the total company knowledge if the knowledge is shared. In every company, there are individuals and groups who may not share knowledge easily so that a system has to be set up to ensure sharing of the combined company knowledge for a project. In incremental product development, as much as 90% of the knowledge required is already available inside the company if there is knowledge sharing. In product innovation, the present knowledge may be as little as 40% of the knowledge required – if any less, then warning bells should sound for the project.

Knowledge can be created from information inside the company – in the files, databases, library and information system. An important information source within the company is staff personal records; people often record detailed information, which is condensed in reports or even not reported. Some information may not have been significant in one project, but is in a later project. Some companies try to collect this into a central computer system, but even then it may be difficult to retrieve the information without the individual's interpretation of it. Within the company it is always difficult to balance the costs and usefulness of stored information. Between the two extremes of 'wiping the slate clean' and 'paralysis by information and analysis', there is a balance for every company that depends on the knowledge level of their staff and the costs of storing information. One persistent problem is to ensure systems are in place to record all significant data and events.

The company's capabilities or expertise can be described as a combination of the company knowledge, the company skills and the availability and relevance of the company information, as shown in Fig. 4.6. The collective body of

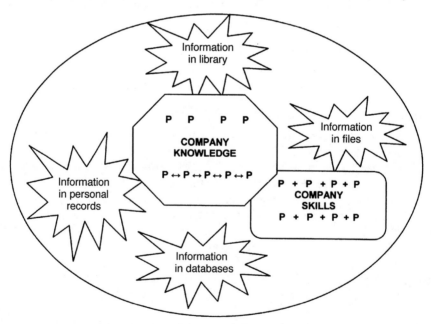

P is personal knowledge, in people's heads; and personal skills.

Fig. 4.6   Knowledge and information in the company.

company knowledge (core memory) plus the company's skills (core competencies), plus the information system within the company combine to form the company's core capabilities. The knowledge can be the separate knowledge of people (P), but much more important is interactive knowledge between people (P ↔ P ↔ P ↔ P). In the case of company skills it is an additive effect of individual skills (P + P + P + P); in the company, people always have to work together and must not work separately. The information system may be separate sources or combined on a network in an information system.

Fleck (1998) identified expertise as knowledge (philosophy, technical specialities), power (sociology) and tradability (economics). Power is part of the organisational structure, within the company but also in professional and other organisations, and knowledge is seen as embodying social relations within which power is mediated and reproduced. There is no question that power in the company can influence the direction of knowledge in product development, and also the resources for that knowledge, whether the power is held in finance, marketing, production or top management. Tradability is determining if a particular knowledge is better at carrying out a task than a competing knowledge. This is often an argument in industry against using more complex knowledge – will the product development be more efficient and effective or just more costly? Power and tradability emphasise that expertise is not 'pure' in a company; the selection of knowledge is influenced by the people with and without power inside and outside the company, and its marketability both internally and externally. Technological knowledge is often influenced by both of these factors. Communication is also important for the company's capabilities (Court, 1997), and this has been shown in Fig. 4.6 by interrelating the people P ↔ P ↔ P, but individuals, P, have also been shown who are in the company's capabilities but are not interrelating.

It is important to differentiate between information and knowledge, although there are certainly grey areas where they mix. One can consider them as weighted:

Data = d
Information = d × c = data × content
Knowledge = d × c × e = information x experience

Databases have been shown as a particular type of information. There is a grey area between databases and information, but data are usually considered the raw facts, information is data that have been worked on to give a meaning that can be understood (Court, 1997). Usually databases are the lists of product sales figures or the demographic information on population. There needs to be recognition that information only becomes active knowledge in product development if it is linked either with past experience or experience in the present PD project.

Sources external to the company are important for knowledge not available in the company. Depending on the state of the company's core capabilities, it is important that necessary additional knowledge and skills are transferred from external sources to the company. In particular the knowledge of customers and suppliers provides important sources, where there is close contact between the company and the outside sources of knowledge. There is knowledge and information outside the company that can be used to fill the gaps in the knowledge for specific projects. This is illustrated in Fig. 4.7.

The place of consultants in the knowledge/information spectrum is varied; there can certainly be skills transferred, but unless the consultant is intertwined in the company, only information can be transferred and not knowledge. This is similar for universities and research institutes; skills and information can be transferred, but knowledge can only be directly used if it is relevant specifically to the company. Company staff and company management need to go outside the company to identify new tacit knowledge from consumers, customers, or scientific and technological centres. Many ideas can come from outside the company but there needs to be the internal tacit knowledge to recognise them and relate them to the company product development (Lenzner and Johnson, 1997). It is not enough to read the information coming into the company, it may be out of date and not easily related to the company, and so its effective incorporation has to be organised and ensured.

4.3.4   Sources of information
The sources of information are both internal and external to the company. They can be grouped as tacit, mix of tacit and explicit, and explicit sources.

- Tacit – company staff, personal experience.
- Mix of tacit and explicit – business consultants, customers, exhibitions/trade material/conferences, family and friends, government agencies, other companies and competitors, suppliers/sales representatives, trade associations/professional bodies.
- Explicit sources – in-house databases/reports, information brokers, libraries, media, on-line sources, patent information, trade journals.

In the Italian industry, Evangelista (1999) showed that for technological information, the internal departments were the most important channels for information into the manufacturing companies as shown in Table 4.4. Internal sources were not as important in the service companies as in manufacturing. Among the external sources for information, clients, customers, suppliers of equipment, materials and components were the most common sources. Information flowed from both the upstream and downstream user/supplier interactions. Consultants were more important in the service industries than in the manufacturing companies. Other sources – universities and higher educational institutes, private research institutes, public research institutes, agencies for technological transfer, patents, licences and other external sources – were very important to less than 5% of the companies.

Campbell (1999) also found in New Zealand that customers and company staff were the important and most used sources for information. The heavily used sources were personal experience, customers, company staff and in-house sources; the moderately used sources were exhibitions/conferences, other

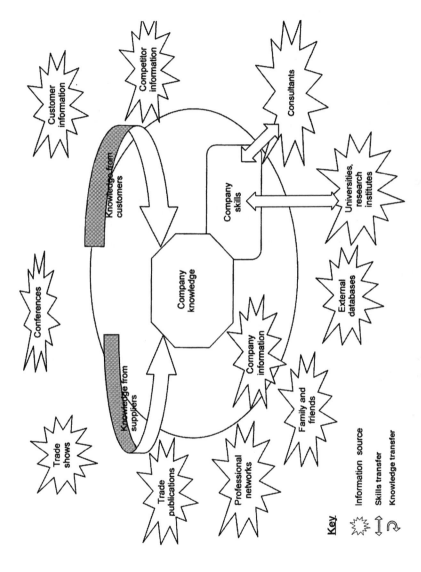

Fig. 4.7 Knowledge and information from outside the company.

Table 4.4　Sources of technological information in manufacturing and services

Sources	Innovating firms for which the source is very important	
	Manufacturing % of total (rank)	Services % of total (rank)
Internal sources (Production/delivery, R&D, marketing department)	63 (1)	37 (1)
External sources		
Clients or customers	44 (2)	34 (2)
Suppliers of equipment and components	36 (3)	30 (3)
Fairs and exhibitions	33 (4)	14 (7)
Competitors	23 (5)	21 (5)
Consultancy firms	15 (6)	27 (4)
Conferences, seminars, spec. journals, etc.	13 (7)	17 (6)

Source: From Evangelista, 1999, by permission of Rinaldo Evangelista and Edward Elgar Publishing Ltd.

companies, suppliers and trade journals. Campbell also enquired if the sources were tacit or explicit. Of the heavily used sources, personal experience and company staff were tacit, customers were tacit and explicit, in-house sources were explicit. The moderately used sources, which were external to the company, were a mixture of tacit and explicit knowledge. Most of the infrequently used sources were explicit, but some professional bodies, business consultants and government agencies provided a tacit component.

Campbell found that highly innovative companies used information more than the least innovative companies, as shown in Table 4.5. The sources where there was no real difference between the three groups were professional bodies, media, trade journals, information brokers and in-house sources; apart from trade journals and in-house sources, these were infrequently used. Overall the highly innovative companies used a greater range of information sources in relation to their product development activities. The highly innovative and moderately innovative companies made use of both formal and informal acquisition methods, the least innovative companies were more likely to gather information informally. The moderately innovative companies tended to use more formal information acquisition methods than the highly innovative companies, both internally and externally.

In looking at the stages in product development, Campbell found differences between the three stages: pre-development analysis, product design and testing, product commercialisation as shown in Table 4.6. There was a surprisingly low use of external information sources. In the pre-development stage, only personal experience for initial screening, and customers for preliminary market analysis, were used by over 80% of the companies. Customers in initial screening and detailed market research, and personal experience in financial feasibility, were

Table 4.5 Information usage for product development

Source	Companies					
	Highly innovative		Moderately innovative		Least innovative	
	Use	Importance	Use	Importance	Use	Importance
Customers	4.3*	4.7†	3.7	4.9	3.9	4.3
Company staff	4.0	4.2	3.5	4.1	3.3	3.5
Suppliers	3.5	3.6	2.5	3.3	2.5	3.0
Exhibitions/ conferences	3.3	3.6	3.0	3.5	2.4	2.6
Other companies	3.2	3.8	3.2	3.7	2.2	2.5
Business consultants	2.6	2.5	2.5	3.0	1.7	2.6
Family and friends	2.4	2.4	2.0	2.4	1.9	2.2
Libraries	2.0	2.2	2.3	2.8	1.7	2.1
Govt agencies	1.9	2.5	1.9	3.1	1.5	2.2
Patent information	1.8	2.1	2.6	3.6	1.9	2.0

* Use scale 1 = not at all to 5 = all the time
† Importance scale 1 = not important to 5 = vitally important.

Source: From Campbell, 1999.

used by over 70% of the companies. In product design and testing, only personal experience in prototype design and detailed design, company staff in trial production, and customers in test marketing, were used by over 70% of the companies. In product commercialisation, the use of information was the lowest of the three stages. Only company staff and customers in production start-up, and customers in market launch, were used by over 70% of the companies. Overall in product development, highly tacit information transfer was used. Only customers'

Table 4.6 Information sources in three stages of product development

Sources	Percentage of companies		
	Pre-development analysis	Product design and testing	Product commercialisation
Personal experience	64	57	44
Customers	57	46	42
Company staff	44	57	48
In-house sources	43	38	32
Other companies	30	–	–
Suppliers	28	24	–
Business consultants	19	–	–
Exhibition/conference	–	–	17

Source: From Campbell, 1999.

information incorporated an explicit content. New Zealand companies are small by international standards, so some of these uses of information may not be true for large multinational companies. But in product development, there does appear to be a strong reliance on tacit information and less on explicit information.

Think break

1. List the information sources used by your company in product development. Which sources give you vital information, useful information, interesting information, useless information? Would you drop some of these information sources? Do you need to include other information sources?
2. Because of the increasing volume of information, companies have set up systems based on information technology to receive, store and distribute information to the various functional departments, R&D, engineering and product development (Graef, 1998). Describe the important knowledge areas to be included in an information system for product development. What are the important factors to be considered in building an information system as a basis for effective and efficient product development?

4.4   Tacit knowledge in product development

Tacit knowledge is the opposite of explicit knowledge, which is knowledge that can be expressed in words or numbers, in a formal, systematic language, and easily stored and communicated. Tacit knowledge is much more difficult to define (Madhaven and Grover, 1998). It is essentially personal knowledge and therefore is communicated person to person or within a group of people (Nonaka *et al.*, 1996). Tacit knowledge in the product development team and the company is a combination and reinforcement of the many individuals' tacit knowledge caused by their interactions. Madhaven and Grover (1998) called this *embedded knowledge* in product development. Evangelista (1999) combined the tacit and explicit knowledge in the group and called it *disembodied technology*. The results of using disembodied knowledge is *embodied knowledge*, in this case the product and the other outcomes of the product development. So knowledge is brought into the product development project, knowledge is generated as the project progresses, and finally the knowledge passes from the project as a new product, production specifications, marketing strategy, financial predictions. This knowledge is not only used in the production and marketing of the product, but can be stored for future projects – either as tacit knowledge in people's heads or preferably stored as an explicit knowledge base.

Often tacit knowledge is specific to a context or area; for example some product developers have tacit knowledge of consumers, products and processes, but may have no tacit knowledge of product testing techniques or of process reactions and would need to rely on explicit knowledge in books and manuals. An important tacit knowledge is the understanding of the defining company situation: where the company is, what it wants to achieve and what are its restrictions/limits on product development, and how does it want to achieve product success.

4.4.1   Individual knowledge in product development

In the product development project, the team therefore is relying on its tacit knowledge augmented with explicit knowledge to provide the required knowledge for the project. Sometimes the project starts with only the tacit knowledge of the team members and further knowledge is generated in the project. This often happens in the small company which lacks information sources, but the larger companies can also do this if there is a perceived need for a fast start to development. Under these circumstances, product success is very dependent on the tacit knowledge and the creative ability of individuals.

Product development is dependent on the dynamics of people and their networks. The individual in product development has knowledge developed from education and experience, but they are also involved in knowledge networks that may be in professional organisations, industry groups, company staff, customers and suppliers. So the individual is dynamically exchanging knowledge while also building up their own knowledge base. Very often, the relevant knowledge for the product development project is general knowledge in this particular group or even in a wider society (Senker, 1998), and only the 10–20% created is really new knowledge. Companies who have great secrecy barriers are sometimes losing this general knowledge and are making knowledge creation more difficult for themselves.

Very often there are individuals in companies who because of their strong connections with external sources are able to gather information that is both relevant and up to date. They may be thought of as information gatekeepers. With their understanding of internal needs and communication systems, they can translate this information into a form that is useful to the organisation (MacDonald and William, 1993). These information gatekeepers can be very useful in product development. Information gatekeepers can also be important within the company, transferring information between groups and translating the information into knowledge for the project.

It is important that the company supports creative individuals, and also provides them with the environment to create knowledge. Creative people are not always the easiest to manage in the traditional power-down method and organisational structure. In some cultures such as in Scandinavia, they are given an important status and given the resources to work together. Scandinavians often wonder why, in American and Canadian research to improve product development, communication is identified as difficult to achieve; they communicate all the time!

Every person in the company can contribute tacit knowledge to product development. Product development is re-creating the company, in a small way or even dramatically, therefore all staff are involved, as also are the customers. It is important in recruiting not to have closed minds, for example recruiting only from certain universities; the company will have too many people with the same way of thinking and not be able to create and tap the necessary new knowledge. A variety of knowledge, skills and motivations are required. And it needs to be combined with experience. There is a need to have a combination of people who have built up their experience in the company and others who have outside experience in the food industry or even in another industry. People with tacit knowledge from other companies will bring new tacit knowledge into the company, which can revitalise the tacit knowledge in the company.

4.4.2   Using tacit and explicit knowledge in product development

Sometimes tacit knowledge has been defined as knowledge from experience that cannot be explained, but this is not usually true in product development except for very simplistic product design. Product development is essentially problem solving, and therefore basic principles are often combined with the results of experience to find solutions. But essentially product development is a defined process, which can be written down in explicit knowledge; as can many of the activities, techniques and decision making. The tacit knowledge is often used to choose the activities and maybe the techniques that can give the necessary outcomes for the decision making. The skills for the techniques are often explicit, being taught and written down in manuals and textbooks. But they may not rely on a thorough understanding of the scientific principles involved, and some would say that they are therefore tacit knowledge. In the food industry, this is a significant point for discussion. The tacit knowledge of the craftsman who could feel bread dough and say it was at optimum fermentation, led to the tacit knowledge of the technician who tests the dough with an empirical instrument that states it is correct for baking. But is the explicit knowledge on bread dough based on scientific principles? In some ways the consumer and market researchers are further advanced as they are using explicit consumer knowledge based on social science research methods. It is based on statistical analysis and not mathematical models, but is explicit for a certain population. It is the change from tacit knowledge to explicit knowledge, which is important for the future – how far is it necessary to make the change so as to have knowledge capabilities for product development in the future?

Although there is written information on problem solving, this is really one area where tacit knowledge is important. In academic education this tacit knowledge is difficult to achieve, because there is often neither the time nor the resources to allow students to develop problem-solving skills under guidance from people with a great deal of tacit knowledge. Very often one can see the process design and product development projects being dropped or never included in food technology courses. These are two of the areas where problem solving can be taught by experience and advice. Indeed in companies, the acquisition of tacit knowledge to support innovation is a purposive activity of much industrial development, design and testing of prototypes and pilot plant (Senker, 1998). This is illustrated in Fig. 4.8 showing how a barrier that stops an idea moving directly to an innovative product may be overcome by intermediate steps such as making

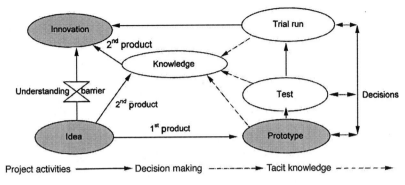

Fig. 4.8   Experience (tacit knowledge) building.

up prototypes and testing them (if needed, with internal recycling), and so establishing sufficient knowledge for implementation of that innovation. When this has been done once or twice new tacit knowledge is created which may enable some or all of the intermediate steps to be cut out in broadly similar innovations.

Technological change has been tacit knowledge-based because it is so much dependent on the knowledge within the company. Especially in incremental changes, 'doing' mostly creates the minor improvements – it is easier to make up the formulations and see if they work than look up scientific information on the processing changes. Innovations are more often based on scientific knowledge, especially in large companies. But they have to be brought into product successes by using the tacit knowledge in combining the product and production, and the product and the consumer. In emerging technologies – biotechnology, advanced engineering ceramics and parallel processing – the knowledge on particular fields is from education and literature, but the tacit knowledge developed in the company, which builds on the formal knowledge, is essential for developing the innovation (Senker, 1998).

The food system uses knowledge for product development in the different parts of the system as shown in Fig. 4.9. In the early stages, it is animal, fish and plant growing or catching, physiology, effects of feeding, nutritional value, sensory properties, uses; this is followed by preservation, cleaning, extraction, treating and packaging. As already noted, the later stages of the food industry divide into two parts – the food processors making food ingredients, which are scale-intensive companies that produce a high proportion of their own process

Production	Raw materials
Agronomy, horticulture, animal production Breeding, physiology, feeding, protecting Properties – nutrition, sensory, uses	**Harvesting, killing, cleaning, grading** Agricultural engineering, animal welfare
Fish catching, fish farming Fish stocks and control, catching methods Breeding, physiology, feeding, protecting Properties – nutrition, sensory, uses	**Size reduction, extraction** Chemical/process engineering **Preservation and packaging** Physiology, microbiology, packaging protection, refrigeration, drying

Fig. 4.9   Technological knowledge areas in food production, raw materials.

technology, and the food manufacturers whose product development is largely directed by their ingredient suppliers (Senker, 1998). The food manufacturing companies continue to use tacit knowledge and skills because this is the only way they can cope with the complexity of food systems using the scientific and technological skills that are available to them. The food processors have acquired the scientific and engineering skills of process engineering and are therefore able to use a greater amount of explicit knowledge. But even they are still using tacit knowledge to analyse and plan their product development. Some of the knowledge and skills in the total food system is shown in Fig. 4.9.

In studying engineering designers' use of knowledge and memory in new product development, Court (1997) found that the most prominently accessed information sources were those based upon locally stored information. The engineering designer's personal experience and knowledge, and in particular memory, were constantly used. In many cases, the designer relied solely on recalling items of information and data from their memory rather than spending a large amount of time searching for it. One-third of information accesses were based on memory usage, with higher figures for many individuals. Knowledge formed within memory is of great importance to the engineering designer.

It is important that in product development people not only have skills and knowledge in specific areas but also, through experience, the knowledge and skills to integrate other areas into their particular activities in product development. Someone may be a product designer but they need to be able to integrate both the consumer needs and production needs into their design. People in the product development team also need to have knowledge of the complete product development process and in particular the decisions to be made and the outcomes needed both from their activities and from the activities of the team as whole. There needs to be a shared understanding of the project and its problems on which the team is working, including a shared common language, and a shared organisational memory. The shared memory can be used to solve the present problems, and will affect the outcomes. But of course the whole team can become rather conservative if it has been together for a number of projects and can see only one way to solve problems. Their effectiveness is then reduced. This is important – teams improve with being together but if they are together too long then their product development becomes less effective. On the other hand, one should not keep changing teams too often, as they do not learn to combine their knowledge and develop group knowledge.

The benefits of shared models of the PD Process and the activities in it are true for all projects. The choice of members of a team and its organisation does depend on the level of innovation of the product. For radical innovation there is a greater need for creativity and often for specialist skills, but there is always a need for a wider knowledge of the different activities in product development. The aims of effectiveness and efficiency are always to be remembered:

- **Effectiveness** in a product development team relates to the degree to which the product meets the targeted need of the customer.

- **Efficiency** is defined as a measure of the resources (including time) used for a given output, often compared with some target or ideal (Madhaven and Grover, 1998).

Think break

1. List your tacit knowledge related to product development at the beginning of your working in product development, and then after every five years.
2. What are the most notable areas where your tacit knowledge has grown?
3. How would you help a young person coming into your product development group to develop their tacit knowledge? In which product development areas do you think they most need to develop tacit knowledge for efficient and effective product development?

4.4.3   Changing tacit knowledge into explicit knowledge

Companies are increasingly trying to change the tacit knowledge held in people's heads into recorded or codified knowledge so that this knowledge is less vulnerable and is not lost to the company. This is made possible by the increasing scientific and technological knowledge in the food industry but also by the availability of simple, cheap computer systems such as expert systems linked into internal and external networks. Protection of intellectual knowledge has also pushed this trend; although in the food industry it is often difficult to receive patent protection except for equipment and agricultural plants, the advent of new types of foods in the future, such as nutriceuticals, could lead to more patenting. The other push to more recording of the tacit information was the introduction of quality assurance and management systems, which included careful recording of processes, products and systems.

There is an increasing amount of scientific literature in food science, which could be the basis for innovation. At the present time, it appears extremely varied and much of it is without the focus needed to build it into a new technology, but this explicit knowledge will gradually filter, often via tacit knowledge, into the food industry. As the present food science is adopted into the industry, we can expect greater support for food research and hopefully rapid advancement in food product and processing knowledge. It is a rather chicken-and-egg situation. With knowledge, models of food processes will be accessible that can be used to develop new processes as well as controlling the present processes.

Knowledge can be embodied in theories, equipment use, people and organisations (Fleck, 1998), as shown in Table 4.7. It is useful to recognise these different types of knowledge in a product development team, both the amount of each embodied in the team and also the amount of each needed in a specific project. In the innovation using new knowledge there is often formal knowledge, which has come from research, which has to be developed to fit with

Table 4.7   Components and contexts of knowledge

Key components of knowledge

Formal knowledge: embodied in codified theories
Instrumentalities: embodied in tool, equipment use
Informal knowledge: embodied in verbal interaction
Contingent knowledge: embodied in a specific context
Tacit knowledge: embodied in people
Meta-knowledge: embodied in the organisation

Contexts of knowledge

Domains: the areas to which the particular expertise applies
Situations: assemblage of components, people, domains, etc., at any particular time
Milieux: the immediate environments in which expertise is exercised

---

Source: After Fleck, 1998.

the other types of knowledge which are all fundamental to product development. The informal knowledge is important in product idea generation. Uniting the various types of knowledge with their contexts is one of the important decisions in product development. The instrumentalities, that is equipment knowledge, are important in building up both product testing and process development; but the actual equipment knowledge depends on the situation – equipment available and the expertise of the people to use it. Equipment knowledge also depends on the place of processing and testing – the laboratory or the factory floor, the small company or the large company.

Knowledge can also be divided (Court, 1997) into:

- general knowledge, gained through everyday experiences and general education;
- domain-specific knowledge gained through study and experience within the specific domain that the designer works in;
- procedural knowledge: gained from experience of how to undertake one's task within the enterprise concerned.

Knowledge in product development is needed for the design of the product, production and marketing, but is also needed on the PD Process. There is a standard framework for the PD Process, and the company needs to define the decisions to be made at specific stages. These actually do not differ very much from project to project, but the outcomes needed may vary and the activities vary with the domain in which the product developers are working. The techniques used will vary according to the situations and the milieux.

The knowledge-based innovation usually has the longest lead time of all product development, not only the time span between the emergence of the new knowledge from research but also on long periods before the new technology turns into products, processes and services in the marketplace (Drucker, 1985). It also is usually based on several different kinds of knowledge, which have to be integrated to produce the new technology and the new product. This is why early products sometimes fail, because the product developers may have some of the knowledge but not enough, for example they may know the scientific knowledge but do not have adequate knowledge of consumer behaviour. Food irradiation is an extreme case of this: there was a great deal of scientific and technological research, before anyone thought about the consumer reactions. In the case of microwave ovens, the food industry had done little research on the effects of microwave heating on food ingredients and food, and given little thought to educating the consumers on how to use microwave ovens with their products. It took some time before the food industry caught up with the innovation from another industry. Other examples of lack of holistic knowledge are early freeze-dried foods, and an early attempt to provide unsaturated fats in the meat from ruminant animals, in particular lamb, to give 'healthier' meat. Early freeze-dried products were specific meats, vegetables and fruit, and only when they were incorporated and marketed as convenience meals did they become accepted. The lamb with higher unsaturated fats had health advantages, but tasted like pork and was not acceptable. So it is a case of having all the knowledge threads at the right time, and the milieu of the company that will take the risk, and the domain with all the features to support the knowledge including cost structures.

Think break

In changing tacit knowledge to explicit knowledge, two important storage systems for knowledge and information in the company are the company library and the computerised information technology system.

1. Compare the two systems, listing the advantages and disadvantages of libraries and information technology systems for storing knowledge collected during product development projects.
2. Is it more useful for future projects to have a project's knowledge on a CD in the library than in the information technology system?
3. One of the problems in knowledge/information storage is the age of the material. For how many years should project material be stored? Should resources be made to collect together the information/knowledge from several projects before this time limit?
4. Design a knowledge/information system that will suit your company's need for internal knowledge/information in product development projects.

4.5   Creating knowledge in product development

Creating knowledge is an integral part of product development, and the ability of the individual and the group to create knowledge is important for both effective and efficient product development.

4.5.1   Creating knowledge in the company

Nonaka *et al.* (1996) suggested that knowledge is created in organisations by a process that amplifies the knowledge created by individuals and crystallises it as part of the knowledge system of the organisation. It is hoped that it would not crystallise into a museum piece but act as a dynamic force to move the knowledge system to both encompass wider multidisciplinary knowledge and more detailed explicit knowledge of the present technology. Knowledge can be created within the product development team, the company, and by interaction with the external environment between individuals and groups. This interaction can be between tacit and explicit knowledge, and can cause conversion of knowledge as shown in Table 4.8.

The whole development from the product description to the product design specifications is a conversion of the tacit knowledge of consumers, market researchers and designers, by the development team to an explicit knowledge at a

Table 4.8   Knowledge conversion

Knowledge conversion	Place in product development	People, group
Tacit to tacit	Brainstorming, focus groups, discussion, concepts comparison	Consumer/designer Consumer/market researcher Designer/market researcher/process developer
Tacit to explicit	Product concept creation, product design specifications, modelling, feasibility reports, evaluation reports, production plan, market strategy	Consumer/designer Designer/process developer Development team/ functional depts/ management
Explicit to explicit	Business strategy/product development strategy, unit operations/new process, measures/testing techniques	Management/ product developers Engineers/ developers Quality assurance/ designer/developer
Explicit to tacit	Raw material specifications/ use in product, basic science/technology, reported experience/new problem, product problem/research for solution, product development project report/tacit model for organising future projects	Supplier/developer Researcher/developer Files/developer Consumer/ designer/developer Reports/product development manager

certain level. The design of the product and the process development then converts this often tacit description to a total product and process that can be described explicitly. Modelling is also a way of taking a tacit description of the product into an explicit description – the model can be verbal, physical, computer-based, or indeed mathematical. Explicit knowledge is exchanged in manuals, reports, papers, expert systems. Product design and production specifications, processing and quality assurance manuals, are commonly used for explicit knowledge transfers. It is important that the business, innovation and project development strategies, PD programmes and the aims and plans for individual projects are explicitly recorded for guidance of product development. Marketing can also be using point-of-sales data to develop their launch plan. The passing from stage to stage is often explicit going to explicit; for example the details of the product prototype going to the product specifications in production. It is important that this explicit knowledge is in a form that can be changed to tacit knowledge in a future project, projecting it with safety and confidence into a new area.

Organisational knowledge creation is a continuous and dynamic interaction between tacit and explicit knowledge (Nonaka *et al.*, 1996), and between individuals and groups and also the company as a whole as shown in Fig. 4.10. In product development, the product development strategy develops from the business strategy, gradually increasing the tacit knowledge to explicit knowledge, until individuals form from this an explicit product strategy. Note that even here is tacit knowledge, which has to be communicated to the product development team. Creating a product strategy involves a community of interacting individuals with different knowledge and skills. They have started with the explicit knowledge in the business strategy, and with group discussions are alternately increasing the tacit knowledge and building up explicit knowledge. In developing the new product strategy, there is a need to use not only the traditional analysis and experience, but also to have the strategic imagination to turn the input into new strategic scenarios (Ellis, 1999). There is a need to look in new directions, develop strategies that are innovative, unexpected, original and effective. At least once a year, board and key executives and selected product developers/designers need to

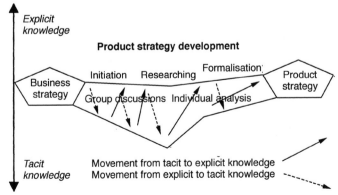

Fig. 4.10   Interaction of tacit and explicit knowledge in product strategy development.

meet to develop ideas for strategies in what Ellis calls serious play. Then management needs to build aggregate project plans, and create the development strategy. To make these activities work, the people charged with their completion need to be educated in the fundamental principles. They also need to enhance their knowledge regularly of the organisation's market position, technologies, production processes, suppliers and competitors (Clark and Wheelwright, 1993).

This process of knowledge building will be repeated in building up the product concept and the product design specifications. This time consumers will be included in the discussion groups with the product development team, marketing staff and technical staff. Product development teams move through these patterns of alternating periods of steady progress in knowledge creation, punctuated by sudden breakthroughs and sometimes changes in direction in the interacting periods. In creating knowledge, individual and group knowledge is interchanged, learning from each other, and then the combined knowledge for the problem is built. The combined knowledge then becomes the company knowledge in the future (Nonaka *et al.*, 1996).

Some of the important factors in using knowledge in innovation identified in the 3M company are shown in Box 4.3. All of these factors are related to knowledge, its communication and the empowerment of staff to use the knowledge. Perhaps the most important is the recognition that product development is a discipline with its own knowledge base. An important company capability in product development is having a thorough understanding of the market by acquiring, disseminating and using market information. But in many companies there are barriers to gaining market understanding, especially of new markets

---

Box 4.3   Important factors in innovation

- Vision for product development which is understood and accepted by management and the product development team, and is also related to the customer.
- Foresight which predicts the customers' articulated and unarticulated needs.
- Stretched goals, setting targets which cause the company to make a quantum leap.
- Empowerment, selecting the right people and then trust them enough to let them have the initiative to work on their own.
- Communications, the free exchange of information, staff understanding that combining and transferring knowledge is as important as the initial innovation.
- Recognition of the importance of innovation as a discipline in all parts of the company.

Source: After Ellis, 1999.

(Adams *et al.*, 1998). In acquiring market information, people focus on either technology design aspects (here is a new product, do you want it?) or business aspects (here is a product, what are predictions of sales revenues?). They ignore product concept development with consumers, identifying target markets and their needs and wants, because the researchers think these are ambiguous. Dissemination of market information is hindered because people focus on their own goals, often departmental instead of the project goals. Cross-functional approaches are needed to give interactive communication so that the market information is incorporated in the product design and also in the development of the marketing strategy. The learning barriers of compartmentalised thinking are overcome by:

- developing common goals that are specific to the product, not to separate departments;
- clarifying each person's role in the product innovation activity so that each knows their part in the larger whole and can help one another;
- learning to appreciate both the contributions from, and the constraints, in the various departments.

In knowledge use, the effort is to try to overcome the inertia to change. People tend to proceed as they always have, maintaining the status quo rather than adjusting actions to capitalise on market learning. Especially with incremental products, it is assumed that the product is just like the present products and there is no need for extensive market research; sometimes the research is done but ignored because it does not fit with preconceived ideas. Managers should enable teams to develop market data. Managers also must help people to extend the usual routines into new practices and promote trust between themselves and the team members, and also within the team. The product developers can make their product familiar to the manager by providing useful information about the product and its market. 'Useful' means that the managers could use the information to follow the development effort and evaluate the product's potential.

An important factor in the product development group is connectivity. People at one time worked in close-knit departments or teams where knowledge would be shared and exchanged routinely and easily. But today there is the problem of not only maintaining contact in a large building but maintaining contact internationally. The product development project team may have no physical contact and often work for different managers, and they may have never met – their only contact is by e-mail (Ellis, 1999). It has been shown that trust in team orientation, that is team members having reciprocal faith in others' intentions and behaviour to work towards team goals, rather than narrow, individual or functional goals or agendas, is essential. As well, trust in team members' competence is important – that they are competent to handle the complex and unknown problems that can occur (Madhaven and Grover, 1998). How does one trust someone through e-mail, far less work cooperatively with them when you have no idea of their knowledge, skills and personality? There needs to be recognition of the team and members do need to meet – not just for the one-day quick meeting but to work together on part of the problem, over several days and weeks. Team members who are able to interact face-to-face will be more effective and efficient at creating new knowledge. Management needs to understand that there are costs in running international product development teams if they are to be effective and efficient – both in having operational, interactive networks and also in having joint working times. Fostering an environment where people share information and knowledge because they know they will get appropriate credit for it, is an extremely important way to create intellectual capital within a company and keep it there.

The company needs to create an environment where individuals are encouraged to preserve and grow their own knowledge, and where they have the mechanism to develop personal relationships so that they share this knowledge with others in an informal interaction. They need to be encouraged to take risks together, and to actively seek knowledge to decrease the risks. The relationships should not be static but should be moving like a kaleidoscope to form new patterns of relationships and new groupings but with basically the same people. People will be lost from product development teams but, if they are properly run, not too often and not the ones who have high knowledge and/or the greatest ability in creating knowledge.

4.5.2   Managing creation of knowledge
In the management of product development Madhaven and Grover (1998) recommended the following:

- Selection of team members with specific knowledge and skills but also an appreciation of other areas of product development from education and experience, and with a shared vision of product development and its procedures. This can be difficult to recognise.
- Selection of the product development manager with multidisciplinary knowledge from education and experience.
- Using a product development process that is used in all similar projects, but could have variations for different levels of innovations and types of products. The decisions and outcomes for each project set out for each stage as well as the project overall.
- Ensuring that the members of the team are familiar with their intended activities, both through experienced team members and well-organised information sources.
- Education of team members on knowledge creation and storage. Also on how to share knowledge and create knowledge by team knowledge sharing and cross-functional development.

Development of people, values and culture in the product development team is very important. Investments in developing knowledge and skills, for technologies, marketing, consumer research and financial analysis, as well as the overall discipline of product development, can be made by employing suitable staff and by educating present staff. As Rouse (1992) said:

Such investments make sense if the people involved have the aptitudes and abilities to gain the knowledge and skills, *and* if they will have opportunities to utilise the newly gained knowledge and skill. Without these prerequisites, well-intended investments in developing people can result in much frustration and not much else.

It is unfortunately fairly common in the food industry for this to happen, and a great deal of talent is lost because people are not allowed to use their knowledge but are tied to a bench doing routine work and not allowed any decision making. Product development is a risk-taking area and people must be allowed to engage together in setting the major decisions and outcomes, and then allowed to make the minor decisions themselves. Again they need to be involved in the discussion and choice of the major activities, and select their own activities and the techniques to be used in them. Techniques especially depend on the knowledge and skills of the people doing the work and if they do not have the major say in choosing the techniques within the constraints of the outcomes needed and the resource constraints, they will have less commitment. This is the way for people to develop their skills and problem-solving abilities. It means that managers have to take risks, because people may fail with poor outcomes or going over the time for the activity. But there are always failures and successes and managers have to increase their own knowledge to reduce chances of failure without reducing people to automatons. Although there are inevitably penalties for shortcomings and failures, it is very important that they be commensurate and not too severe. Managers need to recognise that they need by education to increase their professional knowledge regularly as well as their management skills. A manager needs knowledge across a number of technological areas to lead a product development team successfully.

There is also a need for everyone to recognise the culture and values in the team. If the values are human-centred both in the team and their development of products, then it matters little if a team is laid-back and casual, or conservatively dressed and formal. Different societal cultures outside the team, and indeed the company, affect this aspect of the team. Some cultures encourage communication at the personal level, others do not; and the problem is to give the team itself the values that encourage the sharing of knowledge and the working together. Values have to be realistic reflections of the general society, but they must also encourage effective and efficient product development. The company's values do come down from the Board and the top management, and it may in some cases be difficult to reconcile these with the values of product development. Apart from encouraging people to go to another company with values that are more consistent with product development, what can be done? Company values do change as was seen in the acceptance of total quality management – quality control was thought as only for technicians, until quality assurance and then quality management was developed and sold to management, mostly by outside public relations and sometimes even by government regulation. Product development has to be presented as a discipline and as a system that can produce dividends for the company, and all benefit if both management and the team see it this way.

Technical, organisational and commercial skills and knowledge required for improving product development are shown in Table 4.9. Three groups of abilities are essential for creating product development capability: technical, to achieve product and process integration; organisational, to create the capability of the team; and commercial, to develop effective product concepts and link customer requirements and unmet customer needs to the details of product planning and design (Clark and Wheelwright, 1993).

Some important knowledge seeking and knowledge communication areas in innovative companies (Souder, 1987) are as follows:

- Ability to sense threats and opportunities in a timely fashion, using environmental scanning, technological forecasting and competitive analysis.
- Study of risky opportunities, and accurate assessment of the degree of risk in a project.
- Well-developed project selection systems which effectively communicate the company's needs to the idea generators and foster decisiveness in goal-setting.

Table 4.9 Skill and knowledge requirements for improving development performance

Development participants	Skill/ knowledge requirements		
	Technical	Organisational	Commercial
Senior corporate managers	Understand key technical changes	Recognise importance of creating a rapid learning organisation, lead and provide vision and values	Identify strategic business opportunities
Business unit general managers	Understand depth and breadth of technology	Select and educate leaders, champion cross-functional teams, have career pathing for staff	Target key customer segments, architect product families and generations
Team leaders	Provide breadth of capabilities Comprehend depth requirements	Select, train and lead development team, recognise importance of attitudes and secure functional support	Champion concept definition, competitive positioning
Team members	Use new techniques, apply technologies, develop new technologies	Integrate cross-functional problem solving, create improved development procedures	Operationalise customer-driven concept development, refine concept based on market feedback

Source: Reprinted with permission of the Free Press, a Division of Simon and Schuster, Inc., from *Managing New Product and Process Development* by Kim B. Clark and Steven C. Wheelwright. Copyright © 1993 by Kim B. Clark and Steven C. Wheelwright.

- Interdepartmental debate focused on confronting and resolving conflicts to produce new ideas and a cooperative climate.
- Individuals who play reciprocal roles – persons who generate ideas, who champion these ideas and who link these ideas to the existing organisational goals.
- Organisational structures and climates that foster the development of collaborative roles.
- Long-term commitment to foster technology.

These qualities combined with a willingness of the company to accept change are fundamental to successful new product innovations.

Think break

Technological knowledge is organised and structured in ways that reflect application in product development. The product development team constructs its knowledge around the subsystems in the stages of the product development process. In this way its accumulated tacit and explicit knowledge is organised in the most effective manner for systematic product development and for the activities in each stage (Gawith, 1999).

1. Describe how your company has identified subsystems in your product development processes, and built up knowledge in these subsystems.
2. How do the product development team identify a problem, relate it to past problems and their solutions? Then decide on their method(s) for solving the present problem.
3. How does the team collect together its tacit and explicit knowledge to select the activities and techniques for solving the problem?
4. How does the company management ensure that the whole knowledge system is capable of producing efficient and effective product development?

Overall it is hard to overemphasise the central importance to product development of knowledge and its availability to the individuals and to the team who develop the new products. Some of this knowledge is explicitly written down and codified, but a great deal still lies with the particular people who do the creative work and collectively with their groups. From the viewpoint of the company's continued operation and success, and avoidance of risk from shifting employees, efforts are being made to maximise codification of knowledge. Modern information technology can do much to help with the machinery. Transfer to the record is also helped by the increasing understanding of the knowledge scene and of the philosophical issues on which it rests. But in the long run the knowledge, acquired skills, and powers of analysis and synthesis lying in the individual will always be the key resource. Without it, creativity will stumble, if not founder; with it, will come new products and commercial success relating strongly to the overall skill of the product developers.

4.6   References

ADAMS, M.E., DAY, G.S. & DOUGHERTY, D. (1998) Enhancing new product development performance: an organisational learning perspective. *J. Product Innovation Management*, 15, 403–422.

ALSTON, J.M., PARDEY, P.G. & WALLACE, T.L. (1997) Research policy challenges, in *Government and the Food Industry: Economic and Political Effects of Conflict and Co-operation*, Wallace, L.T. and Schroder, W.R. (Eds) (Boston: Kluwer Academic Publishers).

APFELL-MARGLIN, F. (1996) Introduction: rationality and the world, in *Decolonizing Knowledge: From Development to Dialogue*, Apfell-Marglin, F. and Marglin, S.A. (Eds) (Oxford: Clarendon Press).

BELCHER, J. (1999) Role of packaging in new product development. *IFT Product Development Newsletter*, 5.1, 4.

BETZ, F. (1998) *Managing Technological Innovation: Competitive Advantage from Change* (New York: John Wiley & Sons).

CAMPBELL, H.C. (1999) *Knowledge Creation in New Zealand Manufacturing.* M.Tech. Thesis, Massey University, Palmerston North, New Zealand.

CARDWELL, D. (1994) *The Fontana History of Technology* (London: Fontana Press).

CLARK, K.B. & WHEELWRIGHT, S.C. (1993) *Managing New Product and Process Development* (New York: The Free Press).

CLARKE, P. (1998) Implementing a knowledge strategy for your firm. *Research – Technology Management*, March–April 1998, 28–31.

COOPER, R.G. & KLEINSCHMIDT, E.J. (1988) Resource allocation in the new product process. *Industrial Marketing Management*, 17, 249–262.

COURT, A.W. (1997) The relationship between information and personal knowledge in new product development. *International Journal of Information Management*, 17(2), 123–138.

DRUCKER, P.F. (1985) *Innovation and Entrepreneurship: Principles and Practice* (London: Heinemann).

EARLE, M.D. & EARLE, R.L. (1997) Food industry research and development, in *Government and the Food Industry: Economic and Political Effects of Conflict and Co-operation*, Wallace, L.T. and Schroder, W.R. (Eds) (Boston: Kluwer Academic Publishers).

ELLIS, J. (1999) *Doing Business in the Knowledge Based Economy* (Amsterdam: Pearson Education/Addison Wesley Longman).

EVANGELISTA, R. (1999) *Knowledge and Investment: The Sources of Innovation in Industry* (Cheltenham: Edward Elgar Publishing).

FLECK, J. (1998) Expertise: knowledge, power and tradeability, in *Exploring Expertise: Issues and Perspectives*, Williams, R., Faulkner, W. and Fleck, J. (Eds) (London: Macmillan Press Ltd).

GAWITH, J. (1999) *Total Technology Practice: Preliminary Study for Application in New Zealand Schools.* M.Phil. Thesis, Massey University, Palmerston North, New Zealand.

GRAEF, J.L. (1998) Getting the most from R & D information services. *Industrial Research Technology Management*, July–August 1998, 44–47.

HOOD, L.H., LUNDY, R.J. & JOHNSON, D.C. (1995) New product development: North American ingredient supplier's role. *British Food Journal*, 97(3), 12–17.

LENZNER, R. & JOHNSON, S.S. (1997) Seeing things as they really are. *Forbes*, March 10, 122–128.

MACDONALD, S. & WILLIAM, C. (1993) Beyond the boundary: an information perspective on the role of the gatekeeper in the organisation. *Journal of Product Innovation Management*, 10, 417–427.

MADHAVEN, R. & GROVER, R. (1998) From embedded knowledge to embodied knowledge: new product development as knowledge management. *Journal of Marketing*, 62 (October), 1–12.

MARGLIN, S.A. (1996) Farmers, seedsmen and scientists: systems of agriculture and systems of knowledge, in *Decolonizing Knowledge: From Development to Dialogue*, Apfell-Marglin, F. & Marglin S.A. (Eds) (Oxford: Clarendon Press).

MARTINEZ, M.G. & BURNS, J. (1999) Sources of technological development in the Spanish food and drink industry. A 'supplier-dominated' industry? *Agribusiness*, 15(4), 431–448.

MITCHAM, C. (1994) *Thinking Through Technology* (Chicago: The University of Chicago Press).

NONAKA, I., TAKEUCHI, H. & UMEMOTO, K. (1996) A theory of organizational knowledge creation. *International Journal for Technology Management, Special Publication on Unlearning and Learning*, 11 (7/8), 833–845.

QUINN, J.B. (1992) *Intelligent Enterprise: A Knowledge and Service Based Paradigm for Industry* (New York: The Free Press).

ROUSE, W.B. (1992) *Strategies for Innovation: Creating Successful Products, Systems and Organisations* (New York: John Wiley).

ROWE, R. (2000) The right formula. *Chemistry & Industry*, No.14, 465–467.

SENKER, J. (1998) The contribution of tacit knowledge to innovation, in *Exploring Expertise: Issues and Perspectives*, Williams, R., Faulkner, W. and Fleck, J. (Eds) (London: Macmillan Press Ltd).

SOUDER, W.E. (1987) *Managing New Product Innovations* (Lexington, MA: Lexington Books).

TISSEN, R. (1999) Sharing knowledge does not mean giving away power: it offers a road to success, in *Doing Business in the Knowledge Based Economy*, Ellis, J. (Ed.) (Amsterdam: Addison Wesley Longman).

WALLACE, L.T. & SCHRODER, W.R. (1997) in *Government and the Food Industry: Economic and Political Effects of Conflict and Co-operation*, Wallace, L.T. and Schroder, W.R. (Eds) (Boston: Kluwer Academic Publishers), 107.

5

The consumer in product development

Consumers are the centre of product development in the food industry, directly in the design of consumer products and indirectly in the design of commodity products and industrial products. In industrial product development, the emphasis is on the immediate customer, but consideration needs to be given to the acceptance of the final product by the consumer. It is important in product development to understand basic consumer behaviour and food choice as well as the individual product/consumer relationship (Earle, 1997). Differences among the individual consumers and variations in their environments influence their buying, preparing and eating behaviour. These differences cause variations in food choice and in the degree of acceptance of individual foods. The consumers' total concept of a food is related to their individual characteristics and to the environment in which they buy and eat food.

Consumers are, and will remain, the final arbiters on food product acceptance. The consumer of the new food product gives the ultimate decision on the product development project and therefore it is crucial that the consumer is a major player in critical evaluation throughout the project. But it is even more important to incorporate the consumer in the creative processes in product development. It is wrong to assume that the product designer, the process developer and the marketer know who are the target consumers and what they want and need. The consumers have to identify themselves, and help to create a product that fits into their life styles and also leads them into their desired future.

The consumer needs to be involved in all stages of the product development project. At one time, this seemed an expensive, theoretical and time-consuming activity, and was often ignored in the company. Today, consumer research is more easily coordinated into product development, with the use of modern consumer research techniques to study behaviour and attitudes, develop product concepts and attributes, test product prototypes; and with the use of information technology to set up consumer databases and analyse consumer data.

5.1   Understanding consumer behaviour

Consumer behaviour can be defined as 'those activities directly involved in obtaining, consuming and disposing of products and services, including the decision processes that precede and follow these actions' (Engel *et al.*, 1995). There are environmental influences affecting this behaviour such as ethnicity and culture, social group, regional preferences, as well as food availability and household technology. There are also differences among individuals, not only their age and sex, their education, their standard of living, but also their physiological and psychological make-up. Individuals have their own food choice, which to a greater or lesser extent overrides preferences defined by culture or religion. In the last 30 years, multidisciplinary social science research has increased knowledge of food consumer behaviour and food choice.

Consumer behaviour occurs in sequential stages and at each stage there is a use of knowledge to make decisions. General consumer behaviour has six action stages as shown in Fig. 5.1 (Engel *et al.*, 1995). This sequence can be followed by a further divestment stage where, with food products, the consumer chooses between the options of disposal or recycling of the waste and the packaging. Parallel to these seven consumer actions is the information processing conducted by the consumer. When the consumer recognises the need, there is an internal search in their memory and may be an external search of the supermarket shelves, the menu, and information from other people, media or consumer reports. They may also have been exposed to TV advertising or to promotions in the supermarket; or even to the aroma of bacon sizzling or bread baking in the retail outlet. Engel *et al.* identified five steps in the use of information by the consumer for knowledge building:

1. Exposure to information, communication, the product.
2. Attention given to the information.
3. Comprehension of the information, as it is analysed against the knowledge and the attitudes stored in the memory.
4. Acceptance or rejection of the incoming information.
5. Retention of the new information in the memory as knowledge.

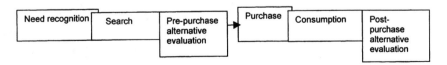

Fig. 5.1   General consumer behaviour in buying and consumption (Source: After Engel *et al.*, 1995).

This knowledge is used to judge the different products; the consumer builds up criteria to judge the products and to compare the different brands and products. These criteria are an important basis for product development. The consumer then decides whether or not to buy the product. After preparing, serving and eating the food there is satisfaction or dissatisfaction, and the decision is made to stay with this product/brand or to look again, as nutrition is a continuing need and the questions for consumers are differential ones related to choice.

Individuals have to be considered in their society: the culture, the social norms, the social structures, and also as part of a group – the family, the home group, the work group, the leisure group. Food eating, even in the case of the individual eating alone, is strongly influenced by other people, indirectly by social influences or directly with the type of foods available. Different cultures and social groups have different values that are recognised in designing products for different markets. Perhaps the reason why American products can be easily accepted internationally is that it is a new country combining many nationalities and food products are designed/promoted so that they are generally accepted by this wide variety of people.

Think break

The core American values have been identified in two textbooks as:

(Engel et al., 1995)	(Peter and Olson, 1999)
Material well being	Achievement and success
Good/bad moralising	Activity
Work more important than play	Efficiency and practicality
Time is money	Progress
Effort and success are related	Material comfort
Mastery over nature	Individualism
Egalitarianism	Freedom
Humanitarianism	External conformity
	Humanitarianism
	Youthfulness
	Fitness and health

1. Study these two versions of American core values, and identify the values that are similar and those that are different. From this develop what you think are your core values. Are your core values different from these lists?
2. Choose two major markets for your company and identify the core values of the consumers in these markets.
3. Compare the core values for the two markets and identify the similarities and differences.

4. For each of these markets set up a core value checklist to be used throughout future product development projects.
5. Discuss how the differences in the core values could lead to different strategies for future product development in the two markets.

When buying food, the consumer usually wants to keep the decision process as simple and quick as possible. Food is consumed two or three times a day, and may be bought every day, so the consumer does not want to spend a great deal of time in buying and today even in consuming food. It is only the special occasion, the special meal or the special food that is given detailed analysis. But it is important for the food designer to recognise that there is detailed and critical thought at certain points of time. Consumers receive a great deal of information on food through the media and advertising as well as by word-of-mouth; gradually and often imperceptibly this information changes their knowledge base and therefore their food behaviour. If there is a great deal of information on saturated fats in the diet, they will consider and may gradually change to low-fat foods. Sometimes their behaviour is changed by a jolt; this could be a food poisoning scare, or it could be food poisoning affecting them directly. These can cause long-term changes in food behaviour, for example reduced meat-eating triggered by reports of BSE ('mad cow disease'), and complete rejection of shellfish caused by a bout of shellfish-related food poisoning.

In the past ten years, food brands in basic food areas have had very similar features and competed mainly on price, so that they have degenerated into commodity products – no brand being distinctive. Also there has been a proliferation of products with very little difference between them – for example in the small New Zealand market, there are about 157 breakfast cereals under the national and retailer brands. Can the consumers differentiate adequately between these, even if they read a consumer magazine? No wonder they choose the easy way out and buy on price, choosing the specials; or keep on buying their familiar product. Foods can be bought on impulse to relieve food boredom or as a treat. A new snack or a new takeaway can be bought to see if it lives up to its promotion; if it satisfies the consumer it can become a regular food.

The food designer needs to be aware, in the target market, of the general consumer behaviour towards foods and eating, and how this is slowly changing with time, but also needs to recognise the sudden change. This can be caused by either new information giving an attitude change, or new foods giving the consumers some greater advantage for safety, nutrition, convenience or attractiveness. Companies that have a long-term relationship with their target consumers build up knowledge about the trends in changes of their behaviour, which is invaluable in product development.

Food consumer behaviour can include the growing of the food but usually in the urban environment it concerns obtaining food from the supermarket, restaurant or takeaway. In the future it may be more distant with food being

Consumer buying sequence   **Consumer thought process**

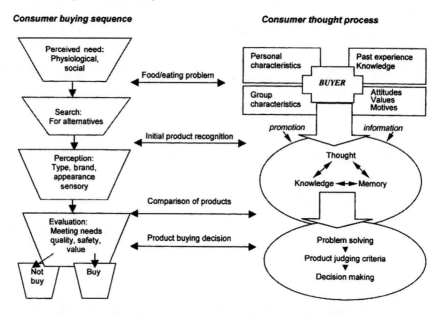

Fig. 5.2 Consumer food purchase patterns.

ordered through the Internet. Food companies must realise that although their food behaviour often starts with the selection of food at the retailer, consumers have an interest in how food is produced and they have their concerns about production and processing. Stages in the consumers' buying sequence and the related consumer thought processes are shown in Fig. 5.2. Recognition of the stimuli that start the buying sequence, the product judging criteria that are used in comparing products during the buying decision, and the level of consumer involvement with the product throughout the purchase and post-purchase sequence, are important in product design.

5.1.1 Stimuli to buy and eat
Stimuli to buy and eat are many, for example the demands of the family or home group, individual hunger, the array of products on the supermarket shelves, the dishes on the menu or even the weather. The influence of family members on the buying of cereals is illustrated in Table 5.1. This shows the strong influence of the child on all stages of the buying action, and in particular on what kind and brand to buy – long a basic premise in promotion by the breakfast cereal manufacturers.

The stimuli to buy and eat can occur before and after entering the supermarket or the restaurant. When buying bread baked in the supermarket, the stimuli may be hot bread aroma, or the known great flavour of the bread. Consumers can identify other bread characteristics such as ease of cutting, calories, free of additives, but may not be stimulated to buy by them. Because

Table 5.1 Breakfast cereals: family-member influence on buying

	Stages in buying		
	Initiation	Search and evaluation	Final decision
Husband	2.64	2.51	2.60
Wife	3.64	3.66	3.88
Child	3.91	3.42	3.62
	Buying decisions		
	Husband	Wife	Child
What kind to buy	2.60	3.81	3.95
What brand to buy	2.42	3.90	3.68
What size	2.16	4.20	2.84
Where to purchase	2.07	4.43	2.29
When to purchase	2.14	4.37	2.75

Each score represents the average of the husband's, wife's and child's perception of family member influence, on a scale where 1 = no input and 6 = all of the input.

Source: After Lawson et al., 1996.

there are so many stimuli the individual does not react to them all on a conscious level, and probably has a basic set used for each type of product. The depth of study in comparing food products and buying is usually not very great; the consumers do it everyday and they want it to be simple and not take time. The product stimuli to buy or not buy include:

- strong 'not buy' factors such as the smell of deteriorating fish, bruising of fruit, unusual colour of bacon;
- strong 'buy' factors such as value for money, sensory attractiveness;

Table 5.2 Consumer actions after buying the food product

Action	Sub-actions	Decisions for and against
Preparation	Transport, store, prepare, cook, serve	*Easy/difficult* *Quick/time-consuming*
Eating	See, feel, smell, bite, savour, swallow	*Enjoy/neutral/dislike* *Easy/difficult* *Clean/messy* *Quick/takes time*
Post-eating	Digest, general feeling, feeling in stomach	*Comfortable/indigestion* *Well/sick* *Pleasant/unpleasant after-taste*
	Dispose of waste Compare with other foods	*None/large, clean/messy* *Like/dislike* *Repurchase/never buy again*

- important 'buy' factors such as the size of a loaf and the thickness of the slices may not stimulate buying because they accept quite a wide variation in them;
- weak 'buy' factors which do not stimulate such as the nutritional value and the ingredients list on the label.

In developing the product concept, it is important to recognise these aspects of stimuli for the new product – strong buy/not buy, range of acceptable variation in important factors, and the low importance factors.

The consumer actions after buying the food, in preparing, eating and post-eating, are important in building up long-term attitudes and behaviour. The decisions that can be made, shown in Table 5.2, can lead to strong acceptance or dislike of the food.

Think break

Compare the complete consumer behaviour from the initial perceived need to buy food to the post-eating actions for the following:

1. A teenager feeling hungry and deciding to buy a takeaway snack.
2. A person buying the week's food for a household of adults and children under ten years.
3. A wealthy person deciding to go out to a high-class restaurant.

5.1.2   Product judging criteria

Product judging criteria during the buying and use of the product are important; for example for bread, they may be judging on: colour of the crust, shape of loaf, fibre content and price. When a person is faced with a food, they perceive its physical and social attributes through the senses of sight, feel, smell, hearing and taste. These in turn arouse the central control unit (the brain) to make a comparison between the perceived sensory properties and the acceptable criteria for the food based on personal preferences and past experience. The result of this comparison is acceptance or rejection of the product. This can occur at any stage of the food behaviour process. The product may be rejected at the search stage, because it does not fit the cultural pattern, someone in the household dislikes it, or it does not suit the eating occasion. It can be rejected during the buying stage because of the pack appearance, the nutritional information, the price, or because the product appears soft to touch, has an unpleasant odour. Similar judgements will take place throughout the preparation, cooking and eating stages.

The **level of involvement** that a consumer has with a product varies with product and environment. Involvement has several facets: perceived importance of product and buying/eating situation, perceived symbolic or sign value, perceived pleasure value and perceived risk (Laurent and Kapferer, 1985). Consumers' product knowledge is based on a chain (Peter and Olson, 1999):

$$\begin{array}{ccccc} product \\ attributes \end{array} \rightarrow \begin{array}{c} functional \\ consequences \end{array} \rightarrow \begin{array}{c} psychosocial \\ consequences \end{array} \rightarrow \begin{array}{c} values \end{array}$$

Some attributes are related to strong core values of the person; others are unimportant and get little response from the consumer. Consumers can believe that product attributes are strongly related to their goals or values, for example that diet foods will help them to achieve their goal of losing weight, and therefore they feel strongly about the low-calorie attributes of the product. Values include **instrumental values**, the preferred modes of conduct, and **terminal values**, the preferred states of being. Consumers also recognise functional product attributes, which are important but not related strongly to either their goals or values; for example that dried soup powders mix easily with water. Finally there may be product attributes that are of no importance to them and these attributes will not gain their interest in the product. These three levels of attribute involvement by the consumer can occur in one product, and lead to the hierarchy of attributes used in product design. The consumers' product knowledge can recognise a number of product attributes, a number of product benefits and also their value satisfaction from the product.

5.1.3 Consumer/food relationship

The consumer and food relationship is important throughout the food behaviour process; both the food and the consumer have attributes and it is the compatibility of these attributes that determines acceptance or rejection of a food product. In product development, consumers' needs and the related products' attributes need to be considered together at each stage of the food behaviour process as shown in Fig. 5.3 (Schaffner et al., 1998). The consumer decision in the post-action to not buy the new product, if widespread, will necessitate a redesign of the product. If it is decided just to drop the product, it is important to determine what caused the failure in the consumer/food product relationship and to store all the information for a later time. The consumer/food product relationship is the basic relationship in food product development and it is important that it is considered in all the steps in the food behaviour process.

Think break

In studying consumer/product behaviour for the development of new product concepts, there are four important stages:

1. Identify the consumer/product relationships for each action stage of the complete consumer food behaviour process.
2. Relate the specific consumer needs/wants with specific product attributes.
3. Rank the specific product attributes for consumer importance in three levels: important, functional, not important.

4. Identify the most important product attributes on which to base the total product concept for the design of new products.

Now study the following buying situations using the above four stages:

- Middle-aged man buying chocolates for a partner's birthday.
- Vegetarian buying a frozen convenience meal for their own consumption.

Identify the important product attributes for the product concepts of a new chocolate product and a vegetarian convenience meal.

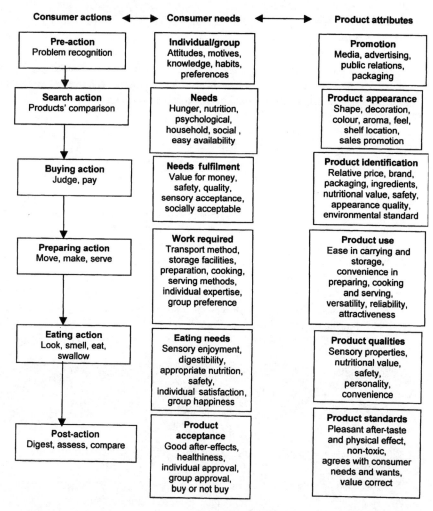

Fig. 5.3 Comparing consumer needs and product attributes in the food behaviour process.

There are three levels of understanding the consumer/product relationship: as an individual product, as a meal and as an eating pattern. The bar of chocolate could be eaten alone, but many food products are eaten together. For example the hamburger is in a bun with lettuce, tomato, a sauce, and it is sold with French fries and a soft drink. Sometimes this juxtaposition of foods is ignored in product development. The consumers also have eating patterns, which do change with time, and the foods have to fit into this eating pattern. So the consumer behaviour is more complex than the single product action model, as each model is interrelated with other product models. The success of the takeaway industry is based on its understanding of these interrelationships. In some cases it has also been used in the supermarkets, for example relating pasta and meat to sauces.

5.2   Understanding food choice

Food choice is an area of research that has expanded a great deal in recent years and whose findings need to be brought into product development. Food choice is caused by the interaction of the person and the buying or eating environment, both the state of the environment and the individual affecting the choice (Bell and Meiselman, 1995). Buying fish and chips served on fine china in a high-class fish restaurant, or buying them wrapped in newspaper from a fish and chip shop changes the interaction between the consumer and the product. Consumer food choice is complex; some of the variables are shown in Fig. 5.4.

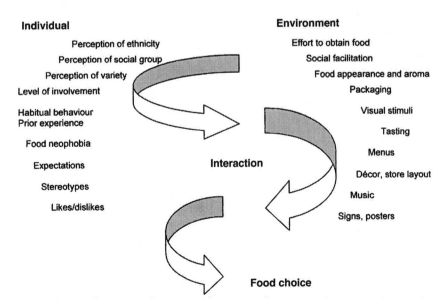

Fig. 5.4   Interaction of the individual and the environment in food choice
(Source: After Bell and Meiselman, 1995).

Food choice can be broken down to the pattern of purchase or rejection of a product, the needs and wants underlying this choice, the psychological attitudes underlying the needs and wants, the effects of the society and the culture. Some important factors for the individual consumer are:

- perception of ethnicity and social group;
- involvement with food;
- habitual behaviour;
- food stereotypes, expectations, likes/dislikes.

5.2.1   Perceptions of ethnicity and social group

Individuals' perceptions of ethnicity and social group, and the relation of these perceptions to the environment and the food, is a basic consumer variable. The term 'situational ethnicity' is used to show that there can be a change in food choice according to the environment except for strongly held taboos. In Box 5.1, the changes occurring in food eating in China are described to illustrate how the change in the environment changes the food choice. Changing the proportion of ingredients but in particular changing the spicing and sauces and the ethnic name can vary a product's ethnicity. This can be used for simultaneous food product development for domestic and export markets. For example, Thai and Australian consumers examined 18 Thai and 18 Australian meat products to determine if meat products acceptable in both Thailand and Australia could be developed (Nantachai *et al.*, 1992). They were asked to identify the context in which each product was used to determine common perceptions of meat products. The common dimension for Australians and Thais was 'social occasions'. Two groups of products satisfied social occasions for both the Australians and Thais – fermented sausages, such as Australian salami, pepperoni and Thai nam, and emulsified sausages such as Australian cabanossi and Thai frankfurter.

5.2.2   Involvement

Involvement, the importance of the food to the individual, can also affect their food choice; for example one consumer may have tea just as a warm drink, and will buy the cheapest tea; another person is very interested in the flavour and will buy on type and brand. There are consumers who are happy with their present range of foods with only an occasional change; other consumers are not strongly involved with the present products and are seeking variety. Much of food product development in the food industry in the 20th century was based on the premise that people want variety, but some products have lasted over 80 years. There is a need for more research to confirm the level of consumer involvement in different types of food products, and the need for variety in all product areas.

Box 5.1   Culture and fast food in the People's Republic of China (PRC)

Chinese cuisine has a long history, a rich culture, and enjoys worldwide popularity. Because Chinese food takes time to prepare even when cooking time is short, most work units (danwei) serve three inexpensive, relative low-quality meals each day. With economic reform and rising living standards, it has become fashionable and affordable for average PRC consumers to consume more time-saving services and to demand food that is different in taste, culture and quality. As happened earlier in Hong Kong and Singapore, demand for time-saving services is increasing faster than income. 'Face' ('mianzi', reputation, prestige obtained through one's efforts or conduct) is related to tangible and intangible personal success. 'Face' makes the Chinese risk-adverse and slower to accept new products, and more loyal than Westerners once brand image is established.

Culture can influence consumers' food choices. The Chinese diet contains more rice, noodles, chicken, pork, vegetables and fewer sweet desserts compared with the American diet of bread, beef, cheese, dairy products and sweet desserts. Therefore, chicken and beef noodle fast-food restaurants are more popular in PRC than pizza and burger restaurants. Beef is scarce, and considered very nutritious in traditional Chinese medicine. The older a person is, the more difficult it is to adapt to the new diet. Therefore older PRC consumers eat burgers for nutrition, and younger consumers eat burgers for taste. Younger persons are more likely to try new foods. Many young, one-child families in urban Beijing take children to McDonald's about once a week. Young people seek novelty and material progress. Although they do not like pizza, Chinese teens sit at Pizza Hut to be seen, older Chinese like low-fat food; all go to McDonald's to be served, enjoy friends and listen to music.

Source: From Anderson and He, 1999, by permission of Haworth Press.

5.2.3   Habitual behaviour

Habitual behaviour is common in food buying – people go to the same supermarket, the same restaurant; shop in the same way – only buying specials, buying at the same time, buying the same brand. Over the years there are trends in buying foods, sometimes because of the availability of new products, and sometimes because of changes in life style. It is important to follow developing trends; for example the sales of frozen foods changed over 15 years in the UK with a gradual decrease in the pioneering frozen products, peas and beans, and the growth of potato products and convenience foods as shown in Table 5.3.

These trends of course are caused by many factors such as availability, lifestyle changes; but they do show how consumers change their habits gradually

Table 5.3   Frozen food consumption in the UK 1974–1989

	1974	1979	1984	1989
	ounce per person per week*			
Convenience meats	0.73	1.31	1.85	2.26
Convenience fish products	0.68	0.81	1.02	1.02
Peas	1.29	1.75	1.70	1.63
Beans	0.44	0.56	0.47	0.49
Chips/potato products	0.48	0.80	1.87	2.82
Other vegetables	0.45	1.01	1.15	1.76
Fruit/fruit products	0.05	0.08	0.03	0.03
Convenience cereal products	0.19	0.44	0.78	1.19

* 1 ounce = 28 g.

Source: From *Consumer Behaviour in the Food Industry* by J. Bareham, 1995. Reprinted by permission of Butterworth Heinemann.

over time in the types of food they buy. These trends, for example the buying of frozen potato products instead of fresh potatoes, have occurred in many Western countries such as the USA, Australia and New Zealand. Habits in food choice do change and the important thing in new product development is to identify what causes them to change – poor products, boredom, new foods which better satisfy consumer needs and wants or new information.

5.2.4   Food stereotypes

Food stereotypes, expectations and liking/disliking, are very important personal factors in food choice. Imagine the surprise on an aircraft when a snack of yoghurt, fresh fruit and muesli is served – it seems wrong because the stereotype for air travel food is 'not fresh and over-processed'. Is your product a stereotype? The important thing in new product development is to break the stereotype and tell the consumer the new product is different, just as British Airways did with its new healthy menu. Consumers also have an expectation of the food. If the actual food is in the direction of their expectation, then the expectation is reinforced for the next time they consider the food – if they expect it to be very sweet which they dislike, and it is very sweet, then they will dislike it more. If they expect the drink to be refreshing and it is, then this will reinforce their expectation for the future. Information is important in expectation – when soups are branded for sensory testing, consumers will have expectations for different brands and their scores will vary from their scores if the soups were not identified. These expectations are an important consideration in product development, and emphasis in developing the total product concept is a basis for product design. And of course there is the basic liking/disliking of products – people have their preferences and it has long been a part of food product development to identify the likes/dislikes of the target consumers.

Think break

A dairy company has developed a line of new nutriceutical products, a range of biologically active dairy-based drinks, and wishes to market through supermarkets, where it is already selling dairy products.

1. Discuss the consumer needs and problems that could be met by these new products.
2. Discuss what prior experience and previous practice would lead to easy acceptance of the new products.
3. What are the types of socio-economic conditions and also social norms that would help the consumers to accept these products?
4. From the knowledge that you have found in 1, 2 and 3, describe the consumers for these products – their food expectations and liking/disliking. Also discuss the food stereotypes that they could have for yoghurts and how the new products relate to these stereotypes.
5. Discuss how you would design the packaging and the in-store promotion to give a total product concept for the new products so that the consumers have the knowledge to trial the products.
6. Where would you position the new products in the supermarket and display them to make the consumers aware of the products?

5.3   Consumers' avoidance and acceptance of new products

Consumers tend to avoid new foods – a phenomenon called *food neophobia*. This is the reason for free samples or in-store tasting of new products; and also a great deal of information with new products. In a Swedish study (Koivisto-Hursti and Sjoden, 1997), fathers avoided new foods more than mothers, and children more than their parents, generally younger children avoided new foods more than older children. Neophobia is related to the marketing classification of consumers in relation to new products as adopters and non-adopters, or innovators, influentials, followers and diehards. *Innovativeness* is the degree that a consumer will try new products, that is how venturesome they are and how prepared they are to take risks. They recognise a perceived risk in the new product as well as the product attributes and benefits that relate to their needs and problems. Other consumer conditions affecting adoption of new products are previous practice and prior experience with products. Prior experience has an effect on new food choice, leading to early acceptance or rejection of the new product. Some people may take the risk of not liking the food or even of safety, and try the new food; other people wait to hear reports on the product. Some consumers are actively looking for new foods and try many new products. In product development, it is important to identify the different types of consumers and to provide the knowledge in the total product concept to overcome their perceptions of risks.

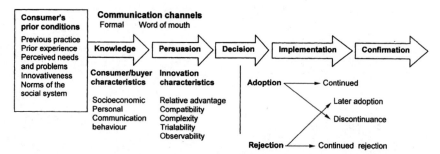

Fig. 5.5 Rogers' model of the innovation–decision process. (Reprinted with permission of the Free Press, a Division of Simon and Schuster, Inc., from *Diffusion of Innovations*, Fourth Edition by Everett M. Rogers. Copyright © 1995 Everett M. Rogers. Copyright © 1962, 1971, 1983 by the Free Press.)

Consumers' prior conditions are an important basis for the diffusion of new products through the target market. If the company understands the different types of people and how they can be encouraged to try new products, then it can plan the product launch to give the optimum diffusion of the new product through the target market. The stages in the adoption process can be summarised as (1) awareness, (2) interest, (3) evaluation, (4) trial and (5) adoption or rejection (Schiffman and Kanuk, 2000). A more detailed innovation–decision process (Rogers, 1983) is summarised in Fig. 5.5.

The total product concept can be designed to give controlled diffusion by understanding the diffusion variables (Engel *et al.*, 1995; Schiffman and Kanuk, 2000):

- Innovation type: continuous innovation, dynamically continuous innovation, discontinuous innovation; related to the increasing level of disruption of the consumers' behaviour patterns.
- Characteristics of the total product concept to the consumer: relative advantage, compatibility, complexity, trialability, observability.
  - (a) Relative advantage is the perceived superiority to existing products.
  - (b) Compatibility is the relation to present needs, values and practices.
  - (c) Complexity is the difficulty in understanding and using the product.
  - (d) Trialability is the degree to which the product can be tried on a limited basis.
  - (e) Observability is the ease that the product and its benefits can be observed, imagined or described.
- Communication: from marketer to consumer through media, public relations, opinion leaders, sales promotion, shows, Internet; and from consumer to consumer by word of mouth.
- Time for adoption: the time for problem recognition, knowledge, persuasion, implementation (buy and use), confirmation and re-buy.
- Social system: ethnic, social, education, literacy, upward social mobility, commercial, size of units, technological level.

5.4   Integrating consumer needs and wants in product development

Food product preference is based on consumer needs and wants. There are a large number of consumer needs and wants, but four important areas to be understood for product development are functional, cultural, sensory and aesthetic. In the past, consumer research and sensory evaluation were two separate areas of research in product development, but today they are combined in an overall analysis of consumers' product preferences (Meiselman, 1994).

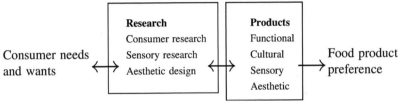

Consumer behaviour and food choice are basic research areas for product development. In the product development project, research is focused on identifying and evaluating specific needs, wants, sensory properties, cultural/ social values and aesthetics, so that specific product attributes can be identified in the product concept and used as a guide in product design.

5.4.1   Identifying consumer needs and wants

As we have seen, there are many factors influencing how and what people eat and buy from the basic hunger pangs to the need for prestige, health. They can be summarised as shown in Fig. 5.6, which has been adapted for food eating from Maslow's hierarchy of needs (Bareham, 1995). In Fig. 5.6, the basic physical needs are shown in the two blocks, holding up the two wants of belonging and love, and esteem. Above these are values or intellectual needs. What is defined as a need and a want varies from person to person, once the physiological, safety and convenience needs have been fulfilled. In Fig. 5.6 a division is shown between the physical needs and the psychological wants, which may not be true for either individual products or individual people. Some people's wants are other people's needs, for example some mothers may feel a need to prepare/serve rich food that indicates care and loving for others; other mothers may see no need for food that gives energy as the children are already overweight and have a sedentary life style. Deciding what are the relevant needs and wants is a difficult decision in building the product concept. Also having satisfied the basic needs, some people may want belonging and love and some esteem. These are shown side by side in this diagram because neither is more important than the other in food eating. The top need is based on acquiring knowledge so that both the eating pattern and the selection of specific foods is based on knowledge of all the needs and wants in Fig. 5.6. This is gaining in importance as people acquire more nutritional and health knowledge.

Lindeman and Stark (1999) studied with young and middle-aged women in Finland:

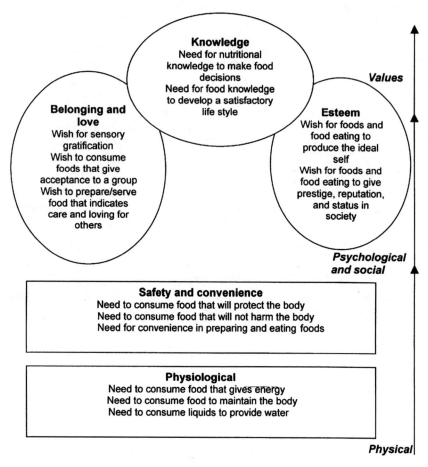

Fig. 5.6   Consumer needs and wants in foods (Source: After Bareham, 1995).

- food choices (health, weight control, pleasure, ideological reasons);
- personal strivings (understanding the world, ecological welfare, slimness, appearance);
- magical beliefs (lay concepts on food contamination and other everyday food problems);
- appearance and weight dissatisfaction, and symptoms of eating disorders.

These consumers clustered into six groups: gourmets, health fosterers, ideological eaters, health dieters, distressed dieters and indifferents. Health concerns were of moderate importance for gourmets and indifferents, whereas the remaining four clusters rated health as a very important factor in food selection. Ideological reasons were high for ideological eaters, health fosterers and distressed dieters. Ideological food choice motives were best predicted by vegetarianism, but also included magical beliefs about food and health, and strivings for self-understanding and environmental welfare. Pleasure was high for the health fosterers and the gourmets. The health fosterers are an interesting group for the future as they are combining pleasure with health and ideological reasons. In a study of adult men, Tepper *et al.* (1997) found that dietary restraint was a consistent predictor of food choice. Restraint influenced the reported consumption of all food groups except desserts; nutrition and food beliefs played only modest roles in food choice. Men showing high restraint were less likely to consume whole-fat dairy foods, eggs, beef and cured meats, fast foods, fats and oils, and regular soda. There have been and will continue to be changes in consumers' concerns about food and health (Ruff, 1995), and it is important to differentiate between the long-term changes and the fashions stimulated by the media.

In the case of the Asian consumer, the top three needs are affiliation, admiration and status. Personal needs in Asia are subordinate to social needs; as a consequence the highest level of satisfaction is derived not from the actions directed at the self but more from the reactions of others. Affiliation is the acceptance of an individual as a member of a group, admiration is earned through acts that demand respect of others, and status comes from the esteem of society at large (Schütte and Ciarante, 1998).

An example of all levels of needs and wants is shown in Table 5.4, which outlines the reasons given for not eating meat. This is an interesting list as it covers practically all the general reasons why people do not eat specific foods. In product development, the meat industry has attempted to reduce these reasons for not eating meat by dealing with the following problems:

- Animal welfare – change from caged to free-range chickens.
- Environment – reduction in forests cleared for animal rearing.
- Health – fat-trimmed lean meat; organic meat.
- Social priorities – barbecue steak, frozen turkeys.
- Displeasure – absorbent pads in meat trays, plastic wrapped trays, meat tenderisation.
- Metaphysical – halal killing for Muslims, small pigs for Samoan celebrations.
- Expense – reducing costs of chicken production and therefore price.
- Inconvenience – easy-to-cook products such as stir-fry chicken, minced pork.

This list shows the variety of consumer needs that trigger changes in the total product concept. The factors in food choice are complex and vary according to consumer and product. In developing a model for food choice, Furst *et al.* (1996) found that ideals, personal factors, resources, social contexts and food context were major influences on food choice. These influences led to the development of personal systems for making food choices that incorporated value negotiations and behavioural strategies. Sensory perception, price considerations, health and nutrition, convenience, social factors and quality were all considered as part of value negotiations, and strategies were developed to simplify food choice. The variation with different product types was shown in a retail study in Britain (Beharrell and Denison, 1991) which found the attribute importance from highest to lowest was as follows:

Table 5.4   Reasons for reduced meat consumption during the 1980s and 1990s

Animal welfare	Moral reasons associated with the view that modern animal production is ethically unacceptable.
Environment	Moral concern that certain features of animal production harm the environment and have undesirable ecological consequences.
Health	Concern about one's own health (1) avoiding natural compounds: cholesterol and saturated fats, (2) avoiding added compounds: hormones, antibiotics, pathogens, (3) avoiding specific health problems: cancer, hypertension.
Social priorities	Conform or adapt to the life style or standards of friends, relations or other influential people. One's own body image is an important example.
Displeasure	Rejection from sensory qualities: sight of meat and associated blood or blood-like drip; sticky texture of meat; taste and elastic mouth-feel of meat when eaten.
Metaphysical	For spiritual, religious, doctrinal or ethnic reasons.
Expense	Abstain because of cost.
Inconvenience or inappropriate presentation	Do not fit into 'light' informal meals, difficult or slow to cook at home, inappropriate for takeaway trade.

Source: After Gregory, 1997.

- Preserves: brand and quality; price; variety and size.
- Bakery products: health; freshness, brand and price; variety.
- Dairy products: health; brand and quality; price; variety.
- Cereals: health and brand; price, size and variety.
- Soups: brand; variety; price; health.
- Fresh meat: quality; presentation, variety, health and price.

Health was of major importance for bakery, dairy and cereal products, but very low for soup and not at all for preserves. Fresh meat was different from the other products – quality was supremely important, other attributes were considerations with little between them. But as described in Table 5.4 in reasons for not eating meat, there are other factors not included in the retail study, which do affect meat buying.

Introducing completely new foods can present problems. For example, Buisson (1995) stated that consumers do not understand functional foods, and they needed to be led gently into such products and the medical benefits not stressed. 'The relative naiveté of the consumer over the links between diet and health is a major impediment to product development of functional foods.' He stressed that great care is going to be necessary in involving consumers in such developments. This is true of all major technological developments in the food industry; the new products based on it need to be developed so that the consumer sees a major health benefit without any major worry about safety.

These various studies show some of the complexity of identifying the consumer needs and wants as a basis for product development.

Think break

In the USA and some other Western diets, the level of fat is too high. Much product development in the past 20 years has been to reduce fat in the diet either by developing low-fat products (Nestle *et al.*, 1998) or by fat replacement (Mela, 1996).

1. Compare the needs and wants of the consumer when buying and using butter and low-fat margarine. What attributes in low-fat margarine could cause the consumer to change or not change their food choice from butter?
2. Butter is often used as flavouring in foods, for example vegetables and sauces. Discuss why it is a popular flavouring, and what attributes a new flavouring would need to replace butter.
3. Some meat is being replaced with tofu in a hamburger to reduce the fat content and the meat content for consumers who are decreasing their meat consumption. Identify the needs and wants of the consumers and relate these to the attributes of the new style of hamburger.

5.4.2   Cultural needs and wants in foods

Broad consumer characteristics such as nationality, religion, race, age, sex, education and socioeconomics are the basis for consumer attitudes, motivation and behaviour. The relationship of some basic consumer characteristics with food may change only slowly if at all, for example religious taboos and requirements such as Hindus not eating meat, Jews not eating pork. Other foods can be replaced by new foods quite quickly, for example fried potato chips replacing rice for young people in SE Asia. With the internationalisation of the food industry, it is important to study these relationships in some depth in introducing new products. There are products that have penetrated the international marketplaces, such as instant coffee, Coca-Cola, Kentucky Fried Chicken and McDonald's hamburgers, which are not related to local foods, but have been very successful introductions. This has been achieved by marketing activities and also by an understanding and use of the market systems to reach the target consumers. Very often new distribution systems have been introduced which may have revolutionised the local food market, in particular bypassing wholesalers and delivering products directly to the retailers, or indeed setting up a new retailer system such as Kentucky Fried Chicken and McDonald's. Some food manufacturing companies with a wide range of products have cooperated in the building of supermarkets and convenience stores, which then have the facilities to accept the companies' new products. This again emphasises that the new product is not just a food but the total product and must be regarded as this in product development.

Acceptance can also be obtained by using a long-established and popular local brand name. But there is still often a need for modification of the products, particularly as regards flavours, and sometimes the packaging and the brand so that the new product agrees with the preferences, attitudes and habits of the new consumers. The size of packaging can be important; for example, the introduction of small pet food packs for the Japanese market where both the houses and the pets are small and the standard can was too large. It can also be a question of price; the pack size can be reduced to a size that is affordable to the target market. New products must relate to the economic level of the target consumers. The culture can also dictate the type of raw materials and processing methods acceptable for the new product. The effects of the culture on product development are summarised in Fig. 5.7 (Bareham, 1995).

Sometimes, the cultural resistance to new product development is too great. Culture is the foundation that underlies all food choices and people use the rules or habits of their specific cultures, and also ethnic groups, to decide what are acceptable and preferable foods, the amount and combination of foods, and ideal foods and improper foods (Nestle *et al.*, 1998). The growth of ready-to-eat cereals in Thailand was slow because the cereals did not fit into the breakfast eating pattern and also because milk was not generally available at breakfast. Steamed rice, fried rice and rice porridge were common at breakfast, often eaten with pork, fried egg or a savoury sauce. Cold cereals did become accepted as snacks for children between meals. This shows the need to understand consumer behaviour in food eating.

To have a food product accepted by a number of cultures, it can be designed with sensory characteristics that are generally acceptable but not distinctive. This often works but if there is an expectation in a culture for specific sensory properties it can be unacceptable. For example in the introduction of instant noodles into Thailand, noodles with international flavours were very slowly

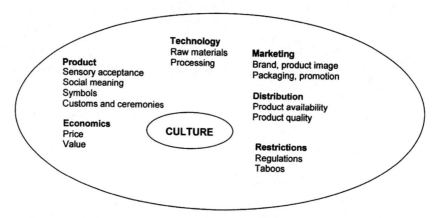

Fig. 5.7 Cultural effects in new product development.

accepted, but when noodles with sachets of Thai sauces and chilli powder followed, sales grew rapidly. When going international there is also a need to understand the geographical situations, climatic conditions and the food industry in the country including processing, marketing and distribution. The local situation must be understood, particularly through cooperation with local people. Raw materials can influence acceptance and there may be a need to replace the foreign raw material with local products. New products need to be seen as superior in some way to the local products, and the marketing has also to be creative. The advertising and promotion needs to be related to the target consumers; for example it may be useful to have cooking demonstrations and video films to show people how to use the product.

The unique functions of the product need to be stressed but also the advertising must fit with the cultural beliefs and practical needs of the consumers. To achieve a 'universal' product, the product must serve a unique but common human need that is seen to improve the life of people in general. Coca-Cola serves a need to have a cheap, safe, refreshing drink, but it is also related to the American life style that is economically higher than in many other nations. So consumers quench their thirst but also hope to raise their living standard to what they see in American films and TV shows. Coca-Cola is adjusted to meet different sensory preferences such as sweetness and acidity in different countries.

Think break

Bring together a group of six to eight people to form a focus group. These people should eat potatoes as part of a meal and also buy potato products sold in the supermarket. Conduct two discussion groups with them:

1. When, how and why do they eat potatoes?
   (a) When do they eat potatoes in a meal and as a snack?
   (b) How often do they eat potatoes?
   (c) Where do they eat potatoes – at home, in a restaurant, as a takeaway?
   (d) What form of potatoes do they eat – boiled, roasted, fried, etc.?
   (e) Why do they eat potatoes?
   (f) What are their general perceptions of potatoes, their experience with potatoes, the stereotype of potatoes, their expectations, their likes/dislikes, good things/bad things about potatoes?
2. When, how and why do they buy potato products in the supermarket?
   (a) What potato products do they buy in the supermarket?
   (b) For what occasions do they buy potato products?
   (c) What are the reasons that they buy them?
   (d) What do they like/dislike about the products?
   (e) What are the advantages/disadvantages of the products?
   (f) What new potato products have they bought?
   (g) What kinds of people buy the different products?

From these discussions, build up a picture of the buying and eating of potatoes and how this is changing. Clearly identify the consumer needs and wants in buying potatoes and buying potato products, including any cultural factors in potato eating. What directions for potato product development does this indicate?

5.4.3   Aesthetics, foods and consumers

There have been over many centuries discussions on the relationship between taste and Taste, that is respectively the tasting part of eating and the faculty of discerning and enjoying beauty or other excellence, especially in art. Is the gourmet, the connoisseur of food and wine, appreciating the *artistic beauty* of food and showing their Taste? Korsmeyer (1999), discussing taste, said:

> The objects of taste not only are fleeting, they participate in the necessary repetition of the practical world of daily life. Eating and all the work that is required to make it possible is a repetitious and perpetual exercise. But this practical fact does not mean that when eating is conducted with reflection and grace it manages to be only pleasant, nor does it mean that its pleasures do not reach beyond themselves to anything more profound.

Eating has always been a symbolic part of culture and society. Particular foods are symbols of:

- religions, for example bread and wine for Christians;
- countries, for example artistically cut vegetables of Thailand;
- celebrations, for example Easter with hot cross buns, Easter cakes;
- marriage, for example wedding cakes.

But the symbols change, their significance changes and they even move their culture. Example of developing symbols, the pretzel and the croissant, are described in Box 5.2.

Food can be representational symbols such as the pretzels, croissants, hot cross buns, but it can also be abstract symbols, for example fresh vegetables can be the symbol of health, chocolate a symbol of indulgence. Some modern symbols are the golden arches of McDonald's, which in many countries symbolises the affluent American society, and Coca-Cola, which for many teenagers symbolises fun and socialising. There can also be products that are symbols of part of the culture, for example beer is related to rugby, an important part of New Zealand culture. Foods can also be related to art, for example in former times a still life with game birds and fish or baskets of fruit signified good living; and in the 20th century the Campbell's soup can became a well-recognised art icon.

Gourmet chefs of course regard their craft as art, both for the visual effect and the eating qualities. They create food for high-class restaurants and at ceremonial dinners, with a great deal of thought and knowledge, to give an aura of

Box 5.2   Pretzels and croissants: changing symbols

The pretzel is said to have been invented by a monk in the early seventeenth century, who twisted a string of dough and baked it into the curved outline of a brother at prayer, to dispense as a reward for his pupils who recited their catechism correctly. This was called bracciatelli, which translates as 'folded arms'. In some parts of Europe, pretzels are a Lenten food, again the arms resembling the folded arms of the monk. But what are pretzels symbolic of today – the USA and casual living, as snack foods for that social occasion? By understanding the symbols of the pretzels, they take on a new expressive dimension, and the aesthetic apprehension of the pretzel expands.

Croissants were invented in Vienna in 1683. In celebration of the successful defence of the city against the Ottoman Turks, Viennese bakers crafted little buns in the shape of the crescent moon on the flag of their enemies. In this case, not only the crescent shape was recognised as denoting the foreign enemy, but also the fact that one devours the crescent re-enacts the defeat of the invaders, and perhaps also represents Christianity overcoming Islam. But today with its expansion through many countries it is seen as representative of France and French breakfasts.

Source: Reprinted from Carolyn Korsmeyer: *Making Sense of Taste*. Copyright © 1999 by Cornell University. Used by permission of the Publisher, Cornell University Press.

wealth, sophistication as well as unusual eating qualities. They of course do create for vision alone in ice sculptures and sugar confectionery. Their creations can be related to the current art climate or earlier art such as art nouveau. Chefs are increasingly being integrated into the product development process by food manufacturers, ingredient companies and supermarkets to develop complete meals and also to develop new food experiences and taste sensations (Hollingsworth, 2000).

Food can mean beauty. It can be identified as outstanding, for example the gourmet and the wine buff have built up their knowledge so that by looking and tasting they can identify the attributes that identify outstandingly beautiful food and wine. At this level, food has an aesthetic value which can be described and admired. Often the product has been developed over many years, such as French wines; but it can also be developed by new technology as has been seen in the wines, developed in the last 20 years in Australia and New Zealand, which have won many medals and can command high prices. Food products can also be associated with beauty; for example all the diet and special foods which are claimed to give beauty to the eater.

Like all art, food does go through fashions, but may not be so directly connected with fashions as clothing, houses and furnishings. A Victorian menu for an important event is very different from the menu at a ceremonial dinner today; the three tier, iced Victorian wedding cake is very different from some of today's wedding cakes, shaped in the form of two hearts, which have soft icing on carrot or chocolate cake. Food fashions do not change very quickly, unlike clothing; except for the gimmick foods associated with a TV programme or a film. But today there seems to be much faster changes, for example in coffee shops, ethnic restaurants and special foods where promotions and advertising are persuading the consumers that there is a new fashion in eating places and they need to change. An example is the recent expansion internationally of coffee houses, which have changed the plain cup of coffee into a range of coffee types, and made it fashionable to go to a coffee house for that casual hot drink. Fashion changes thrive in a group that accepts change (Zelanek, 1999); so with the fast-moving international changes occurring today, one would expect fashions in food to become more important.

But what does aesthetics mean in product development? The design of food packaging through the past 50 years has been very much related to developments in art. The graphics on the can of soup have developed from the simple, plain, printing, through the cartoon type illustration, the representational picture of the food, the inclusion of the consumers with the food, to the showing of raw materials and the ready-to-serve. There have also been some abstract designs during the years. However, food products in general have not tended to go along the aesthetic path and there have been some disasters where main-line food companies have tried to branch into gourmet foods. Perhaps this is the time to join aesthetics with modern technology in designing what are called more customised new products, identified more clearly as foods with Taste and not just taste. The success of wines and cheeses as aesthetic products could be expanded to other product types and in particular to future new products.

Think break

1. Identify five foods that are religious symbols.
2. Study five paintings, one from each of the 16th, 17th, 18th, 19th and 20th centuries, that show foods and decide what they are saying. How has food changed over the years?
3. How has art changed during the past ten years? What direction could this give to food design?
4. Identify foods that symbolise today: health, fun, luxury, good company, wickedness. How could you build up a new diet food to symbolise the ethics in your market?
5. Contrast the aesthetics of a wine with the aesthetics of a fruit juice, using two products with which you are familiar. How could the juice be given an aesthetic value the same as the wine? Who could be the target market for the new fruit juice?

5.5   Sensory needs and wants in food product development

The importance of the sensory properties of the food in acceptance or rejection of the food has long been recognised. The appearance, colour and sometimes aroma of the food are influential in buying; aroma, flavour and texture in eating the food. However, in buying behaviour, taste is not the only crucial determinant, and in some cases is well down the priority list (Raats *et al.*, 1995). Therefore, it is important not only to recognise the sensory properties but also the interaction between them and other product attributes.

5.5.1   Sensory product attributes

A food can be defined at different levels (Cardello, 1996) when considering the sensory properties. There are basic properties of the food that can be recognised by the individual's sensory system and then, by using learning and memory, these sensations are changed into the sensory product attributes perceived by the individual. Taste and aroma are combined in an overall flavour, e.g. an acidic taste with a citrus aroma are combined into an orange flavour; and the mouth feel, biting and chewing are combined in an overall texture – for example consider sticky toffees and hard toffees. Having identified the sensory product attributes, the individual can score them on a liking (hedonic) scale from dislike very much to like very much. Bringing the like/dislike scoring together with other properties of the product brings the consumer to acceptance or rejection of the product. The stimulus, sensation, perception and the response are combined in the individual and the product as shown in Fig. 5.8. The core is physical and chemical properties of the product, which are the basis for sensory properties. But in a food these sensory properties interact with each other, and the consumers have perceived sensory reactions.

Sensory product attributes have to be firstly identified before measurement and hedonic testing. Classifying the sensory product attributes is complex when one moves beyond shapes, sizes, basic tastes of sweet, salty, bitter, sour; and the colour standards of lightness, hue and chroma. Texture can include the finger feel and the mouth feel; the finger feel including firmness, softness, juiciness; mouth feel including the mechanics of chewing such as hardness, cohesiveness, viscosity, elasticity; the geometrical characteristics such as particle size, shape and orientation; and other mouth feels such as moistness and greasiness. These are in a simultaneous or sequential effect during eating, and give a total final reaction on swallowing. For example in testing the texture of black beans (Watts *et al.*, 1989), the descriptions in Table 5.5 were used by a trained texture panel. The magnitudes of these biting and chewing attributes, and also the duration of chewing were determined.

Developing terms for flavours and aroma is much more complex. For example one can take four or five representative commercial samples of the product type (category) or some of the product prototypes with different levels of ingredients, and ask a sensory panel to list the flavour characteristics, then in discussion try to

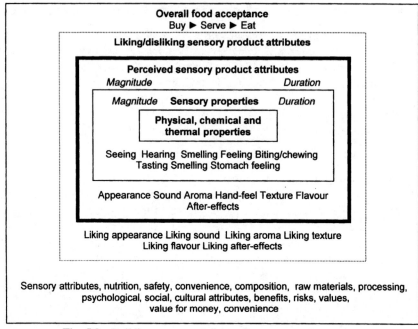

Fig. 5.8   Building sensory attributes for consumer acceptance.

organise the different descriptions into flavour types and use a reference list to group them into categories, as in Civille and Lyon (1996). A reference substance (usually a chemical compound but can be a simple substance) is found for that particular flavour category and used by sensory panels in the future. Flavour terms have been built up as shown in Table 5.6 for some specific foods. These terms have standard descriptions, for example *burn* is 'chemical feeling factor associated with high concentrations of irritants to the mucous membrane', *heat* is 'chemical burning sensation in the mouth and throat'. To explain the difference, the reference sample for burn is vodka and for heat is red pepper.

Table 5.5   Measuring the texture of black beans

Hardness: bite down once with the molar teeth on the sample of two beans and evaluate the force required to penetrate the sample.
Particle size: chew the sample (two beans) for only two or three chews between the molar teeth, and then rub the cotyledon between the tongue and palate and assess the size of the particles which are most apparent.
Seedcoat toughness: separate the seedcoat from the cotyledon by biting the two beans between the molar teeth and rubbing the cotyledon out between the tongue and palate. Then evaluate the force required to bite through the seedcoat with the front teeth.
Chewiness: Place a sample of beans (two beans) in your mouth and chew at a constant rate (one chew per second), counting the number of chews until the sample is ready for swallowing.

Source: From Watts, Ylimaki, Jeffrey and Elias, 1989, by permission of IDRC, Ottawa, Canada.

Table 5.6   Descriptions of flavours of vegetables

Corn	Cucumber	Eggplant (aubergine)	Red peppers	Parsley
Cardboardy	Astringent	Bitter	Bitter	Barny/barnyard
Grassy	Bitter	Cooked	Burn	Bitter
Legumy	Cucumber	Heat	Heat	Fishy
Metallic	Green	Metallic	Pungent	Grassy
Salty	Sour	Mouth numbing	Raw	Green
Starchy	Sweet	Raw	Sweet	Hay
Sweet	Watermelon	Sweet		Salty
Woody				Sweet

Source: From Civille and Lyon, 1996, copyright ASTM, reprinted with permission.

Other sensory characteristics are temperature, pain and sound. A product such as ice cream has its own temperature effect, and there are effects of temperature on the sensitivity of the consumer to flavours and aromas as well as on the volatility of the aroma materials. Consumers expect a certain temperature of foods for eating. Although a canned stew is safe and edible if eaten from the can, can designers have gone to great lengths to design methods of quick heating to make the stew acceptable. Some pains are expected for example in eating very hot curries, drinking a 'straight' whisky, licking an ice block, but some are unacceptable such as finding a hard nut that hurts your teeth. Sound affects the acceptability of many products, for example the snap of raw celery and the crack of a biscuit indicate freshness. Some important factors in the overall sensory character of the product are the order of appearance of the attributes, the magnitude of the attribute and the duration of the attribute.

5.5.2   Interactions of sensory product attributes

Clear definition of the desired sensory properties by the consumer may be difficult because of their interaction in consumer acceptance. The consumer may not be able to separate the sensory attributes that influence their liking or disliking of the food. It is this interaction of even simple tastes such as acidity and sweetness, colour and fruit flavours that can make sensory testing and product design complex. An example of interaction was shown in a study by a trained Canadian sensory panel which showed that the perceived intensity of sourness in lemonade decreased with increasing levels of sucrose added as in Table 5.7.

Table 5.7   Sourness of lemonade with increasing sucrose

Sucrose (%)	0	2	4	6	8
Mean sourness score	13.5	11.0	4.5	3	2.5

Source: After Poste *et al.*, 1991.

Another problem is the interaction of the sensory testing with other consumer characteristics; for example consumers may have a stereotype of the product that disagrees with the sensory properties of the new product, or they may be slimming and therefore will downgrade creaminess and their acceptance of a new ice cream. Also there are differences in national preferences for sensory properties, even between what one would think are similar groups – such as Australians and New Zealanders (Cooper and Brown, 1990). Australian products (potato chips and canned soups) were found to be saltier than New Zealand products; New Zealand products (chocolate biscuits, fruit juices) sweeter. There appeared to be genuine differences in the types of products desired by consumers on opposite sides of the Tasman. Cooper and Brown suggested that one should always regard the consumers' 'norm', what they are familiar with, as a very powerful factor in product acceptability. For example 'tropical' flavour – a mixture of pineapple, passionfruit and orange and having a strong yellow-orange colour – is very popular in New Zealand in many products because it was one of the earliest juices introduced onto the New Zealand market in 1981 and became very popular. But it is much too sweet a flavour for Australians; they like their juices more sour and often lighter in colour.

Sensory acceptance and its relation to overall product acceptance is not predictable from a straightforward simple sensory panel, but needs a more complex research of consumer needs, wants and behaviour (Cardello, 1996). The total product has a strong effect on acceptance. The advertising, the information given, the appearance, the aroma or smell all lead the consumer to expect certain sensory qualities in the food and they will rate their acceptance against this. If it reinforces their good expectation, then the sensory properties will be acceptable; if it is better than their bad expectation, they may even score the sensory properties more highly. In Table 5.8 are shown the consumer scoring in Germany of one sample of Dutch tomatoes with different labels, to show the effect of labels and information on quality determination by consumers (Vesseur, 1990).

Another important factor is the amount of a food product eaten and how often it is eaten. Consumers may become tired of a product and stop buying it – they

Table 5.8   Taste of one sample of Dutch tomatoes with different labels

Label	Average judgement*
German organic	2.5
German open air	2.5
Dutch open air	2.6
Mediterranean	2.8
German greenhouse	3.0
Dutch greenhouse	3.0

* Based on a scale 1 = very good, 6 = inadequate.

Source: Adapted from Vesseur, *Acta Horticulturae* 259, 1990 by permission of ISHS (International Society for Horticultural Science) Leuven, Belgium.

may find it boring or may find some sensory aspect overwhelming when eating a large quantity. Köster (1990) tested two kinds of tomatoes with 200 families over four weeks, one group using product A and one using product B. Product B had a higher liking score than Product A at first sight, but from their diaries:

- More of Product A was used than Product B.
- More of Product B was thrown away than Product A.
- More outside tomatoes were bought when they had Product B than when they had Product A.

In developing a new product, it is important to test for several times with the consumer and with the amount that they normally eat. In home-testing it is useful for them to keep diaries to show how much they use.

Think break

Contrast the sensory product attributes when consumers buy, serve and eat

- fruit juices;
- fresh porterhouse steak.

1. List the sensory product attributes that you identify from buying to after eating.
2. What sensory attributes interact in buying, eating and serving?
3. Identify all the sensory attributes that affect your sensory liking/disliking.
4. What other product attributes would you include in determining your acceptance of the two products?
5. Are there any interactions between these and the sensory attributes?

The consumers' liking and disliking of the sensory properties are complex because they are bringing their product stereotypes, expectations, past experiences, and some of the other product attributes that influence their acceptance of the product.

The problem in product development is how to relate future consumer and market needs, wants and behaviour to product, production, processing and marketing technologies; and also how to relate new technologies to future consumer and market needs. This interrelationship is important in every stage of the PD Process. Therefore the consumer needs to be involved at every stage either directly or through their recorded needs and wants. The product development is focused on the consumers or in the case of the industrial and food service product development on both the customers and the consumers.

5.6   Consumers in Stage 1: Product strategy development

In developing the product strategy, there are seven steps (Linnemann *et al.*, 1999):

1. Analysis of socioeconomic development in the target markets.
2. Translation of consumer preferences and perceptions into consumer categories.
3. Translation of consumer categories into product assortments.
4. Grouping of product assortments in product groups in different stages of the food supply chain.
5. Identification of processing technologies relevant for specific product groups.
6. Analysis of the state of the art in relevant processing technologies.
7. Matching the state of the art of specified processing technologies with future needs.

This study sets up a new product strategy based on the consumers. As Linnemann *et al.* stated, the beginning is a study of consumer trends. For example, is the life style of consumers changing? What are the trends in food buying? How are their economic standards changing? This gives the background to the choice of new product areas. In category appraisal, consumers study the company's and competing products in a product category to investigate the acceptance and standing of the company's products against competitors. Product prototypes can also be studied with the commercial products. Consumers can describe their concepts of the products, and the key product attributes leading to consumer acceptance. They can give opinions on their 'ideal' products. From this information, opportunities for new products in the category can be identified. It may also lead to changes in positioning of the company's products.

After the product strategy development, the consumers need to be involved in the generation of product ideas, screening of product ideas, development of the product concept and finally the product design specifications, so that their needs and wants are brought into the product concept which is the basis of product design, as shown in Fig. 5.9. The activities in Fig. 5.9 show the importance of the consumer involvement in Stage 1 of the PD Process, which lays the basis for the product development project. The most important activities are product idea generation and screening, consumer survey, product concept development, product acceptance predictions and development of the product design specifications.

5.6.1   Product idea generation and screening

Consumers have taken part in product idea generation for 50 years. Initially the individual in-depth interview was used in motivation research to find out why people bought particular products, and product features were built up based on this information. The interviews were usually conducted by psychologists or psychology-trained interviewers. Later consumer discussion groups became more common as the new product ideas were found to be more creative because of the synergy between the members of the group. These small consumer panels are representative of the target market segments; they can be a focus group with

Opportunity identification

Consumer analysis of opportunity
Identifying the issues
Generating ideas for product types
Identifying the consumer segments
Ideas for target consumers and products

Studying the consumers
Behaviour
Perception
Preference
Food choice
Preliminary screening of product ideas

Evaluating the product ideas
Building the product descriptions
Selecting the consumer segments
Identifying consumer needs, wants, benefits, risks
User/distributor requirements
Cost/value relationships
Market potentials
Further screening of product ideas

Product ideas concept development
Factors in product acceptance
Likes/dislikes
Identifying product attributes, benefits, risks
Rating of product concepts

Final product concept development
Target market segments
Product attributes
Product benefits
Product risks
Packaging attributes
Packaging benefits
Product design brief
Final evaluation of product concept

Product design specifications
Quantitative product metrics
Quantitative packaging metrics
Processing method
Raw materials
Costs/prices
Sales forecasts
Market share
Estimation of acceptance, success
Final acceptance of new product for design

Fig. 5.9   Some consumer activities in Stage 1: Developing the product strategy.

free discussion using techniques such as brainstorming and lateral thinking, or a nominal group with a more formal session using questionnaires, which consumers answer individually followed by general discussion. There are usually 6–10 people in each group discussion, but this can be repeated several times to give a total of 60 consumers or more. These consumer group

Table 5.9   Observation form for making choux paste in the home

---

SETTING
Baker: male ..... female .....   **Date ....**
No. in family .....   **Day ....**
Quantity of choux paste per mix ....   **Time to start ....**
Quantity of choux paste per month ....   **Observer ....**

INGREDIENTS
Selection: please state type or brand of ingredients used
Flour ..... Eggs: fresh ..... frozen pulp ..... chilled pulp ..... dried ..... other .....
Fat: butter ..... margarine ..... branded fat ..... other .....
Salt ..... Sugar ..... Flavourings ....
Other ingredients ....

Preparation: please describe
Flour: sifted ..... not sifted ..... other .....
Eggs ....
Fat ....
Water ....

METHOD
First mixing ....
Order of adding ingredients: 1st ..... 2nd ..... 3rd ..... 4th ...... 5th ......
Equipment used: hand whisk ...... electrical whisk ..... cake mixer .....
Method of mixing ....
Time of mixing ....

First heating
Mixture heated .... Mixture not heated ....
Equipment used: steamer ..... bowl over pan ..... bainmarie ..... other .....
Method of heating ....
Temperature of heating ....
Time of heating ....
Any other comments ....
Second mixing ..... as above, note any other ingredients added
Second heating ..... as above, note any other ingredients added
Third mixing ..... as above, note any other ingredients added
Third heating ..... as above, note any other ingredients added
Depositing
Piping bag ..... Spoon ..... Other ....
Baking
Type of oven: electric ..... gas ..... other .....
Temperature of baking .....
Time of baking ....
Time out of oven ....
Complete time for making the choux paste ....
Products made
Eclairs ..... Cream puffs ..... Other .....
Any other comments ....

---

discussions are faster than individual interviews, cheaper, more flexible and reduce distance between company and consumer. They can develop ideas when little is known about a product area, and investigate the trade-offs the consumers are making. The most important aspect of consumer group discussions is that the results are in a ready form for developing the product concept and the whole basis of the product design. The criticisms are that the groups are small and not statistically representative of the target market, and that some consumers are influenced by other members and the group leader. At this time in the project, quantity of information is important in building up the ideas and the product concept.

Observing the behaviour of the consumer from buying to disposal of waste can also generate ideas for new products. Table 5.9 shows an observation form on the baking of choux paste by a home baker to aid the development of a complete baking mix for choux paste. Observation is a useful technique to study consumer behaviour, but it must be used carefully and wisely. An unusual happening in the environment or even the observation itself can lead to unusual behaviour. It is very useful to study how the consumer prepares/serves the product and the method of eating. Observation can provide a first hand, authentic picture and is the best method of studying consumer behaviour, but usually it can observe only public behaviour. The consumers can record their own behaviour, if it is not possible to observe their actions.

Think break

1. Making choux paste can be a long and difficult procedure, not always successful for the new baker. Can you suggest new products that could make this process quicker and easier and would guarantee success for the household baker?
2. Using a focus group develop ideas for new products to solve the following problems for consumers:
   (a) improve the nutritional value of ice cream,
   (b) increase the safety of oysters.

The consumer groups also screen the new product ideas, in combination with technical and company screening. The initial stages of idea generation and screening (Roberts, 1997) for a food service new product are shown in Table 5.10. In this case there are three groups – the supplier, the menu planning decision maker and the consumer. Roberts explored all three in the initial stages. Table 5.10 shows the interactions of the suppliers (meat processors), the menu planners (usually chefs) in the motels and hotels, and the consumers (the customers of the motels and hotels). It is important to combine the needs and wants of the consumers, the technical staff and the company in building up the product descriptions.

Table 5.10 Multistage idea generation and screening for a meat product for hotels and motels in Melbourne, Australia

Activities	Participants	Techniques
Preliminary idea generation	Food technologists, caterers	Nominal group technique Brainstorming Synectics
Preliminary screening	Researcher	Qualitative: fulfil consumer requirements, offer benefits for menu planner, value for money, competitive advantage
Outcome: 30 product ideas in categories		
Development of ideas	Menu planners	Focus group on list of product ideas
	Consumers (beef eating)	Focus group on a list of product ideas
Outcome: 30 ideas with benefits, attributes needed by consumers and menu planners		
Technical feasibility screening	2 meat technology experts	Interview on technical feasibility, technology availability, competition, demand volume
Outcome: 32 technically feasible products		
Checklist screening	4 food technologists	Individual scoring on marketing and technical factors
Outcome: 14 product ideas in 5 categories		
Development of product descriptions	Menu planners Consumers	Focus group Focus group
Outcome: 14 product descriptions		
Suppliers' company specific screening	Managers Food technologists	Individual scoring Individual scoring
Outcome: 5 Agreed product descriptions for development		

Raw beef product untreated – tender beef in thin slices.
Raw beef product treated – flavoured pickled beef.
Prepared ready-to-cook product – fricadelle (new beef burger), coated beef product.
Pre-cooked beef product – precooked meat loaf.

Source: From Roberts, 1997.

5.6.2   Consumer survey in the early stages of product development

When more quantitative data are needed, for example in determining the target market and predicting the sales to the target market, a consumer survey using a randomly selected sample of the population is needed. Consumer surveys are usually personal interviews using a formalised questionnaire, but sometimes a qualitative unstructured interview can be used where there is little information about the new product. Methods of organising a consumer survey in product development projects are described in West and Earle (1987).

The consumer survey usually compares three or four new product ideas that have been generated. The information sought can be past and intended behaviour; general opinions and attitudes on eating characteristics, nutritional value, safety, cooking/serving/eating needs, size of packs and related cost; and demographic data such as socioeconomic characteristics and level of knowledge. It is useful to identify the usage patterns for products at present on the market and to assess consumer attitudes and opinions on particular types of products, as well as seeking information on the new products. The information can be analysed to give market share by consumer classifications, method of purchasing products, frequency of purchase and ways of using products. The researcher defines the market segments and forecasts the market potential. From consumers' opinions, attitudes and general comments, assessment can be made of product needs and the inadequacies of present products.

With information from the consumer survey, new product descriptions can evolve to product concepts, with definition of basic product attributes such as size, storage life, function, price range, ingredients, desired eating characteristics and cooking method. The target market segment(s) are identified, so that the choice of representative consumers for developing the product idea concept for the product design can be made. Consumers in the original focus groups are probably chosen as representative of the segment that the researcher believes will accept and buy the product. But there is a need to confirm this is correct in a larger survey. Target consumers can be identified on demographic factors such as age and education, but target segments based on such factors as usage of product, life style, personality and social groups can be more directly related to the product.

5.6.3 Product concept definition and optimisation

The product concept is built up in stages – attributes identification and screening, attributes measurement, complete product concept, product concept evaluation. Saguy and Moskowitz (1999) said 'Innovative products possess spatial and temporal limits' and what is attempted in product concept development is to outline these limits and in the product design specifications to give them quantitative values. Development of the product concept and the product design specifications are outlined in Fig. 5.10, showing the consumers

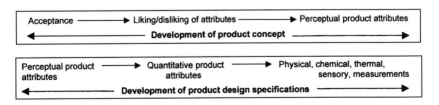

Fig. 5.10 Attributes in product concept and product design specifications.

after their acceptance of a new product description, identifying the product attributes important to them in the product. The product designers with consumers and analytical technologists build up metrics, that is quantitative measurements of the product attributes, as the basis for the product design specifications. Often, these activities are not sequential from consumers to food technologists but they are cycling backwards and forwards between the three groups, as the concepts of the product characteristics identified by the consumers gradually become the physical and chemical testing of the food technologists.

The consumers who take part in developing the product concept are category users or, if it is an innovation, the predicted category users. The consumer focus group is invaluable for building up the product concept. Usually 30–60 but sometimes up to 200 consumers take part in small discussion groups of six to eight people. The discussions are usually free ranging so that the consumers can discuss their own attitudes and behaviour towards the products and identify their needs in the product. The consumers are using as models the company's present products, competing products and early product prototypes.

In identifying product attributes for the product concept, it is important to discover from the consumers everything they recognise in the product so as to discover all the product attributes. The consumers combine what they identify as similar attributes into one attribute; then they develop a description of this attribute. The attributes are grouped by the consumers into core values, functional attributes and unimportant attributes. The core values are what consumers wish to feel/achieve when they buy/eat the food and after the food is eaten, for example feeling healthy, happy, not hungry. The functional attributes are the qualities of the product needed for use. The essential attributes, the 'benefits' that consumers identify to differentiate the products and also the 'risks' that they identify with the product, are recognised. Included are all the different types of attributes – basic product, package, use, psychological, social, cultural and environmental.

Think break

A simple product description given to a focus group of women with children was:

A new fruit salad topping is to be produced, using fruit, and containing no synthetic flavours and colours. It is to be used like other toppings on ice cream and other desserts.

The focus group identified some important product attributes:

- Target consumers: bought by families, used by children.
- Functional: packed in 300 ml 'squeeze', plastic container, same viscosity as present toppings on market, used on ice cream, pancakes, etc.
- Values: natural, real fruit, low calories.
- Economics: price £1–1.50 for 300 ml.

1. Study the product attributes identified by the focus group. Can any of these attributes be combined? What other product attributes can you identify in the new product? Are they unimportant or could they have significance to the mother, the company, the retailer?
2. This type of product is eaten mainly by children. What product attributes would be important to children? If possible discuss this with a child who eats this type of product.
3. Develop a product idea(s) from all the information.
4. Write a product idea concept to be given to new focus groups. This time there will be separate focus groups for mothers and children. How would you organise the focus groups to give more detailed attribute information for the development of the final product concepts?

It is important that the consumers examine the attributes together, as they often interact with each other in the product. Consumers may not be able to describe new products, especially radical innovations that they have never seen, but they can compare different combinations of product attributes and select what suits them. The next step in the development of the product concept is to test different combinations of the identified attributes. The attributes are brought together as a variety of combinations in separate product concepts and assessed for acceptance (or purchase intent) by the consumers. The different product concepts can also be compared with competing products to see how they perform competitively. Using statistical modelling such as conjoint analysis, the product designer can identify the crucial attributes, recombine them (adding any new attributes that the consumers have identified as missing) and gradually optimise the product concept. Concept screening not only helps to select the best concept and determine the contribution of individual attributes, but also shows how concepts can be restructured.

The multi-attribute approach for product concept generation and evaluation has led to a systematic method instead of the old 'try it and taste' method. It has increased the basic knowledge of food products, and their relationships to each other both on product platforms and positions in the market. It has identified the common attributes related to types of products, and also the differences between specific products. The use of statistical techniques with their associated computer software has given a quantitative base for product and attribute identification. The techniques include factor analysis, clustering methods, multidimensional scaling (MDS), conjoint analysis (Shocker and Srinvasan, 1979; Martens et al., 1983; Green et al., 1988), and in sensory studies, descriptive sensory analysis and principal components analysis (Gacula, 1997; Meilgaard et al., 1999). These methods have been widely used in the food industry (Schutz, 1988; Moskowitz, 1994; Saguy and Moskowitz, 1999).

Multivariate analysis is used in grouping attributes. Ninety-two New Zealand consumers compared 45 meat cuts, including beef, lamb, hogget, mutton, pork,

Table 5.11 Grouping of product attributes for meat quality

| On buying | Meat quality | |
	On cooking and serving	On eating
Stringiness	Prestige value	Acceptability of flavour
Spicing requirements	Number of dishes	Acceptability of colour
Convenience of meat	Cooking time	Value for money
Wastage	Expense of meat	Preparation time
Bone content of meat		Flavour strength
Fat content of meat		Nutritional value
Tenderness of meat		
Juiciness of meat		

Source: After Wilkinson, 1985.

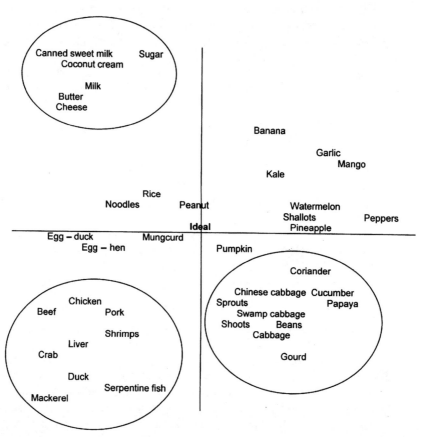

Fig. 5.11 MDS for Thai raw materials and dairy products (Source: From Anderson, 1974).

ham, bacon, sausage products, offal cuts and white meats using 18 product attributes that had been identified by consumers. Wilkinson (1985) grouped the attributes, using factor analysis, into three main groups – on buying, on cooking and on eating. The consumers regarded each meat cut as having a unique 'blend of appeals' under these three general groups shown in Table 5.11. He concluded that the individual product profiles he developed for the 45 meat cuts using these attributes had very important implications for new meat product development. To produce a more desirable mutton cut, one objective could be to improve the negative aspects of mutton while not affecting the attributes that identify the product as a mutton cut. To produce a beef-like product from mutton, the new mutton product would have to mimic the specific attributes of beef if the consumers are to be convinced that the new product, based on mutton, is indeed like beef.

MDS and clustering techniques can be used to place the product concepts with products already on the market, to confirm the positioning of the product concepts with the competing products. Anderson (1974) in developing a quantitative model for the design of nutritious and acceptable foods in Thailand, used MDS to compare dairy products with Thai food raw materials in Fig. 5.11. Groupings of vegetables, meat and fish and dairy products with coconut cream and sugar can be seen. The position of an ideal product can also be determined using MDS.

Principal components analysis can reduce the number of attributes, as it identifies the smallest number of latent variables, called 'principal components' that explain the greatest amount of observed variability in products. It can show the relationships of the products to each other and also the associations among the attributes. In testing a number of orange drinks, the original attributes were sweetness, pulpiness, colour, thickness, natural colour, sourness, bitterness and after-taste, and these were represented by two principal components. The branded orange drinks were in roughly four groups, with four orange drinks grouped around the ideal (Cooper *et al.*, 1989).

As the product attributes are identified and the product concepts developed, there may be a need for further evaluation, a comparison of the final product concepts. For the specific product concepts, there should be a comparison by the consumers with:

- ideal product on attributes;
- competing products on attributes;
- competing products on buying predictions at different prices;
- competing products on relative positions in the market.

There could be a need for more detailed analysis of the product concept for the feasibility study, with predicted sales volume, sales revenue, market share, probabilities of product success and market success. For the general food type these can be predicted through reference to statistics of food distribution, sales records and supermarket data. For example the sales of a product that has been accepted by consumers, and has been on the market for some years, may be used to predict the sales of a new product with similar properties. Consumer diary

Table 5.12   The product concept for design

- Product – category, image
- Consumers' market segment(s)
- Behaviour of consumer with product – buying, preparing/serving, eating, disposal
- Consumers' perceptions of the product type and new product
- Consumers' positioning of the product in the market
- Consumer-identified product attributes
- Consumer-identified product values, benefits, functional attributes
- Packaging – colour, symbol, style, information
- Pricing, value for money – consumer pricing, competitive prices, price position
- Availability – distribution and retailing for consumer
- Consumer concerns about raw materials, production and processing

---

Source: After Earle and Earle, 1999.

records and pantry surveys may also be used to predict consumer acceptance of new products. But there can be a need for a consumer survey on the new product concept, particularly for the innovative product with no related products on the market. Predictions of market behaviour are made under a range of possible future environments predicted from trends in consumer and social changes.

The product concept is then detailed for the product design specifications and the product design. The sections in the product design concept (Earle and Earle, 1999) are shown in Table 5.12.

5.6.4   Developing the product design specifications from the product concept

An adaptation of the profile test – the optimum location profile – can be used by the consumers to identify firstly the important product attributes, and then define their ideal point for the product on each attribute. The consumer panel needs to be representative of the diversity among potential customers in the targeted segment. If there is more than one segment, then panels are organised for each segment. Of course the consumer cannot score some imaginary product but can only compare their ideal with some real products. So they have to be presented with either the company's present product, competitors' products, a home recipe or early samples of the new product. There is a need to choose samples with differences in the magnitudes of the attributes, so that the consumers can identify their ideals (Booth, 1990). The product designer can try variations of the new product, altering ingredients to give variations in some attributes, so that the consumer is relating the changes to their preference at the same time that the designer is starting the creation of the new product. This is a valuable interaction, and cuts down a great deal of time later in the development. The consumer response is directly linked to the stimulus (McBride, 1990).

The optimum product profile identifies the 'strength' of each attribute in the product concept. The consumers score the product concepts usually with competing products, on each attribute using scales, for example, for viscosity:

Thin, runny		Very thick

It is difficult for the consumers to do this in the abstract and prototype or commercial products are tested. The mean scores of the consumers are used to derive an attribute profile, as shown in Fig. 5.12 for three baked beans products, a standard, one with increased tomato and one with increased beans. Other attributes for the bean products could be availability (difficult/easy), convenience of use (difficult/easy) or product image (everyday food/special food). The consumers rank the attributes from most important to least important, and note any attributes that are missing. There is constant checking throughout the build-up to the product design specifications. The consumers can also score their overall acceptance for each product to find the effects of the product changes on acceptance. Product profiles are normally identified by product type, but it is important to realise that the brand and the packaging can have an effect on the rating of products. The next step is to ask the consumers to score their ideal product on the same scales. This gives an ideal product profile for the product design specifications.

The consumers' definitions of the attributes and their magnitude can be related to an expert panel's sensory analysis of the product (Moskowitz, 1997) or to measurable physical properties such as viscosity and colour, microbiological standards, nutritional value and chemical composition. It is this correlation of the consumers' attribute scaling with either analytical sensory testing or standard product testing (Muñoz, 1997) that has changed the product design and process development method.

In developing the product design specifications, there are difficulties in measuring sensory properties on an objective basis and correlating these with consumer panel data. The complicated interaction of sensory properties, both

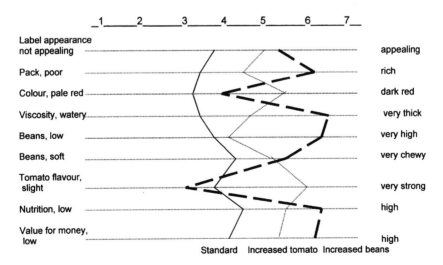

Fig. 5.12 Profiles of three baked bean products.

within the same sensory property, flavour with flavour, or between different sensory properties, flavour with texture, has made the setting of design specifications difficult. Sometimes, it is not possible to identify objective measurements for the consumers' ideals and it can be necessary to have the consumers test the experimental products. As consumers may have problems remembering their ideals, they can score on how near the new product prototypes are to their ideal; this consumer testing during design can also identify any unknown factors that may be affecting their acceptance of the product.

The product qualities are directed mainly by the consumer but of course they must conform with any legal regulations and with defined company policy. The product design specifications also include marketing, production, distribution and environmental factors (Earle and Earle, 2000); these are mainly technical but there are also consumer needs and wants on raw materials and processing methods which are incorporated.

Think break

Develop an optimum product profile for your new product ideas in the following product categories:

- extruded, shaped, snack food,
- canned sardines,
- refreshing drink.

1. Write a description of the new product.
2. Identify the product attributes for the new product.
3. Develop scales for each attribute.
4. Score two competing products on the scales.
5. Score your ideal product on the scales.
6. Compare your optimum product profiles for the three products.

5.7   Consumers in Stage 2: Product design and process development

During the early stages in developing the product prototypes, consumers may not test the product, but whenever the prototypes are coming near to the standards set in the product design specifications, consumers' needs are brought into the product testing.

5.7.1   Including the consumer in product design and process development

It is important that the experimental techniques and objective tests related to the consumer attribute standards are used to connect the study of the different levels of the processing conditions, or different proportions of raw materials, with the rating of the product attributes by the consumer (Saguy and Moskowitz, 1999). Sometimes, it is not possible to identify objective measurements for the consumers' ideals and it is necessary to have the consumers test the experimental products. As consumers may have problems remembering their ideals, they can score on how near the new product prototypes are to their ideal; this consumer testing during design may also identify any unknown factors that may be affecting their acceptance of the product.

There is consumer involvement in the packaging design, and in studying the relationship of the product prototypes to the food behaviour – what are often called 'use' tests. As the designer is reaching the stage of optimisation, the product prototype (or two to three product prototypes) is tested by a larger number of consumers in central location tests. From this the product is optimised and the final prototype developed. This is usually tested for acceptance in a large random consumer test. This building up of the consumer testing in product design and process development (Earle and Earle, 1999) is shown in Table 5.13. This scheme is only an indication of the techniques to be used at each stage.

Table 5.13   Consumer testing in product design and process development

Steps	Activities	Techniques
'Getting the feel' Consumer panels	Ideal profiles	Profile tests Descriptive sensory analysis Multivariate analysis
	Product 'Mock-ups'	
'Screening prototypes' Consumer panels	Product comparison	Difference testing Ranking
	Elementary product prototypes	
'Ball park studies' Consumer panels	Acceptability of attributes Acceptability of products	Attribute scoring to ideal Preference panel Hedonic testing
	Acceptable product prototypes	
'Optimisation' Central location test Consumer panels	Product improvement Competitive comparison Food behaviour study Packaging testing	Acceptance testing Hedonic comparison In-home use tests Ergonomic testing
	Optimum product prototype	
'Scale-up' Random consumer test Small buying experiments Consumer panels	Buying predictions Commercial product concept	Acceptance testing In-depth interviews
	Semi-commercial product	

Source: After Earle and Earle, 1999.

Generally the numbers of consumers taking part increase as the development progresses; there may be only 30 at the 'mock-ups' but gradually building to 200 or more at the last step of scale-up. At this time, samples can be made on pilot plant or semi-production plant so there is product for large-scale consumer testing. It is important that the knowledge about consumers is being built up with the design so that the large-scale testing is a confirmation rather than an unknown. It is possible to measure the design's impact on consumers' product perceptions as the design is developing, so that the product/consumer relationship is known and is optimised. As the product attributes are being built into the design, the consumer is reacting to them and is conveying their perception of them – which may be quite different from that of the designers (Veryzer, 1997)!

5.7.2   Product attributes evaluation

The problem in product development is identifying what are the critical attributes of a food to the consumer and then measuring them in the design process. Three questions can be asked:

1. Is there a difference between two food samples? Used when trying to duplicate a product or to see if there is a difference between product prototypes. Difference tests such as triangle tests, paired comparisons can be used.
2. Is the product acceptable? How acceptable? Used for the optimum prototype products when testing by large consumer groups or smaller, representative panels.
3. What are the characteristics of the products? How strong are they? Used when building up the product concept and also in designing the product prototypes. This can be called descriptive analysis or product profiling. The profile method is designed to give a profile of the overall sensory properties by describing and determining the relative magnitudes of the attributes.

The sensory attributes of the product are designed to find not only the ideal magnitude ('*bliss point*') for the individual sensory attribute, but also the combination of sensory attributes and their magnitudes for the optimum combined sensory attributes, which gives a high hedonic acceptance (Moskowitz, 1994). After optimising the sensory attributes, there is still the acceptance test to see if the total product concept has been achieved. Formerly in product development, there were many hedonic tests with consumers in developing the product, but with today's analytical sensory and other techniques, the product design specifications can be detailed so that much testing can be done with physical, chemical, thermal tests or with analytical sensory testing. Perception tests such as ideal product profile tests are conducted with consumers at a set stage in the product design, and then total product testing when all the specified consumer attributes in the product are nearing optimum.

Having identified the attributes, the method of testing must be selected. Sensory science has developed a great deal in the last 30 years and there are many techniques which have been tested and recognised (Meilgaard *et al.*,

Table 5.14   Attribute evaluation in product development

Product stage
- Product description
- Product concept
- Product design specifications
- Basic product
- Product prototypes

- Unidentified final product
- Identified final product
- Commercial product

Type of respondent
- Expert judges
- Trained panels
- Company staff
- Focus consumer panels
- Representative consumers

- Randomly selected consumers
- Target market

Measurement procedure
- Attribute selection
- Quantitative measurement of attributes
- Hedonic measurement

- Total product acceptance
- Buy/use intention
- Sale/use

Source: From Moskowitz, Benzaquen and Ritacco, 1981, by permission of Business News Publishing, West Chester, PA.

1999). Descriptive analysis techniques are used in product development because they describe and measure the multiple attributes in the product, and determine the magnitude of the attribute – in other words they are quantitative. The results from a trained panel can be replicated and therefore can be used in the statistical analysis of data, using linear relationships, which can be used to optimise the product prototypes. The use of ideal profile for following design is useful in determining whether the optimum product prototype is being achieved. 'Just right' scales are also used to determine if the product is nearing an optimum. The selection of type of stimuli, type of respondent and measurement procedure (Table 5.14) is important in ensuring the validity of the sensory results (Schutz, 1993).

Think break

In developing a new tomato pork sausage, a product developer had five prototypes which are being tested by a small 30 member panel of the target consumers – teenagers. The sausages had two levels of salt and two levels of pepper. A simple 2 × 2 experimental design was used with Product A low salt, low pepper, Product B low salt, high pepper, Product C high salt, low pepper, Product D high salt, high pepper, Product E medium salt, medium pepper. The panellists were asked to study the flavour level, whether any flavour is dominant, and the level of salt.

	A	B	C	D	E
	Number of people				
Flavour level					
Too high	3	5	15	18	11
Just right	10	14	8	5	7
Too low	17	11	7	7	12
Flavour balance					
Some very dominant flavours	3	7	10	11	8
Some slightly dominant flavour	2	5	5	10	7
Rounded balanced flavour	25	18	15	9	15
Harshness					
Very harsh flavour	0	3	5	7	3
Slightly harsh flavour	1	5	3	9	3
Mild flavour	29	22	22	14	24
Saltiness					
Too salty	1	2	10	12	8
Just right	12	11	11	10	14
Too low	17	17	9	8	8

They identified the dominant flavours in each product as:

Product A – meatiness, Product B – spiciness, pepperiness, Product C – salt, Product D – salt, pepperiness, Product E – spiciness.

Study these results and discuss what is happening with increase of salt and pepper. What levels of salt and pepper would you use in optimising the sausage formulation? Discuss how you would experiment further with the sausages.

5.7.3 Ball park experiments

In many studies it is important to relate the characteristics of the raw materials to the final product qualities so that in product formulations, raw materials can be varied to achieve the optimum product. For example the sensory properties of the raw materials can be related to the sensory properties of the products which in turn are directly related to the consumer acceptance of the final product. The sensory properties of each available raw material and its competitive and compensatory actions need to be identified either from recorded information or from experimental work. A linear function can then be used, within a linear programming model, to select subsets of raw materials, which can be combined through processing to have attributes acceptable to the consumer (Chittaporn, 1977). Moskowitz (1994) showed how products can be built up and optimised by studying the effects of ingredients in the acceptance of the product prototypes; sometimes the relationships between the formula ingredients and the sensory attribute ratings are non-linear, and there are interactions occurring, so that the relationships can be complex.

During the preliminary design and process development, a wide range of ingredients may be screened and the processing and cost limits are identified. Product testing at this stage is better done internally (Wilton and Greenhoff, 1988), but once the limits of the design are identified and optimisation of the product quality starts, it is useful to bring in consumer testing. Sometimes internal development is completed before consumer testing is started, but this may lead to recycling of the design and accompanying delays. If factorial experiments or mixture designs are used to follow changes in processing variables and proportions of ingredients, then consumers can test the products and not only show how overall acceptance varies but also how critical product attributes change (Schutz, 1993). From their responses can be predicted the optimum area for further development. By regression of the results from the factorial experiment, an equation relating acceptability scores or ratios of attribute scores to the ideal product, to the ingredient and processing variables is obtained, which can be used to predict the optimum processing point. If this is a linear relationship, it can be used in linear programming with other factors such as nutritional compositions and costs.

An alternative method is to use expert optimisation where the consumer sets their ideal points on the product characteristics in the product design specifications and then a small group of trained panellists can follow the changes of the specific product attributes and relate these to the consumer ideals. Physical tests can be used which are corrrelated with the consumer measurements of a product. Use of trained panellists quickens the design and development. Care must be taken to bring in the consumers when the prototype products are nearing the optimum to check if the ideals are still related to acceptance by the consumer and that no rogue characteristics have crept in which are not wanted by the consumers.

In developing a nutritionally balanced snack product for urban Thai children, Sinthavalai (1986) developed a formulation by linear programming based on nutritional needs and costs, and developed some preliminary product prototypes by interacting expert sensory testing with the linear programming choice of ingredient levels. A consumer panel tested the final prototype from this experimentation. The magnitudes of the attributes and also the ideal scores for each attribute were recorded. The ratios of the experimental scores to the ideal score (1.0) are shown in Fig. 5.13. It was found, shown by the distance from the ideal, that the prototype was rather crumbly and gritty, and had a rather strong mungbean flavour and weak fruit flavour. Crumbliness in particular was not ideal. This initial prototype set the direction for further experimentation in processing and raw materials, these new prototypes being tested by large panels of Thai children.

5.7.4   Optimisation

In the optimisation stage, there are several types of activity taking place which need a combination of consumer panels with larger-scale consumer testing in

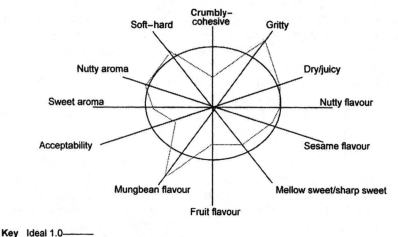

Key   Ideal 1.0———
      Sample, ratio of mean score to ideal score·················
(Ingredients: mungbeans, glucose syrup, banana syrup, deep-syrup cooked banana, roasted white
sesame seeds, skinned roasted peanuts, ascorbic acid,riboflavin, vitamin A, folic acid)

Fig. 5.13   Thai snack bar profile, ratio to ideal by small consumer panel
(Source: From Sinthavalai, 1986).

central location tests and use tests. The product itself is now perceived 'right' by
the designers, but has to be checked by consumers to see if they perceive any
improvements. There is a need to confirm consumer acceptance of the product
and the individual product attributes. Different aspects of the product can be
tested from the basic product to the total product with packaging and sales
promotion.

The unidentified product prototype(s) is often tested in large-scale central
location tests held in a shopping mall or a movable, testing caravan, which can
be taken to a central spot in the city. A single sample or pair of samples or
sometimes multiple samples are used in the central location test to determine if
the product is better than the competitors' products. When increasing the
numbers of consumers testing the product, the first decision is are you testing the
product only or are you testing how the consumer cooks, serves and eats the
product? If you wish to test the product only, then one needs control of all the
'use' factors; these are standardised so that the product preparation, the
temperature and the serving are all the same. Central location tests are used to
test the product and also to gain information from the consumers about their
attitudes to the product and their predicted behaviour towards the product, as
well as identifying the acceptance of different characteristics of the product. The
cooking and presentation procedure is absolutely standardised. The type of
information gained from a central location test is used for directing the final
stages of product design and early stages of product commercialisation. The
response of the consumer is immediate and the results can be analysed very
quickly. They can be used to compare different prototypes to direct further
development. Packaging and information can also be tested and even the total

product with advertising. Central location tests are standardised, informative, speedy and usually cheaper than randomly selected 'use' testing. They give individual responses either in a one-to-one interview or by using a self-completion questionnaire. They can have from 50 to 200 consumers.

In-home, better called in-use tests, are related to how the consumer or the household or the food service chef fits the product into their eating pattern, and into their cooking and serving facilities. If the product has to be prepared, it gives the acceptance of the 'cook' as well as the acceptance of the people eating the product. It gives the marketer and the developer the sure knowledge that the commercial product is acceptable to the target market. It is the assurance before the launch that the product and the package are accepted. It can also show what the designers and their consumer panels have forgotten or ignored! One major defect or asset can influence the acceptance. An important aspect, which is found by in-use tests, is the emergence of changes that occur over time, that is with the usage or the eating of the product over a week or two. This is an important aspect when an unusual product is being tested, which may be rejected at the first test but is gradually accepted. For example in introducing an oil-based mayonnaise to target consumers who were used to eating a condensed milk-based mayonnaise, the oil-based was rejected in the first test but accepted in the second test.

The packaged product is tested in a 'use' situation to see how the combined product fits with consumers' food preparation/eating behaviour. This can range from the basic question 'can the consumer pick up the pack and open and use the product?' to the more complex 'how does the product fit the consumers' need to show they care for their family?' The study is usually made by allowing a panel of consumers to use and eat the product at home or at the place of eating the product, and then interviewing them. It can also be done by an observation test in a central kitchen.

Because of the need for confidentiality, consumer panels often test the total product with brand, packaging design and in-store promotion material. These are usually focus groups that study all aspects of the product to confirm that the different attributes are related to the consumers' needs, wants and behaviour. The aim of optimisation is to discover the best product overall, and to determine if it will beat the market leader or the company's direct competitor.

5.7.5   Scale-up

This is the last time before commercialisation to check if the product is accepted by the target market segments, that it has the attributes wanted by them and that they will buy the product at the suggested price. Acceptance of the final product prototype can be tested in a large randomly selected consumer test, and consumers can be asked the probabilities that they will buy the product at different prices. But buying predictions are more accurate if consumers are presented with the product in a buying situation.

The tests can be by representative panels with in-depth discussion for the initial products from the scale-up experiments. The final product is tested either in a large survey of the target consumers or by a small buying study and sometimes even a town/city market. Acceptance scoring, hedonic (like/dislike) scoring, ranking and scoring of acceptance on different characteristics are all used. Hedonic scores will differ among different groups of consumers, for example in testing a grilled steak some people may like it rare and others like it well done and will score differently for hedonic scores. Therefore it is important also to score the attributes so that the designer can know why the consumers are accepting the products in a different way. It is important to have a homogeneous group of consumers. If there are two groups of consumers with different needs, then the scores can be bimodal, that is split into two groups. It is wise to look at the distribution of the scores and to see if there is clustering of scores, which indicates that there are specific groups within the sample reacting in different ways to the product.

Consumers can be asked their prediction of buying the product at different prices – how much they will buy, and how often they will buy. There are problems in asking the consumers if they will buy the product, which can be overcome by using small buying tests. The product can be sold in a supermarket or in another type of retailer, for example a few restaurants or takeaways; or a pseudo retailer can be developed by the company to sell products, including the new product.

In this scheme, there are some important decisions on timing – when is the product to be identified with a brand or company name, when is the price to be introduced, when are the products to be compared with competitors' products? It is important to introduce these quite early to small consumer panels, although there may be problems of secrecy. It can be a disaster if they are only introduced in the last major tests to confirm buying predictions. In Table 5.15 are shown some effects of branding snack bars on acceptance of the sensory characteristics (Moskowitz et al., 1981). In the same test, one brand, Snickers, had a 124% increase in liking; the other brand, Milky Way, had 54%.

The validity of consumer testing in predicting the outcome of the product development project improves as the product goes from product concept to prototype product to commercial product, the product measurement goes from

Table 5.15   Effect of branding on sensory scores

Product	Inside chewiness	Chocolate flavour	Sweetness	Overall quality
Snickers				
Not identified	60.21	36.85	53.0	65.84
Branded	62.35	53.17	62.2	84.16
Change on branding	4%	44%	18%	27%
Milky Way				
Not identified	59.68	50.59	74.8	88.20
Branded	67.26	64.03	76.1	104.96
Change on branding	13%	27%	2%	19%

acceptance to buying and more consumers become involved in product testing. The involvement of consumers in focus groups in the building of the total product concept, and their acceptance of the prototype product and their predictions to buy and use during product design, increase the validity of the predictions for product success at the end of product design. This can reduce the need for large-scale consumer and market testing during commercialisation.

Think break

A company is developing a new liquid breakfast for adult consumers between 20 and 40 years old. The consumers have identified product attributes that they wish the product to have: nutritional value (high fibre, low fat, low sugar, low salt, calcium, folate, iron and vitamins), health value (sustainable but not weight increasing); value for money; easy to pour; portion about 250 ml; not sticky or sickly sweet; mild roasted cereal flavour.

1. Outline the steps in designing the product.
2. How would you organise the testing of the product prototypes at the different steps of the ball park experiments, optimisation and scale-up so that the consumer needs and wants are evaluated at each step?
3. How would you test the various attributes identified by the consumers during the design steps?
4. How would you test the final product prototype for commercialisation?

5.8   Consumers in Stage 3: Product commercialisation

The consumer is involved in two parts of the product commercialisation: commercial design of product and marketing, and the commercial testing of the total commercial product (Earle and Earle, 2000) as shown in Fig. 5.14. Figure 5.14 shows only the steps in the commercial design and commercial testing where consumers may be involved.

The final design of the commercial product is interwoven with the design of the marketing – the market channel, the promotions and advertising, the selling method and the pricing. The consumers are involved in various aspects of these designs – usually in focus groups or other types of consumer panels such as advertising panels.

5.8.1   Final consumer product concept

The product prototype is built up to the total product concept, which is accepted by the consumer and the society. The company core product with its attributes and packaging has been optimised in the previous stage, but the final aesthetics of product and packaging, the brand, the name, the price, the advertising and the

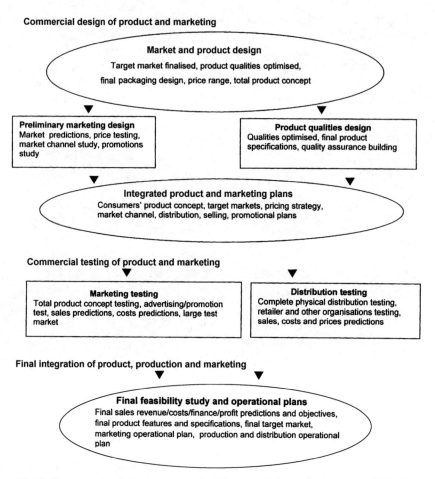

Commercial design of product and marketing

Market and product design

Target market finalised, product qualities optimised,

final packaging design, price range, total product concept

Preliminary marketing design
Market predictions, price testing,
market channel study, promotions
study

Product qualities design
Qualities optimised, final product
specifications, quality assurance building

Integrated product and marketing plans
Consumers' product concept, target markets, pricing strategy,
market channel, distribution, selling, promotional plans

Commercial testing of product and marketing

Marketing testing
Total product concept testing, advertising/promotion
test, sales predictions, costs predictions, large test
market

Distribution testing
Complete physical distribution testing,
retailer and other organisations testing,
sales, costs and prices predictions

Final integration of product, production and marketing

Final feasibility study and operational plans
Final sales revenue/costs/finance/profit predictions and objectives,
final product features and specifications, final target market,
marketing operational plan, production and distribution operational
plan

Fig. 5.14   Steps involving consumers and their needs in product commercialisation
(Source: After Earle and Earle, 2000).

promotion are added to give the commercial company product. In other words
the marketing attributes of the product are added to give the final product image.
The raw materials from the primary production, the ingredients from the primary
processing, and the processing during manufacturing of the consumer products,
are included, as these are important areas of concern for the consumers and the
society. In building the final consumer concept, it is important to include the
consumers' environment, what is happening and predicted to happen. The
consumer's product concept is influenced by:

- competitors;
- social, political, economic, physical environments;
- media and communication;
- consumers' own behaviour towards the product.

The consumer product concept is designed to convey to the consumer that the product satisfies the consumers' eating needs and wants, nutritional requirements, availability, use of appropriate materials, reliability, maximum safety, good appearance, low cost and psychological acceptability. But this has to be located in the environment surrounding the consumers, and has to be related to the correct target consumers. The product is rated with the competitors, showing product leadership, product parity or product loser. The relationship between price and the user-perceived quality is carefully evaluated to fit the consumers' perceived value for money. They may accept a high price indicating high quality, or a low price indicating ease of purchase, or they may have in-between attitudes trading off quality and price.

5.8.2   Consumers in marketing and production design

There can be some involvement of consumers with production as production builds up the product qualities for the production specifications, at the same time as they are involved in marketing design (Earle and Earle, 2000), as shown in Fig. 5.14. The marketing design includes, together with the product:

- market information – market research and analysis, particularly the targets to be set for the launch and post-launch and the methods to monitor these;
- market channels and distribution – choice, control and development of market channels, transportation, storage;
- pricing – price range, relation of price to demand, margins, discounts, specialling;
- promotion – retailer and consumer promotion, advertising, public relations;
- sales – methods of selling, terms of sale, sales reporting, analysis and forecasting.

Consumers can be involved in several of these marketing decisions. There could be consumer surveys to predict the buying of the product, and from these the prediction of sales for the sales targets. The market channel must make the product available to the consumer at the right place, the right time, the right price and the right quality, so there is research on where the consumer buys the product, how often, and how they store the product and for how long. They can be involved in retail surveys and also in shelf-life trials. Lack of shelf-life testing with consumers may show up as unwanted deterioration of the bought product in the household refrigerator. Shelf-life testing is usually done by a trained sensory panel using descriptive sensory analysis. The critical product attributes are measured over time and the changes in the retail outlet and the home storage are measured. The consumers set the acceptance levels of stored products and these tolerance limits are used as a guideline by the trained panel.

Often consumers are involved in promotional design, particularly in development of the visual material for sales promotion and the video/film for TV advertising. Focus groups or promotional consumers take part in developing the product image, slogans and educational material. Promotional material is tested with consumers to compare the various designs using measures for awareness and persuasion to try, and also to examine the clarity of the information about the product.

5.8.3   Commercial product testing

The types of consumer testing on the final product and marketing designs vary according to the type of new product and the amount of consumer research in the previous stages. If it is an incrementally improved product, there is already a great deal known of the market and, if it has already been tested in a consumer test, then there will likely be no need for a test market and it can go straight into a launch. But if the new product is a major innovation, it can justify in-depth studies with consumers on the final consumer concept, and large-scale consumer tests as well as a final test market. There can be new products within these two extremes. It is a case of balancing the risk of failure through lack of knowledge with the costs of time, money and other resources. Delay may cause failure because of launching at the wrong time or loss of confidentiality allowing competitors to launch ahead. Lack of knowledge can also cause failure because misunderstanding consumer perceptions may lead to an uninviting product image in the promotion. Some questions to be answered in the testing of the commercial product are shown in Table 5.16. The questions needing answers lead to the type of testing required.

If a great deal of information on the product and the relationship of the consumer with the product were needed, it would be a consumer use-test of the total product. Consumers would be interviewed, using in-depth questioning, on their reactions to the product, and their predictions of their future behaviour

Table 5.16   Questions in commercial product testing

Marketing
- What will be the consumers' purchasing/repurchasing behaviour?
- What will be the consumers' reactions to the prices, the promotions?
- What are the predicted pessimistic, most likely and optimistic sales units and revenue over the next months, years?
- What are the predicted competitive reactions?
- What are the predicted market shares?

Product
- Is the product what the consumers want?
- Does the product have the benefits wanted by the consumer?
- Does it have the desired attributes wanted by the consumer?
- What are the consumers' concerns about the product?
- Is the package acceptable, right size?
- Are the product and packaging attractive at point-of-sale?
- Are the product and pack ethical, legal?
- Do the brand and the product image relate to the product?
- What is the consumers' total concept of the product?

towards the product. In the consumer test, a statistically representative sample is chosen from the target market(s) so that an indication can be obtained of the opinions and attitudes of the consumers in the market. Usually a single sample presentation (a monadic test) is used and the complete commercial product is tested with appropriate sales promotion and public relations material. The product can be delivered by mail or hand delivery. Information is best obtained by interviewing either personal or telephone, but sometimes self-administered questionnaires are used. Using consumer tests to predict buying behaviour can be inaccurate as consumers have trouble themselves in predicting future behaviour.

If information on the marketing methods and their effects on consumer buying behaviour is needed, a test market would be used. The consumers would have the opportunity to buy the product in a supermarket, restaurant or other relevant retail outlet. This could be in one or two supermarkets or restaurants or takeaways, using only the in-store promotions, or it can be in a market area with the media advertising and public relations. The consumers who are buying the product and perhaps some of the consumers not buying the product are interviewed, to determine the acceptance, competitive difference, uniqueness, aesthetic worth, brand attitude and product worth. It is important also to determine consumer reactions and consumer buying behaviour in the test market, by interviewing consumers about their purchase and repurchase of the product, their use of the new product and their opinions on the new product. The sales data are found from the computer sales records of the retailers and from this, national sales can be forecast. Companies can still be experimenting with different options during the test market, for example different prices and different displays (Hisrich and Peters, 1991). The interactions of the variables in the marketing mix can be determined.

Ethical product testing is an area that is increasingly important today. This is relating the product and the marketing to the ethics of the society. Ethical testing is related to a particular society; but basically, in most societies, people want to be able to trust the company not to harm them or use fraud and deceit against them (Earle and Earle, 2000). In the food industry, this is even more important than in other industries because people consume all the products and their health depends on them. For a mutually satisfactory future, all the product testing must be truthful and encompassing so that the company earns a good reputation in launching new products.

During product commercialisation, not only has there to be testing of the product to build up the knowledge about its benefits and defects, but plans developed for both short- and long-term consumer testing after the launch.

Think break

Chilled salmon is being developed in New Zealand for marketing in Europe. The salmon are grown in the same environment in the clean seawater of the Sounds in New Zealand, eating the same food, so their quality is consistent. They are harvested in a rested state (anaesthetised before harvesting), so the flavour and texture during storage are assured. They are packed in a regulated gas atmosphere in special packs to slow down the ageing process in the chilled fish. The aim is to send the salmon by sea which will take 60 days; but there is a need to test the market before the final development of the package and containers. The fish will be flown to Europe for the test market.

1. Discuss the problems in test marketing this chilled product which is to be sold in supermarkets.
2. Describe how a test market could be set up in one supermarket. Which country and which type of supermarket could be used?
3. How long would a test market need to run to give a reliable prediction of sales?
4. Outline how consumers buying the product could be surveyed and what information would be asked from them.
5. How would you relate the product to the competing products?
6. If the test market is successful, what problems do you identify in organising a product launch?

5.9   Consumers in Stage 4: Product launch and evaluation

The involvement of the consumer can be divided into the consumer launch and the consumer evaluation after the launch.

5.9.1   Consumer launch

The launch to the consumer depends on the type of product, the innovation level of the product, the budget and the general organisation in the company for product launches. The aim is to achieve the diffusion of the product through the target market as set in the launch targets. Products vary in terms of newness and therefore the education needed to make the consumer aware of the product's values and benefits, and the 'learning' needed by the consumers to try to adopt the product. The consumer also sees risks and costs in trying the product, which need to be predicted and the correct reassurances given. At one end of the product spectrum is the line-extension and the improved product, which are easily accepted by the consumer, and at the other end is the absolutely new food of which the consumer has little understanding. The company's launch plan has to consider the needs of the consumer in adopting the product and the time it will take for the consumer to accept the product (Earle and Earle, 1999). There are high-learning and low-learning products. For the high-learning products, the marketing strategy is based on a prolonged market development effort during which special attention is given to sales and distribution as well as constant examination for any product weakness and any unprecedented consumer reactions (Hisrich and Peters, 1991). Advertising and public relations need to make the consumer aware of the product; trialing of the product needs to be encouraged by free samples and tasting of the product. Low-learning products are usually improvements on competitors' products and the introduction there is emphasis on the competitive advantage. There is less time needed for education and it is mostly just awareness by the consumer of the product and encouragement to buy through the extra benefits of the product.

Another feature of the market to consider is the growth rate in the market. If it is a market growing fast with many new products, then it is important that the consumers who are adopting the new products are recognised and targeted in the marketing. If they can be encouraged to buy the company's new product and the product does have the needed values and benefits, then they could stay with the company's products and ensure a growing share in a growing market.

The timing of the launch is also important to the consumer – is it the right time of the year to eat the food? Is it the right time of the month for consumers to have the money to spend on a new product? Are they becoming bored with the competitors' products? Often the timing is dependent on the type of consumers in the target market; if they are highly receptive to new products, then the products can be marketed aggressively; if they are fearful or antagonistic to new products, then the launch is slower with much more education and trialing.

5.9.2   Consumer evaluation of the launch

Whenever the product is launched, the consumer's buying behaviour is studied, who is first-buying and who is re-buying; how much are they buying at each purchase and what is the time between purchases? Buyer diaries can be used, where the consumers record their purchases, not only of new foods but also of other foods, especially in the product category. The company can determine who is buying, the re-buying pattern, the timing and the amounts for each purchase. They can also determine from what particular brands it is gaining customers, to what other brands the product is losing customers.

When a new product enters a market, there will be a gradual shift in consumer perceptions of the product category and the competing products. The factors on which consumers make their decisions may change and the positioning of the product in the market may change. Therefore after the launch, the marketers need to test at regular intervals the whole category and whether the positions of the competing products are changing. This is particularly necessary where the new product is a major improvement or a new product.

There is also a need for information on the consumer's perceptions of the new product and the product values and attributes. It is necessary to ask them how they found it at all stages of the food behaviour sequence from buying to consumption and disposal. It is important to test the product attributes on which the product was designed, and also to see if consumers identify new product attributes that are influencing their acceptance of the product. The consumers can be contacted outside the supermarket, just after they have bought the product, and then interviewed, when they have eaten the food product, either by telephone or by face-to-face interview. This survey will indicate not only how the product can be improved but also how the marketing mix is performing. Two important attributes to discuss with the consumers are nutrition and safety, as well as any environmental reactions to the product.

The product quality is also followed with the consumers to see if they are noticing any variations or defects. Any products returned to the company by consumers need to be examined carefully. Raw materials/ingredients may change after the launch and the product may need to be reformulated; also there may be processing changes to make the process more efficient. The products from these changes need to be consumer tested to confirm that the changes have not lowered the acceptability of the product. Difference testing, in the form of triangle, duo–trio or paired comparison tests, can be used to see if the product change is significantly different. If there is a noticeable difference the product needs to be tested by consumers in an acceptance test. For example in launching an instant drink mix, the full-scale production had temperature control problems so that the solubility of the final powder varied, sometimes taking some time to dissolve. This made the product unacceptable for the market and caused a product failure. Quality audits need to be done after the product launch at various points in the distribution chain. These are usually done by trained panellists using descriptive techniques; but the consumers again need to set the standards for the trained panellists, or consumer acceptance tests need to be done on the retail product. Trained sensory panellists will probably do trouble shooting in the distribution chain, for example for a distinct off-flavour and will be backed up by instrumental analysis to identify the compounds. This may be the basis for an insurance claim or a legal case and therefore has to be a quantitative test.

Despite all the brainstorming, technology, streamlining of operations and large amounts of money spent on promotion that go to make up a new product, and the best efforts of management to predict what the market should want, what really matters is what the customer wants. The gap between hopes and reality reduces as the customer is better understood. Consumer research has made great advances in recent years, and many of the results of the research can be applied directly in the food industry. The progressive and successful food business should look very carefully at them. The potential losses on unsuccessful new foods create a very substantial pool from which even quite major investments in time and effort in consumer research can be profitably funded if they lead to real improvement in predictability. The evidence suggests they do. This chapter has introduced the very active literature, and set out avenues that lead towards increased precision in understanding the needs of the customer who actually buys the food and eats it.

Think break

1. Review the consumer research methods available and discuss their use in product development if your company is:
   (a) a specialised ingredient supplier to the baking industry,

(b) a manufacturer of dried soups in small consumer packs for convenience stores,

(c) a large-scale sugar refiner marketing individual packs of sugar to coffee shops,

(d) a manufacturer of consumer margarine products for supermarkets.

2. Consider your company's consumer research budget in relation to that for advertising and promotion, and set out your view of their relative significance on overall profitability.

5.10 References

ANDERSON, A.M. (1974) A quantitative model for the design of nutritious and acceptable foods. PhD Thesis, Massey University, Palmerston North, New Zealand.

ANDERSON, P. & HE, X. (1999) Culture and the fast-food marketing mix in the People's Republic of China and the U.S.A.: implications for research and marketing, in *Consumer Behaviour in Asia: Issues and Marketing*, Chan, T.S. (Ed.) (New York: International Business Press).

BAREHAM, J. (1995) *Consumer Behaviour in the Food Industry: A European Perspective* (Oxford: Butterworth-Heinemann).

BEHARRELL, B. & DENISON, T. (1991) Food choice in a retail environment. *British Food Journal*, 93(7), 24–30.

BELL, R. & MEISELMAN, H.L. (1995) Food retailing and the food consumer, in *Food Choice and the Consumer*, Marshall, D.W. (Ed.) (London: Blackie Academic and Professional).

BOOTH, D.A. (1990) Designing products for individual customers, in *Psychological Basis of Sensory Evaluation*, McBride, R.L. and MacFie, H.J.H. (Eds) (London: Elsevier Applied Science).

BUISSON, D.H. (1995) Developing new products for the consumer, in *Food Choice and the Consumer*, Marshall, D.W. (Ed.) (London: Blackie Academic and Professional).

CARDELLO, A.V. (1996) The role of the human senses in food acceptance, in *Food Choice, Acceptance and Consumption*, Meiselman, H.L. and MacFie, H.J.H. (Eds) (London: Blackie Academic and Professional).

CHITTAPORN, P. (1977) A quantitative model for the design of a processed infant food product for Thailand. PhD Thesis, Massey University, Palmerston North, New Zealand.

CIVILLE, G.V. & LYON, B.G. (1996) *Aroma and Flavour Lexicon for Sensory Evaluation*, ASTM Data Series DS66 (West Conshohocken: American Society for Testing and Materials).

COOPER, H.R., EARLE, M.D. and TRIGGS, C.M. (1989) Ratio of ideals, a new twist on an old idea, in *Product Testing with Consumers for Research Guidance*, STP 1035 (West Conshohocken: American Society for Testing and

Materials).

COOPER, H.R. & BROWN, S. (1990) Trans-Tasman differences in consumer perceptions. *Food Australia*, 42(10), 494–497.

EARLE, M.D. (1997) Changes in the food product development process. *Trends in Food Science and Technology*, 8, 19–24.

EARLE, M.D. & EARLE, R.L. (1999) *Creating New Food: The Product Developer's Guide* (London: Chadwick House Group).

EARLE, M.D. & EARLE, R.L. (2000) *Building the Future on New Products* (Leatherhead: Leatherhead Food R.A. Publishing).

ENGEL, J.F., BLACKWELL, R.D. & MINIARD, P.W. (1995) *Consumer Behaviour*, 8th Edn (Fort Worth: Dryden Press).

FURST, T., CONNORS, M., BISOGNI, C.A., SOBAL, J. & WINTER-FALK, L. (1996) Food choice: a conceptual model of the process. *Appetite*, 26(3), 247–266.

GACULA, M.C. (1997) *Descriptive Sensory Analysis in Practice* (Trumbull: Food and Nutrition Press).

GREEN, P.E., TULL, D.S. & ALBAUM, G. (1988) *Research For Marketing Decisions*, 5th Edn (Englewood Cliffs, NJ: Prentice-Hall).

GREGORY, N.G. (1997) *Meat, Meat Eating and Vegetarianism: A Review of the Facts*. MAF Policy Technical Paper 97/16 (Palmerston North, New Zealand, Ministry of Agriculture and Fisheries).

HISRICH, R.D. & PETERS, M.P. (1991) *Marketing Decisions for New and Mature Products*, 2nd Edn (New York: Macmillan).

HOLLINGSWORTH, P. (2000) Culinary's expanded role in cooking up product success. *Food Technology*, 54(7), 38, 40, 42, 44.

KOIVISTO-HURSTI, V.L. & SJODEN, P.O. (1997) Food and general neophobia and their relationship with self-reported food choice: familial resemblance in Swedish families with children of ages 7–17 years. *Appetite*, 29(1), 89–103.

KORSMEYER, C. (1999) *Making Sense of Taste* (Ithaca: Cornell University Press).

KÖSTER, P. (1990) Sensory analysis and consumer research, in *Workshop in Measuring Consumer Perceptions of Internal Product Quality*, Acta Horticulturae Technical Communication 259, Jordan, J.L. (Ed.) (Wageningen: International Society for Horticultural Science).

LAURENT, G. & KAPFERER, J-N. (1985) Measuring consumer involvement profiles. *Journal of Marketing Research*, 22(Feb), 41–53.

LAWSON, R., TIDWELL, P., RAINBIRD, P., LOUDON, D. & DELLA BITTA, A. (1996) *Consumer Behaviour in Australia and New Zealand* (Sydney: McGraw-Hill).

LINDEMAN, M. & STARK, K. (1999) Pleasure, pursuit of health or negotiation of identity? Personality correlates of food choice motives among young and middle-aged women. *Appetite*, 33(1), 141–161.

LINNEMANN, H.R., MEERDINK, C.H., MEULENBERG, M.T.G. & JONGEN, W.M.F. (1999) Consumer-oriented technology development. *Trends in Food Science and Technology*, 9, 409–414.

MCBRIDE, R.L. (1990) Three generations of sensory evaluation, in *Psychological Basis of Sensory Evaluation*, McBride, R.L. and MacFie, H.J.H. (Eds) (London: Elsevier Applied Science).

MARTENS, H., WOLD, S. & MARTENS, M. (1983) A layman's guide to multivariate data analysis, in *Food Research and Data Analysis*, Martens, H. and Russwurm, H. (Eds) (London: Applied Science).

MEILGAARD, M., CIVILLE, B.S. & CARR, B.T. (1999) *Sensory Evaluation Techniques*, 3rd Edn (Boca Raton, FL: CRC Press).

MEISELMAN, H.L. (1994) Bridging the gap between sensory evaluation and market research. *Trends in Food Science and Technology*, 5(12), 396–398.

MELA, D.J. (1996) Implications for fat replacement for food choice and energy balance. *Chemistry and Industry*, 329–332.

MOSKOWITZ, H.R. (1994) *Food Concepts and Products: Just-In-Time Development* (Trumbull: Food and Nutrition Press).

MOSKOWITZ, H.R. (1997) Experts versus consumers: a comparison, in *Descriptive Sensory Analysis in Practice*, Gacula, M.C. (Ed.) (Trumbull: Food and Nutrition Press).

MOSKOWITZ, H.R., BENZAQUEN, I. & RITACCO, G. (1981) What do consumers really think about your product? *Food Engineering*, Sept., 80–82.

MUÑOZ, A.M. (1997) *Relating Consumer Acceptance and Laboratory Data*. PCN28-030097-36 (West Conshohocken: American Society for Testing and Materials).

NANTACHAI, K., PETTY, M.F. & SCRIVEN, F.M. (1992) An application of contextual evaluation to allow simultaneous food product development for domestic and export markets. *Food Quality and Preference*, 3(1), 13–22.

NESTLE, N., WING, R., BIRCH, L., DISOGRA, L., DRENOWSKI, A., MIDDLETON, S., SIGMAN-GRANT, M., SOBAL, J., WINSTON, M. & ECONOMOS, C. (1998) Behavioural and social influences on food choice. *Nutrition Reviews*, 56(5), S50–S74.

PETER, J.P. & OLSON, J.C. (1999) *Consumer Behaviour and Marketing Strategy* (Boston: Irwin, McGraw-Hill).

POSTE, L.M., MACKIE, D.A., BUTLER, G. & LARMOND, E. (1991) *Laboratory Methods for the Sensory Analysis of Food*, Publication 1864E (Ottawa: Research Branch, Agriculture Canada).

RAATS, M., DAILLANT-SPINNLER, B., DELIZA, R. & MACFIE, H. (1995) Are sensory properties relevant to food choice? in *Food Choice and the Consumer*, Marshall, D.W. (Ed.) (London: Blackie Academic and Professional).

ROBERTS, L.M. (1997) A new beef production adoption model for hotels and motels in Greater Melbourne. PhD Thesis, Massey University, Palmerston North, New Zealand.

ROGERS, E.M. (1983) *Diffusion of Innovation*, 3rd Edn (New York: The Free Press).

RUFF, J. (1995) Consumer expectation in the food industry: a vision for the 21st Century. *Food Science and Technology Today*, 9(4), 195–205.

SAGUY, I.S. & MOSKOWITZ, H.R. (1999) Integrating the consumer into new product development. *Food Technology*, 53(8), 68–73.

SCHAFFNER, D.J., SCHRODER, W.R. & EARLE, M.D. (1998) *Marketing: An International Perspective* (New York: McGraw-Hill).

SCHIFFMAN, L.G. & KANUK, L.L. (2000) *Consumer Behaviour* (Upper Saddle

River: Prentice-Hall).

SCHÜTTE, H. & CIARANTE, D. (1998) *Consumer Behaviour in Asia* (Basingstoke: Macmillan Business).

SCHUTZ, H.G. (1988) Multivariate analysis and the measurement of consumer attitudes and perceptions. *Food Technology*, 42(11), 141–4, 156.

SCHUTZ, H.G. (1993) Measuring consumer acceptance of flavours, in *Flavor Measurement*, Ho, C.-H. and Manley, C.H. (Eds) (New York: Marcel Dekker).

SHOCKER, A.D. & SRINVASAN, V. (1979) Multivariate approaches for product concept evaluation and generation: a critical review. *Journal of Marketing Research*, 16 (May), 159–160.

SINTHAVALAI, S. (1986) Development of nutritionally-balanced snack product for urban school-age Thais. PhD Thesis, Massey University, Palmerston North, New Zealand.

TEPPER, B.J., YOUNG SUK CHOI & NAYGA, R.M. (1997) Understanding food choice in adult men: influence of nutrition knowledge, food beliefs and dietary restraint. *Food Quality and Preference*, 8(4), 307–317.

VERYZER, R.W. (1997) Measuring consumer perceptions in the product development process. *Design Management Journal*, Spring, 66–70.

VESSEUR, W. (1990) Tomato tasting test and consumer attitude, in *Workshop in Measuring Consumer Perceptions of Internal Product Quality*, Acta Horticulturae Technical Communication 259, Jordan, J.L. (Ed.) (Wageningen: International Society for Horticultural Science).

WATTS, B.M., YLMAKI, G.L., JEFFREY, L.E. & ELIAS, L.G. (1989) *Basic Sensory Methods for Food Evaluation* (Ottawa: International Development Research Centre).

WEST, S.J. & EARLE, M.D. (1987) *Market Research Methods*, IDRC MR151e (Ottawa: International Development Research Centre).

WILKINSON, B.H.P. (1985) Attitudes and Behaviour of Consumers to Meat in Palmerston North, New Zealand 1979–85. PhD Thesis, Massey University, Palmerston North, New Zealand.

WILTON, V. & GREENHOFF, K. (1988) Integration of sensory techniques into market research. *Food Quality and Preference*, 1(1), 33–35.

ZELANEK, M.J. (1999) *Consumer Economics: The Consumer in Our Society*. (Scottsdale: Holcomb Hathaway).

Part III

Managing and improving product development

The effectiveness of product development in a company is determined by the people responsible for product development from the directors to the project managers; by firstly their basic philosophy and understanding, secondly their abilities and thirdly the clear recognition of their roles.

Product development combines people, their individual knowledge and skills. How they collaborate to produce the company's abilities in product development is the basis for product success or failure. Product development needs knowledge and skills in all areas of the company – R&D, marketing, production and finance, and in particular top management. This is why product development management is complex and often becomes swamped in the management for today. Companies appear to have great difficulty in deciding where to place product development management – in marketing, R&D, production or as a separate department; over the years product development management is apt to be reorganised several times because of problems that have been identified. This is no bad thing since product development, because of its nature, is always changing, going in different directions as technology and consumers change. But the core product development knowledge needs to be kept intact and allowed to grow through the product development projects.

It is important to recognise that there are different layers of product development management, layers that are interacting with the management of the functional departments and indeed are sometimes the management of the functional departments. The key issues are in two areas:

1. Management needs in the vertical responsibility from directors to project managers.
2. Interaction between these management people and the functional departments.

The product development activity in a large company can be very complex. In the small company, the director, chief executive, product development manager and project leader can be one person. But in all companies it is important that the product development strategy, product development programme, PD Process and the overall aims of the programme and the individual projects are clearly defined. Then everyone understands their place in the jigsaw, the outcomes expected from their work and the decision-making process.

The other key issue in management is the interrelationship between product development and the functional areas of R&D, marketing, production, distribution, and finance. In looking at the typical activities occurring throughout the product development process, it is clear that the support of these groups is essential. Both the inputs needed for product development and the outcomes from product development are related to the functional departments, which are very much involved in making and marketing today's products (Stockwell, 1985). There are inputs such as product mix strategy and sales forecasting from marketing, product trials and quality assurance development from production, predicted returns on investment from finance, and outputs such as market plan from marketing, production schedule from production and net profit forecast from finance.

In Chapter 6, the basic needs in product development management and the people who are responsible for making product development both effective and efficient – producing the optimum product at the right time and within budget – are outlined. The PD Process is the focus – design the PD Process; establish the key decision points and the decision makers; establish outcomes, budgets and constraints; organise and manage. The chapter ends with a discussion on managing and organising product development in the company, and collaborating with outside agencies.

Chapter 7 illustrates product development at different points in the food system in four case studies. Management of product development is different among primary production, processing of food ingredients, manufacturing and food service, because of the different scientific and technological bases, the different needs of the target markets and the time for development. It is important to understand that there is a common product development framework but the activities can be different.

Chapter 8 studies the searching for best practice in product development and in particular the improvement of R&D management by benchmarking. The changes that are occurring in food product development and what may happen in the future are discussed.

Reference

STOCKWELL, D. (1985) Managing product development as a business activity, an address to the ASEAN Food Conference, Manila, Philippines.

6

Managing the product development process

Product development management in the food system is complex, long term and capital intensive. It is total company management involving every function in the company – so it is managing either a microcosm of the company or an integration of the company functions. For a major innovation, the company may set up a new venture company or division; or a new group of people may form a new company. At this time when many new companies are being formed on the innovations of information technology and biotechnology, it is interesting to speculate on new venture companies in the food industry and the basis of their new innovations. But at the present time, it is the large multinational food companies that dominate product development at all levels in the food industry, and it is management of product development in these companies that is the main basis for innovation in the food industry. There are many small food companies that are also involved in product development on a small scale. Management of product development in the food industry varies from a group in the small company sitting around a kitchen table to the multinational food company with large R&D laboratories, small-scale production development plants and product development teams in many countries. The basic principles of product development management are the same in large and small companies, but often more difficult to apply in the large company because of rigid hierarchies.

The framework for management in the food industry is the PD Process, and the recognition of management at the different stages.

6.1   Principles of product development management

Several principles of product development management have been identified (Souder, 1987; Ganguly, 1999). Relative importance does change but the basic principles are robust and are useful as a basis for product development management. They can be grouped under basic philosophy, understanding, abilities and organisation of the company as shown in Fig. 6.1.

6.1.1   Basic philosophy and understanding

- Belief in product development as a major business strategy.
- Understanding emerging worldwide technologies, in-depth knowledge of technologies.
- Understanding the transformation of technologies into want-satisfying products, intimate understanding of changing consumer needs.
- Developing a creative climate, creating spontaneous teamwork.
- Patience, realising that innovations take time, going through cycles of success and failure, and that management has to aid and direct them to the end of product success.
- Recognising the need for skills in systematic decision making and risk-taking.

These are still essential elements in product development that have continued to demonstrate their significance over a great diversity of situations and times, not to mention fashions! Unless management, especially top management, believes that product development needs knowledge of technology and consumer, and of their optimum relationship, then product development will stumble. Having recognised these basic knowledge needs, they have also to recognise that there needs to be a creative atmosphere and time to reach product success. Lastly they have to believe that the success of product development depends on their decision making, its quality and timeliness (Lord, 2000).

Fig. 6.1   Basic principles of product development management.

6.1.2 Abilities

- Systematic selection of best projects, using information sharing and group decision making, creating idea generation and evaluation with all people involved in product development, setting decision processes based on the product development goals.
- Careful analysis of the customer's level of sophistication and the product designer's level of technical sophistication, creating collaborative roles between product design and consumer/market research, educating product designers on consumer needs and wants, educating marketing on technical possibilities and problems.
- Finding and coordinating the resources and knowledge for product development, upgrading knowledge to make use of new technology, nurturing methods for new technologies, selection of technology with fit to present or planned future company technology, predicting costs of adoption of new technology in finances and company organisation.
- Elimination of disharmony between R&D and marketing groups, making open communication an explicit responsibility of every employee, using joint R&D/marketing task forces.
- Reducing complexity and problems, breaking large projects into manageable stages, identifying and eliminating mild problems before they become major.

Management at all levels needs to have the abilities to recognise the path of the project and to coordinate the knowledge, resources and people to follow the path efficiently and effectively to product success. There is a great deal spoken about multidisciplinary, cross-functional, inter-functional, intra-functional, integrated product development, but basically product development needs to be recognised as a many-faceted process which can only be achieved by collaboration between people with different knowledge and skills. It cannot be enclosed in specialist or functional boxes such as marketing or production (Harris and McKay, 1996). Management needs to understand the meaning of company collaboration and to have the ability to put into action a multifaceted product development project based on collaboration.

6.1.3 Organisation

- Design of product development organisation, ability to set the tone, posture and prevailing attitudes towards product development, creating an organisation to fit the needs of members and of customers, encouraging responsibility and creating multidirectional communication.
- Cost-effective project management, selecting the method that relates to the problem, for example incremental innovation using commercial line management, technical innovation using technical management, major innovation using separate project management or a new product committee.
- Flow management during the project, organising the timely transfer and flow of product prototypes and knowledge, encouraging the skills and knowledge

for the evolving technology and keeping team members involved to greater and lesser extent throughout the project.

- Product development budgeting techniques, understanding the changing cost/time ratios between projects and within projects, the financial analysis of the different stages of the PD Process to identify the costs and their possible improvements, the financial controls needed for the different cost/ time ratios.

Management has to design the organisation for product development in the company, both for the overall new product programme and for the individual project. There needs to be coordination among projects to have the optimum use of people and resources, as well as planning and control for the individual project so that it flows towards the final product launch without stumbling too often. Radical innovations are never straightforward linear progressions through the project; there is often recycling especially during the earlier stages, but these returns to earlier stages in the project need to be managed.

Think break

In your company:

1. What is the basic philosophy of product development management?
2. Describe the understanding of technology changes and consumer needs changes.
3. How are these changes affecting product development?
4. Has your company the abilities to develop new products related to these changes?
5. If not, what new abilities need to be found? How could this be done?
6. What are the organisational methods used by your company?
7. Do they ensure effective and efficient product development?
8. If not, what changes need to be made in the organisational methods?

6.2   People in product development management

It is important to recognise that there are different layers of management. The different levels of management can be identified as directors, chief executives, product development managers and project leaders, although the actual titles of the managers may be different from this in the individual companies (see Fig. 6.2). The directors are at the business strategy level, the chief executive at the product/innovation strategy level, the product development manager at the new product programme level and the project leader at the level of the individual

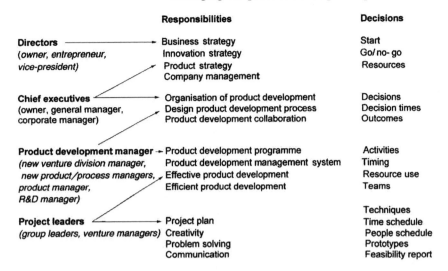

Fig. 6.2   Levels of PD management.

project as shown in Fig. 6.2. All of these different levels have their own basic philosophy and understanding, abilities and responsibilities, but they need to be coordinated into the whole product development management.

The interwoven nature of responsibilities should be noted; people may have individual responsibilities at their level of management, and also joint responsibilities at higher levels. There is no clear demarcation of responsibility in product development because it has to be collaborative management.

6.2.1   Directors

Directors on the Board of the company have their vision for future growth of the enterprise, commitment to product development as a method of ensuring company growth, understanding of the knowledge and skills needed for product development, provision of resources for product development. They need to have the abilities for: incorporating innovation into the business strategy, analysis and decision making, intelligent risk-taking, selecting and directing a chief executive with the knowledge and skills for the innovation strategy that they have developed. They need an understanding of the company's technological and marketing environments, the competitors' innovative strategies and multi-industry evolution, the company's structural and cultural context, and the company's resources and capabilities. The Board sets the overall innovation strategy and the philosophy for product development, gives appropriate allocation of resources and makes the major decisions in the development. As with most business activities, product development is most successful when it starts from the top.

6.2.2   Chief executives

Chief executives have commitment to the organisational role of product development, understanding the needs of product development, recognising the knowledge and skills for product development, recognising the product development process as it relates to their company. They need the abilities:

- to develop the structure in which product development operates;
- to organise a management system for product development;
- to integrate all the functional areas taking part in product development;
- to develop a clear product strategy and a product development programme;
- to set clear goals;
- to indicate the decisions to be made at different parts of the product development processes, and make decisions with careful analysis.

They need to be able to define the long-term company development strategy and assess the strategic importance of new company initiatives and their relation to the present core capabilities (Cooper, 1998). They are responsible for effective portfolio management, making strategic choices of markets, products and technologies that the business will invest in (Cooper *et al.*, 1999). The chief executive develops a positive environment, actively supporting, leading and directing product development on a continuous basis, and providing integrating communication between different groups, usually the functional departments of marketing, production, R&D and finance, with product development. Since product development spans many disciplines, it should not get locked into one image – marketing, production or R&D.

6.2.3   Product development managers

Product development managers have commitment to the company's PD Process and integration of the skills and knowledge for this process, understanding of the customer and consumer needs and wants, knowledge of present and emerging technologies, understanding of the company and the external environment. They need the abilities to:

- identify the outcomes necessary for each stage of the PD Process for the chief executive's and Board's decision making;
- identify the time and other constraints on the project;
- identify and find the resources for the product development;
- encourage the creative and technical achievements of the people involved in product development;
- analyse and make the decisions.

They need to be able, with top management, to obtain/maintain support for new initiatives, to define the company's strategies for the new initiatives, and to cooperate with the project leaders in defining projects. Product development managers integrate the various projects into an overall product development programme. They set with the project managers the timing of stages in the PD

Process, plan and control the resources so that they are available at the correct time and are of the right quality, analyse the results of the development and make decisions for further stages. They need to be aware, guide and be available when necessary, to help the creativity and the problem solving. Every company needs a person who is responsible for new products and is recognised as this. This person must have product development knowledge and skills as well as management knowledge and skills. There needs to be a balance between the innovation development and the management. Over-management can stifle innovation, but uncontrolled product development may lead to inappropriate products, inefficient product development and time/cost overruns – in other words commercial failure.

6.2.4   Project leader

Project leaders understand the consumer and market as well as the PD Process and the product; recognise and foster innovative, creative, problem-solving skills; and understand integration of people with different skills and philosophies. They need the abilities to:

- drive the project to a successful conclusion;
- identify the outcomes for each stage of the PD Process and important sections in the individual stages;
- relate the outcomes to the activities in the project;
- choose the techniques for the activities that relate to the knowledge and skills of the team, and the resources available.

They have the capacity to develop the business strategy for the new product as well as to define the technical/marketing development, and to build the organisational structure for the development. The project manager is leading a team of people who are skilled in different disciplines – consumer research, marketing, product design, processing development, production and finance. Although the project manager may not have an in-depth knowledge in all areas, there is a need for basic knowledge in each area and the ability to see the interrelationships between them. The project leader is responsible for ensuring that the project progresses smoothly, meets all interim objectives and targets on time and within budget, and makes sure that the necessary resources are available when and where they are required. They also are the primary channel of communication between the project team, senior management and external organisations (Jones, 1997).

6.2.5   Important factors in management levels

There are three important factors to recognise in these four different types of management. Firstly there is a need for *championing*: strategic championing at the directors/top management level, organisational championing at the product development manager level, and product championing at the project leader

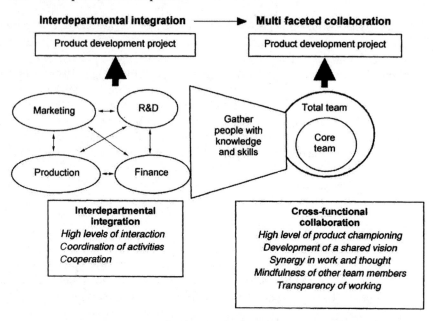

Fig. 6.3 Comparison of integration and collaboration in product development
(Source: After Jassawalla and Sashittal, 1998).

level. Secondly, *decision making* needs to be spread throughout the three groups and to be clearly defined. There is a need for decision making at all levels and not only at top management level. There is nothing more restrictive in product development than all decisions having to be made by one person; also it tends to slow development because there is endless reporting and decision making. Decision making is a collaborative activity between all levels of management. Thirdly, *collaboration* is an active aim of product development management. Collaboration is much more than cross-functional integration, it is the active working together of people from different disciplines, different functional departments, and different levels of management in product development, all with common aims for product development as shown in Fig. 6.3.

Cross-functional collaboration includes an equal stake and responsibility for the outcomes, and a willingness to understand the other people's viewpoints so that they can be blended to give higher levels of combined knowledge in the product development. Such collaboration involves synergy in thought and action, which leads to outcomes from the product development exceeding the capabilities of the individual participants in the PD Process (Jassawalla and Sashittal, 1998). Collaboration can be hard to achieve in radical innovation where uncertainty leads to tensions between people, with different functional groups blaming the others for delays, poor product qualities and increased costs. It is much easier in incremental product development where the risks of failure are much less and many activities become routine. Stage-specific collaboration is more likely to lead to new product success, rather than integrating all functions during all four stages of the PD

Process, with the three functions – R&D, manufacturing and marketing – playing the central role in turns (Song *et al.*, 1998).

Think break

Are you a manager? If yes, answer the questions directly. If no, answer the questions as related to your manager.

1. For what area of product development are you (or the manager) responsible?
2. Identify your responsibilities.
3. How do these relate to the responsibilities of the manager above you?
4. How do these responsibilities relate to your abilities? Do you need more abilities or improved abilities? How could you attain these?
5. How do you relate to other areas of product development, other people in product development?
6. Would these relations improve with a top management decision to have collaboration as a basic company philosophy?

6.3   Designing the PD Process

The PD Process is the framework to build up both product development projects and the product development programme. The selection of activities in the PD Process depends on the knowledge and skills in the company, knowledge easily available outside the company, the risks involved through lack of knowledge, the importance of the product development for the future of the company, and most important the level of innovation.

6.3.1   The effects of knowledge and skills on the PD Process

Knowledge and skills in the company, and also the company's philosophy on product development, affect the choice of activities and therefore the structure of the PD Process. If the company is not able to do consumer research, it may do personal research with the retailer, food service buyer, their family and friends. If the company does not have formulation skills, it will take a formula from the ingredient supplier. If the company does not have process engineering skills, it will buy a turnkey processing line from an equipment supplier. Or if there are no innovation skills in the company, then it will acquire another company! Some company managers are conservative and some want excitement, some are risk takers and some are fearful of risk; these differences cause differences in the PD Process used in the company. Some companies bring new products through to market as quickly as possible, missing activities such as business analysis and test marketing, as they are prepared to live with failures; other companies include all activities so as to reduce their risk of failure in the market. So the PD Process is specific to the company and its knowledge, skills, and philosophy.

6.3.2   Level of innovation and the PD Process

A comparison of the PD Processes for radical and incremental changes is shown in Table 6.1. The sequence of the activities varies, for example in the incremental change there is generally a linear sequence in the PD Process, but in the radical innovation, there is often recycling of activities. The incremental product changes can be developed and marketed according to a standard PD Process with strong involvement of the functional departments such as marketing and production. The radical innovation uses generalised

Table 6.1   Differences in the PD Process for platform and derivative products

Radical innovations (new platform products)	Incremental changes (derivative products)
Stage 1: Product strategy development	
Probing problems with consumers	Consumers setting attributes
Focused project objectives	Clear schedules and time goals
Developed product concept	General product concept
New product design specifications	Standard product design specification
New market probing studies	Market surveys
Stage 2: Product design and process development	
Building product attribute measures	Refining product attribute measures
Product/process interrelationship studies	Product formulation
Pilot plant studies	Process improvement
Frequent product testing	Strategic product testing
Stage 3: Product commercialisation	
Design of new production method	Adaptation of production
Design of new quality assurance	Adaptation of quality assurance
Commissioning of new plant	Minor plant changes
New marketing strategy	Improvement of marketing strategy
Detailed business and market analysis	Setting market and financial targets
Stage 4: Launching and post-launch evaluation	
Rolling launch or pre-launch test market	National launch
Continuous market analysis	Assessing if market targets met
Continuous financial analysis	Assessing if financial targets met
Overall management	
Project milestones to control	High importance of speed
Long-term commitment of capital	Short-term commitment of capital
Long-term commitment of human resources	Short-term commitment of human resources

Source: From Earle and Earle, *Building the Future on New Products*, © LFRA Ltd, 2000, by permission of Leatherhead Food RA, Leatherhead, UK.

activities because creativity and problems in the project are difficult to predict.

For the radical innovation, there is a need to develop technical and market knowledge in the first two stages of the product development process and to include product/market testing and business analysis in the product commercialisation stage. In the initial stages:

- product concepts are developed with the consumers;
- product designers make some models or simple prototypes of possible products and ask the consumers to evaluate them;
- further product concepts are developed; and
- evaluated by marketing and processing technologists to see if any are possible for commercialisation.

Later in the PD Process, early versions are marketed quickly on a small scale, obtaining the user feedback and making modifications before expanding the market. Usually for the radical innovation, the company's resource commitments are made at sequential times in the PD Process, and not at the beginning of the project as for incremental products (Mullins and Sutherland, 1998).

In the incremental product projects, a great deal of the knowledge is already in the company, so there is less need for new research in building the direction of the project (Earle and Earle, 2000). The product concept can be developed by a marketer and a product designer, evaluated by consumers to check that no mistakes have been made, and the product specifications written in the standard form for this type of product. A national launch usually targets the total market.

Between the extremes of radical and incremental changes, major product changes can need different types of PD Processes. If major changes in product, market and production are being made, they can be similar to the PD Process for radical innovations. If the major change is marketing related, for example a positioning change, the PD Process is similar to the incremental PD Process with an emphasis on the marketing change; if production related, such as a new process, it is an incremental PD Process with an emphasis on technical development.

6.3.3 Other factors in designing the PD Process

Other factors to be considered in designing the PD Process framework for the company are: place in the food system, environment, technology, marketing and company resources. Primary production, industrial and consumer food product development have differences in their PD Processes, which will be described in Chapter 7. There may be strong societal and political constraints on product development that need to be included in the PD Process; obviously food regulations limit the processing and the raw materials, but religious requirements are also often important; or it can be recycling of packaging or other environmental problems. If these are not included in an early part of the PD Process, a great deal of time and money can be wasted. Technology in processing, distribution and marketing has also to be considered – what is standard, what is new? (Earle and Earle, 2000). With the large multinational food companies, there is often a requirement for fast processing and large-scale equipment. This means that the product design, and the process development, start in the first stage of the PD Process with the consumer research, and then develop together. In recent years there has been a great deal spoken about concurrent engineering in other industries (Tomiyama, 1998), developing the product and the production methods in parallel. In the food industry it is crucial that the product design and the process development are interwoven from early in product concept development and product design. For the smaller companies using simpler equipment, it can be possible to do a significant amount of product design on 'kitchen' size equipment before building up the process.

6.3.4   Using and changing the PD Process

The company, often through experience, has found the important activities for their industry and business – although sometimes this is not so much by careful analysis after product development projects are completed, but by copying the actions of competitors or the industry in general or the latest fashion. Selecting both the activities and also intensity of work in each activity needs to be based not on 'we have always done it that way' or 'the industry does it that way' but on what is needed to achieve the target aims and outcomes. Selection of the activities involves:

- reason for the activity;
- resources needed;
- outcomes expected;
- timing of activities;
- controls for measurement of progress (Gruenwald, 1988).

The PD Process is not a static framework but needs to be evaluated regularly so that it can be updated for the new knowledge and skills in the company, the new directions chosen for the company and the changing environment. The use of some novel activities in the PD Process leads to the competitive edge of the company. But the PD Process always has a basic framework, which has been built up by experience and is only slowly changed. Top management can instigate new strategies and therefore new PD Processes for product development but this needs exceptional entrepreneurial skills, sustained commitment to new products for company growth and an acceptance of risk taking (Gruenwald, 1988). Without making drastic changes, top management can send the right signals to the organisation for evaluation and necessary change to the PD Process, for communication between different people to recognise changes, and for a dynamic organisation that can cope with change in the PD Process (O'Connor, 1996).

The management of the PD Process varies from company to company. Some companies put the framework on the internal communication network and expect product development staff to consult it; other companies set out a rigid PD Process with the activities and often the techniques identified as to what is to be done in all projects.

Think break

Does your company have a PD Process framework?

1. Discuss how the PD Process framework:
    (a) recognises the company's philosophy, knowledge and skills,
    (b) adapts to innovation level, type of technology, marketing, environment, food regulations,
    (c) identifies the activities for each of the four stages of the PD Process,
    (d) puts these activities into parallel or sequential positions,
    (e) shows the necessary collaboration between people and departments.
2. What changes could be made to your company's PD Process framework that would make product development more effective and efficient in your company?

(If your company does not have a PD Process framework, design one using the above criteria.)

6.4   Establishing key decision points and the decision makers

Many people make decisions in product development at all levels – from the top management and the Board of Directors, who decide on the overall project and its resources, to the process worker, who decides on the detail of production for the new product, and the salesperson, who decides the factors that they think will encourage the customer to buy the product. All the decisions follow on from decisions made by other people, and are linked with other decisions. This interrelationship between decisions is not always recognised in the company and therefore important decision making is not identified. The top management receives a report that is based on the knowledge and decisions of middle management or perhaps of a product approval committee, and then makes its decision on this report combined with other knowledge it may have. The decision made by top management determines the project for the project leader, whose decisions determine how the project is to be organised. Decisions and the knowledge used to make them are the foundation stones for product development; poor decisions based on inaccurate knowledge lead to product failures, and good decision making based on sound knowledge leads to product success.

6.4.1   Top management's decisions

Top management and often the directors on the company Board handle the critical decisions between the stages. Basically there are two decisions:

- Do we go on or stop the product development?
- What resources does the company put into further development?

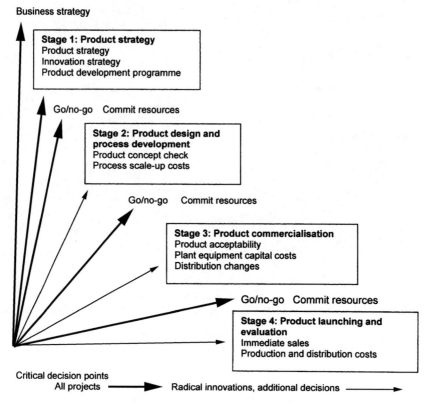

Fig. 6.4 Critical decision points for directors and top management.

During Stage 1: Developing the product strategy, there are another three critical acceptance decisions – does the company accept the product strategy, the innovation strategy and the product development programme?

In the first stages of product development, Urban *et al.* (1987) identified the important top management decisions as: commit company to new products, approve PD Process and budgets, product portfolio and market strategy. Before product design, there was a review of market entry strategy and then after product design, review and commit to test, followed by go/no-go decision and review launch plan. They noted that after launch the important decisions were approve one- and five-year plans and finally approve strategy to 'milk' or 'rejuvenate'. The general critical decision points for top management are shown in Fig. 6.4, and some additional decisions for radical innovations are shown within Stages 2, 3 and 4. Because of uncertainties in radical and some major product changes, there can be changes in the product concept during design, changes in processing which need design and building of new equipment, changes in the storage conditions which alter the distribution method, and changes in the target market which need product or marketing alterations. These uncertainties need to be recognised at the beginning and the additional critical decision points they introduce included in the PD Process. In incremental new products, it is unlikely for top management to be involved in decision making within stages unless there are unusual problems that could for example cause disharmony with company ethics or cause time and costs overruns.

Senior management needs to review the right information at the right time to make the right decision. It also needs to make its decisions quickly, so that the project is not delayed while people wait for decisions. If the project is allowed to proceed to prototype product or to product commercialisation before the top management decision, it may be too late to stop the project even though it will need more development and therefore more resources. But management must not get directly involved in 'fighting fires' on products that are late or have problems. By this time most of the major decisions have been made by other people, and the real impact of top mangement involvement is minimal. The decision makers at this stage are the technical and marketing people who have the knowledge to solve the problems. If major problems occur, the project leader is responsible for reviewing the project and for devising a solution that can then be reviewed by top management, who make the decision to proceed or kill the project. Several important aspects for top management reviews are (McGrath *et al.*, 1992):

- to provide a clear and consistent process for making major decisions on new products and improvements;
- to empower project teams to execute a project plan;
- to provide the link for applying product strategy to product development;
- to provide measurable checkpoints to monitor progress;
- to establish milestones that emphasise a sense of urgency.

For all these critical decisions by the directors and senior management there is a need for knowledge, much of which is in the outcomes of the various stages within the PD Process. If the critical decisions and the knowledge needed to make these decisions are clearly identified by top management, during design of the PD Process framework at the beginning of the individual projects, the product development team knows what information it has to produce and in what form. There also needs to be a warning system in place for resource or time overruns or difficult problems impeding progress in the project.

6.4.2   Product development manager's decisions

The product development manager, the person responsible for the product development programme, also has decisions to make. He or she is responsible for the administrative framework of the product development programme, deciding on the activities, the schedules and the budgets, and most important the integration of different product development projects that are running in parallel. He or she decides on the workload forecasts, the standard of product development performance of the team and individuals, the efficiency expected of the PD Process (Gruenwald, 1988). In particular, a product development manager has to decide how and when to involve the company's functional groups in the PD Process, and very importantly how the consumer needs are related to the product. The product development manager leads the coordination between the consumers and the development team, and helps the project leaders to integrate technical and business perspectives. He or she is responsible for the development of new skills and knowledge, as well as bringing in outside consultants and other sources of knowledge (Burgelman and Maidique, 1988). Some important decisions of the product development manager in Stage 1: Product strategy development are:

- aim, outcomes and constraints for the individual projects;
- depth of activity in developing the product concept and the product design specifications;
- accuracy needed in the market predictions;
- amount of product design and other technical development;
- time schedule for the activities;
- resources available for each of the activities;
- involvement of the functional departments in developing and evaluating the product concept and the product design specifications, and in predicting the market and technical success;
- communication methods with the team members and among them;
- recognition of lack of skills and knowledge for the later stages and planning of involvement of outside knowledge sources and of education of team members.

The product development manager has to decide how to produce the outcomes for Stage 1, but has also to look ahead to later stages so that the necessary basis for them is laid during this stage. The later stages have similar types of decisions and these become more complex as the costs of the development and the opportunity for a large failure increase. It is very important that the product development manager ensures that decisions by the top management and by the project leaders are made at the right time. Otherwise the project can lose its urgency and the time to market extends. It is important when technical and marketing decisions are proving difficult that resources are organised by the product development manager to solve the problems without delay.

6.4.3   Project leader's decisions

The project leader has also decisions to make so as to achieve the outcomes set by top management within the time and budget structure of the product development manager. Firstly the project leader has to decide with more senior management on the aim, outcomes and constraints of the project. The project leader selects techniques for the activities identified by the product development manager, which are within the capabilities of the team members or outside agencies, and which will produce the product with the qualities needed by the consumer. The project leader decides how to do this within the resources and the time allowed so that the project remains on schedule. The project leader decides the balance between the effectiveness and the efficiency of the product development, that is balancing the quality of the product development and the time and resources used (see Fig. 6.10 on page 295). This is very difficult decision making, especially for the young project leader. There needs to be help from more senior management. Nothing is worse than senior management telling the project leader to produce the ideal product but not allocating the resources to attain it, or to give them a project known to be a problem without the knowledge to start solving it. The project leader's most important organisational decision is the product development project plan with the activities, resources, time, and the communication and control during the project. The project plan with its predicted timing and use of resources is the basis for decisions that determine the efficiency of the product development; the project leader using it to make the decisions on overruns of time and resources.

Few product development projects, except for simple incremental product changes, can be predicted accurately – either the direction of the design and development or the outcomes in the results. So the project leader is continuously making decisions during the project on the relationships between:

- product and the consumer needs;
- product and the company;
- process, distribution and marketing;
- production and marketing functions.

These decisions are made not only with the core product development team but also with the wider team in the functional departments and in outside organisations. These decisions are extremely important to the quality of the research and need to be recognised reasonably early before the project becomes confused and disorganised.

Think break

1. For product development in your company, outline the PD Process for each of the following:
    (a) product improvements,
    (b) new product introductions,
    (c) process improvements,
    (d) new process introductions.

If your company does not have PD Processes, design them for the company. Identify the individuals or groups who are responsible for the product strategy and the product development programme, and for go/no-go decision making at the end of Stages 1, 2, 3 and 4 in the PD Process.

2.   Select a recent project, which was a radical product innovation, and identify the critical decisions, where they occurred in the project and who were the decision makers. Identify decisions and decision makers in:
     (a) the go/no-go decisions,
     (b) decisions that led to final product qualities, product image, product features and uses,
     (c) decisions that led to production method, distribution method, marketing strategy, costs and pricing,
     (d) decisions on the project efficiency, in achieving timing and costs.

6.5   Establishing outcomes, budgets and constraints

Decision making is the key framework for product development, but knowledge fills out the framework to give the live project. Knowledge is brought into the framework, and also created within the framework so that the decisions can be made. But knowledge has a cost; increasing the knowledge adds to the cost of a project. In product development, the intelligent and systematic balancing of knowledge and costs is fundamental to successful product development. Others might say it is the balancing of the costs of product failure against project costs. In product development, there is a need to define outcomes (the collected and created knowledge) and the budget (the costs) at the beginning of the project, and also to reconsider these at the completion of the four stages and at any other critical point in the PD Process. Conditions change throughout the individual projects and there is a need for top management to reconsider the outcomes and the budgets at the same time as permission for the project to proceed to the next stage is given.

6.5.1   Defining outcomes
At the end of each stage of the PD Process, the top management requires for its two outcomes:

*   product form to that stage of development;
*   report on which to base its decisions.

The product form will develop in the project from a product concept, to product design specifications to prototype products, to commercial product, to final launched product. The report can vary from many pages with detailed knowledge to a one page executive summary, dependent on what top management feels it needs to know. The areas listed in the reports in Table 6.2 are important areas of knowledge in product development for decision making but the top management may not wish to see any details, especially in incremental product development where there is not a great risk of making wrong predictions. Some of these critical decisions may be made by the middle management before the final executive

Table 6.2 Outcomes (knowledge needs) for decision making

Outcomes (knowledge needs)	Decisions
Stage 1: Product strategy development	
Product/innovation strategies	*Strategies acceptance*
Product development programme	*Programme acceptance*
Project aim, objectives and constraints	*Project acceptance*
	Resources for initial investigation
Product design specifications	*Product idea acceptance*
(or product concept)	*Resources for product design, process*
Product report:	*development*
• technical feasibility	*Timing of programme*
• marketing suitability	*Harmony with business*
• consumer acceptance	
• project costs, risks	
Stage 2: Product design and process development	
Final prototype product	*Acceptance as new company product*
Feasibility report	*Resources for commercialisation*
• target consumers	*Total company involvement*
• product qualities	*Harmony with business*
• processing method	
• marketing strategy	
• predicted sales	
• predicted costs	
• project costs, risks	
Stage 3: Product commercialisation	
Commercial product	*Acceptance as new product into product mix*
Commercial report	*Launch agreement*
• production plan	*Capital investment*
• distribution plan	*Acceptance into company organisation*
• marketing plan	
• financial plan	
• risk analysis	
• capital investment	
• human resources	
• effect on company	
• effect on society	
Stage 4: Product launch and evaluation	
Marketed product	*Long-term acceptance into product mix*
Final evaluation report	*Feedback to future business strategy*
• product quality and position	*Future product development*
• production and distribution efficiency	*Resources for future product development*
• costs against targets	
• sales against targets	
• indicative return on investment	
• effect on company	
• market acceptance	
• society acceptance	

Source: From Earle and Earle, 1999, by permission of Chadwick House Group Ltd.

summary is prepared for the chief executive and the directors. A summary of the knowledge and a decision direction is made by the middle management for top management. In simple incremental product development, the Board may have set an overall budget for the product development programme; the decisions on the product concept acceptance and the resources for the individual projects are made by the middle management.

There is no question that the decisions about the product/innovation strategies and the product development programme are set by top management, but it is also important that middle managers, and even the project leaders in large projects, are involved in deciding on the knowledge needed in the outcomes. Only they have the detailed understanding and skills to know what knowledge is really needed, and what knowledge can be created with the capabilities of the company and available outside sources. The final decision is by top management but the knowledge basis for it is from the collaboration of all the people in the product development group (Jassawalla and Sashittal, 1998). Strategic planning for new products and new product selection are the most critical issues in product development management (Scott, 2000). It is important to align the business strategies with the technology strategies. This can be difficult where the technology strategy may be very long term, say 10 years, and the business strategy is shorter, usually 5 years. So the strategic knowledge outcomes are not set in concrete but are studied yearly, and are designed so that they anticipate, but are also reactive to, change.

The linking at all levels of management of anticipation and reaction are important in the early stages of the PD Process: project plan, product concept, product and process specifications, product design choices and process design choices. There needs to be recognition (Verganti, 1999) of both:

- anticipation capabilities – the capabilities to anticipate information into the early phase of product development;
- reaction capabilities – the capabilities to introduce changes late in the PD Process at low cost and time, to cope with unexpected constraints and opportunities.

The early decisions, and their related desired outcomes identified in the early stages of the PD Process, play a central role in the further development of the product development project. This is why there has been much emphasis in recent years on spending more resources on the 'fuzzy' front end, so that the future development of the project can lead more surely to product success. But, especially in the innovative and long-term projects, it is not possible to anticipate everything that is going to happen. In aiming for integrated product development performance (that is shortest time to market and optimum product quality) with new product lines and new product platforms, Verganti (1999) found that companies used different mixes of anticipation and reaction between the extremes:

- Detailed approach. Companies devote great efforts in the early stages to reduce uncertainty about downstream constraints and opportunities, and tend

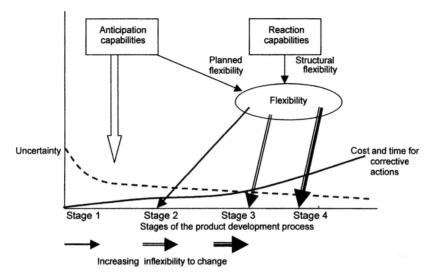

Fig. 6.5   Anticipation and reaction capabilities in product development
(Source: After Verganti, 1999).

to keep reaction to a minimum. They try as far as they can to anticipate what is going to happen.

- Postponing approach. In this fully reactive approach, companies simply start the implementation of product development without anticipating knowledge about downstream opportunities and constraints.

Other companies aim at reducing the probability of running into serious unexpected events, through selective anticipation of critical areas and reactive capabilities that can handle inevitable late corrections at low cost and time. The ratio of anticipation to reaction depends very much on the company philosophy on risk-taking but also on the level of innovation of the project.

There is a need to combine anticipation of decisions/outcomes with planning of the ways to react to new knowledge developed in the PD Process, so necessary changes can be introduced late in the PD Process without too much cost in resources and time. Because making changes in the later stages is more costly in time and resources, there needs to be a balance in anticipation and reaction in the early decision-making as shown in Fig. 6.5.

Flexibility in product development (the cost and time for late corrective actions in a project) may be seen as consisting of two major components: structural flexibility and planned flexibility. Structural flexibility is the reaction capability that unfolds through long-term practice, project after project; planned flexibility is the project-specific flexibility, built through decisions taken in the early phase of the particular project. Anticipation capability is the capability to identify, clearly and early, specific critical parts of the project and plan the trigger reaction measures to manage these critical parts of the project. Decision making at the early stages of the PD Process is a balance of anticipating predicted outcomes and setting up possible reactions to uncertain outcomes later in the project. This balancing of the anticipation and reaction capabilities is related to the level of innovation of the project, and to the company's capabilities and resources.

Outcomes can be set in general strategic terms by the top management, but they have to be developed into much more specific outcomes by the product development manager and the project leader. Designers are given product design specifications and these are the blueprints for their development of a prototype product. Process development engineers are given product qualities from the designer's outcomes and design a process to produce these product qualities. Production engineers are given the production specifications from the process engineers' outcomes and have to design a production system. The different outcomes may be developing in sequence or in parallel. For example the marketing strategy development and the process/production development both start from the product prototype and its qualities, and end at the same time, as the production plan and the marketing plan. This integration of times for outcomes is very important so that there are no waiting periods in the project.

A problem is sometimes caused by ranking tangible outcomes higher than intangible outcomes, for example the process design before the consumer attitudes. Product development often goes astray because a great deal of time and money is spent on developing a technical product and process, only to discover that the product is not what the consumers or customers wanted. The technical product is a tangible outcome, the consumers' concept of their ideal product is intangible, but it is actually the true outcome.

Think break

In Table 6.2 are some of the outcomes required by top management to make decisions at critical points. The product development manager is required to produce these outcomes for top management, in turn the product development manager has to identify for the project leader, the specific outcomes from the activities in each stage.

1. For Stages 2 and 3 in the PD Process, identify the specific outcomes needed by the product development manager from the project leader to build up the outcomes needed by the top management.
2. Draw an outcome 'tree' to show the relationships between the two levels of outcomes – for top management and for the product development manager.
3. In some past product development projects in your company identify the critical outcomes for top management decision making. What outcomes were unimportant to the decision making of top management? Discuss how this could affect the identification of critical outcomes in future projects.

6.5.2   Setting the budget

Setting the budget for the product development programme and the individual projects is related to the outcomes desired and the resources of the company. The central figure in setting the budgets is the product development manager who discusses the project needs with the project leader and then negotiates for the budget with the top management. This is essentially a human relationship although it is clothed in quantitative terms such as predicted financial outcomes, costs and probability of success. In the radical innovation, these predictions can be inaccurate and the budget for the project is secured on, as Tighe said in 1998, the 'selling' of the project to the higher management based on a combination of estimates and confidence. For incremental product development, the predictions are more accurate and the decisions are more pragmatic. Essentially the product development budgets are competing with the budgets for the 'today' functions; either in the main company budget, or if product development is in a functional department/division, within its working budget. The project leader and even the product development manager can be looking for a sponsor, who could be someone on the company Board, in top management, in a functional department, or in a new venture corporation. For incremental product development, increasing a product line or re-positioning a product, the marketing manager is probably the most interested person in the outcome of the project. For process development, it will be the production and engineering managers. If the project is a radical innovation, going across the functional areas in the company, then it is at top management or Board level. In a budget presentation, the project is clearly defined as to:

- aim and outcomes and their relationship to the strategic direction of the company or the department;
- effects on profits, revenues; costs and the qualitative targets for the project.

In the presentation it is important to relate the outcomes of the project to the business and also to show how the resources sought in the presentation are to be used. The project budget needs to relate to the company strategies, plans and budgets, because the resource allocation is not only dependent on the project needs but also on its relationship to the overall company resources and the competing departmental and project needs.

Important resources needed by the project manager are:

- financial;
- expertise (people with the required knowledge and skills);
- equipment;
- raw materials;
- information.

For a company in a given environment, some or all of these resources may be in short supply, expensive or shared; there are limited resources for the project. In companies with a number of projects and project managers, competition for limited resources may add to resources allocation problems. The consequence of these situations, in terms of reaching the project outcomes on time, has to be assessed and problem areas identified as early as possible. Bottlenecks are predicted and resolved before they become serious. Failure to identify such bottlenecks can result in missed deadlines, which are frequently associated with large cost penalties or lost market advantages.

Budgeting should not be complex. A straightforward approach is to take each activity and look at its resource requirement on a monthly basis and so build up a schedule of costs, personnel, equipment and raw materials. It is important for the project leader to analyse each activity to see if other techniques could be used that would be less costly, more within the present capability of the team, and yet still yield the desired quality and efficiency levels in the project. If the monthly expenditure is presented graphically as cumulative expenditure, critical areas for cash flow during the project can be identified. The capital expenditure is treated separately. One aspect to consider is the future use of the capital equipment; if the equipment is to be used only for one project, writing off this capital is an important point in determining the validity of the project.

One of the factors in helping the project team to product success is to give it a secure basis for development; this is largely based on the availability of finance. So all levels of management need to spend some time in the early stages predicting the costs for the project and the relationship of these costs to the company's availability of finance. They need to be involved in the decision to finance the project. The product development manager has to combine the resources detailed in the plans of the individual projects. The projects are integrated so that there is optimal use of resources, the specific resources are within the budgets of the different functional departments – R&D, marketing, production, and the finance required is within the overall budget predictions of the finance department. This can present a problem in some companies where departmental budgets are not specifically related to product development and only R&D is identified as product development. As is well known the main expenses are in product commercialisation and launching, and these costs need to be identified at the beginning of the project so that they are within the financial resources of the company or available from venture capital. The top management and the Board of Directors have to view the budget proposals for the product development programme and judge them within the strategies and resources of the company. It is not sufficient just to agree to the funds for one year, but the funds for future years need to have support. There also need to be contingency funds either to allow for rapid product development sparked by a change in the environment or to solve unforeseen problems that may have arisen.

Think break

In two recent product development projects in your company (one incremental and one radical innovation):

1. Identify the times in the PD Process when the need for resources was identified and when the resources were allocated.

2. What resource needs were identified – finance, people, equipment, raw materials, information? How were the resources found? Were there any other resources needed?
3. Who were the people who developed the budget proposal in these projects and who made the decision to guarantee the budget?
4. Were there any delays in the projects because of poor prediction of financial needs? Were there cost overruns?
5. Compare the resource management and financial control in both projects, and identify the difference between an incremental project and a radical innovation.

6.5.3   Setting the constraints

Product development does not occur in a vacuum! There is an environmental situation that sets the parameters and constraints for the project. All the layers of the environment – society, industry, market and company – limit or constrain the area of the project as can be seen in Fig. 6.6.

Parameters that need to be considered at the beginning and throughout the project are the needs and wants of the consumers, the processing and marketing technology available, the knowledge capability, the time and the resources. Some important company parameters are the business strategy, the innovation level, expertise, management style, location of plants and markets, distribution system, product development organisation and management. Some of the environmental parameters are local government, national government, industry

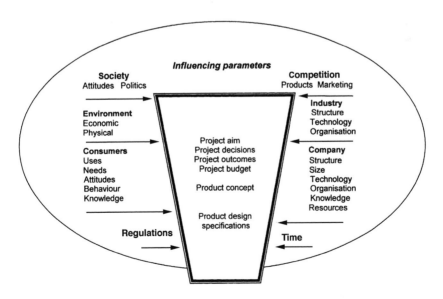

Fig. 6.6   Parameters that develop constraints in Stage 1 of the PD Process.

agreements, farmers' agreements, economic and technological status, business cycle as well as any social restrictions and attitudes.

Constraints caused by the different parameters are identified. Constraints can be set on the product, processing, marketing, finance, time and resources. These constraints are often quantitative, for example on the product there can be chemical composition, microbiological level, nutritional value, and a specific requirement such as the use/non-use of an ingredient. Some of the constraints important in product development projects are shown in Table 3.1 on page 100. Three important areas to develop constraints are the competitors, government regulations, user needs with their related societal attitudes (Holt, 1983). The company selects the important competitors, and the areas for competition, for example product technology, marketing, costs and prices, and the position of the new product as related to the competing products. Functional requirements or some other product quality may then be set according to the qualities of the competing products. These will then be part of the product design specifications and limit the area for design. Another important area to develop constraints on the design of the product is government regulations, both internal to the country and at entry to the country. These must be identified early in the project. There is little point in designing a canned fruit in syrup only to discover that there are high import duties related to the level of sugar in the product; or there is a quota on certain meat or dairy products into a country and these are already filled; or that only certain ingredients are allowed in bread; or there is a basic vitamin content if the label says vitamin-enriched. The government regulations need to be surveyed and the important parameters and constraints identified to avoid developing products that will violate the law. The very important area for setting constraints is the study of the user – what their needs, wants and fears are, and how these relate to the product design specifications; their attitudes to pollution, farming practices, environmental sustainability. Various factors may be constrained, for instance for a nutritional drink:

- product – nutritional, minimum percentage of protein, maximum percentage of fat;
- processing – minor adaptation of present production;
- marketing – existing distribution channels;
- financial – maximum level of investment in the project;
- company – two project members, one technical, one marketing; functional departments as support groups;
- food regulations – no preservatives, nutritional labelling required.

The parameters that can constrain product development are many and it is not possible to study them all when identifying constraints on the product development project. It is important to identify the critical constraints and not to have these constraints any tighter than is necessary to meet the parameters. One always needs to ask is this constraint valid? Is it necessary? If the constraints are very tight, then the opportunity for creativity is reduced. The constraints are important in the product screening and project evaluation, and are used in building up the product concept and product design specifications. It is important that they are clear and as quantitative as possible.

Some parameters for the Mexican and US markets for tortillas are identified in Box 6.1.

Box 6.1   Mexican tortilla firms stage US bake-off

Few things are more Mexican than the tortilla. But when it comes to making money off the ancient disks, most of the action today is north of the border. That's why two of Mexico's biggest food companies, Grupo Industrial Maseca SA (Gruma) and Grupo Industrial Bimbo SA have chosen the US as the main battleground for their fight to control the $5 billion world tortilla market. So they're pitching their mass-produced, packaged tortillas to a foreign audience and honing their marketing skills for the day when Mexico's tortilla market joins the modern world.

Mexico

Mexicans eat 360 billion tortillas per year, 10 times the number of tortillas per capita as Americans. The market is overwhelmingly dominated by tortillerias, small businesses. More than 96% of all tortillas are sold in little shops licensed by the government. These outlets, many grinding tortillas on hand-powered conveyor belts, are virtual monopolies in their neighbour-hoods, with a captive market that so far has resisted modern sales efforts. In part the reason is cultural: Mexicans like their staple fresh, hot off the press. But more importantly, Mexico subsidises small tortillerias with cheap prices on corn flour, making it possible to sell corn tortillas for less than the production cost. 'People stand in line for two to three hours for tortillas. It's worth it because they are so cheap.'

Thus, despite modern baking technology, companies such as Gruma and Bimbo simply are unable to make tortillas cheaply enough to compete with the small businesses, however inefficient they may be. Once tortilla subsidies are phased out, millions of Mexicans will buy their tortillas just like Americans do, in plastic bags in supermarkets. If Mexican companies could raise their packaged-tortilla sales to 20% of the Mexican market from 5%, they would match the tortilla output in the USA.

USA

There is a Mexican immigrant market but tortillas are also making in-roads among Anglo families. The US market is growing 10% per year in dollar terms, compared with just 2% for Mexico.

Gruma purchased Guerrero Foods, an East Los Angeles tortilleria founded by Mexican-Americans, in 1988. It works with Mission Foods, another

Box 6.1 (continued)

Gruma operation. Guerrero's nemesis is La Tapatia, another family start-up, snatched by Bimbo last year.

Mission is the more upscale product, packaged with Mexican recipes in English on the back, for sale to Anglo supermarkets and institutional customers like PepsiCo Inc.'s Taco Bell unit. The play to national sympathies is obvious. Immigrant tortillas are packaged in bulky two-dozen, three-dozen and five-dozen, even 100-count bags, usually with a logo heavy on corncobs. Anglo tortillas are sold in slimmer packs with flashier logos; there are kid-size tortillas with green carton dinosaurs on the package, fat-free tortillas for the health conscious, and 'home-style' with lots of lard.

Bimbo has battle-hardened sales forces in the USA and the best distribution system in Mexico. Gruma's edge is its tortilla technology, with high speed mass production. In Los Angeles, it has the biggest tortilla factory in the world, runs three separate lines, one for institutional clients like Taco Bell, one for retailers and one for snack foods like tortilla chips. Gruma's R&D produce most of the industrial tortilla machinery used today. It also keeps down its raw material costs, because it mills its own flour in the USA and Mexico. Gruma is the biggest US player, with a 15% market share.

Analysts say that the future belongs to the tortilla-maker that can transfer US marketing practices to the home market.

Source: From Millman, 1996, reprinted by permission of the *Wall Street Journal*, 1996. Dow Jones & Company Inc. All rights reserved worldwide.

Think break

In Box 6.1, there is a comparison between two markets and two companies in the tortilla industry.

1. Compare the parameters in the USA and the Mexican market affecting the development of 'commodity' tortillas, specialised consumer tortillas and institutional tortillas.
2. Identify the constraints that limit product development in the three product areas.
3. If the two companies were to consider entering the Mexican market when the corn flour subsidy was reduced by 50%, what constraints would each company have in developing tortillas?
4. What would be the major innovations for each company in product development?
5. What would be the major differences in the product development of the two companies?

6.6   Organising the PD Process

The PD Process, the decision-making, the outcomes, the budget and the project constraints are set and now someone has to start creating the product! There are two dimensions in carrying out the product development activities:

- PD Process capability;
- functional/technical knowledge and skills (expertise).

Companies have various degrees of functional excellence and technical know-how; but independent of these is the skill in organising, carrying out, maintaining and improving the complex business process of product development. McGrath *et al.* suggested in 1992 that companies could be graded on their product development capability according to these two dimensions. Companies tend to focus in one direction – technological superiority or business process capability, but the world-class companies are trying to increase their capabilities in both dimensions.

6.6.1   Identifying activities, knowledge and skills

The choice of activities is not only determined by the knowledge needed in the outcomes but also the resources and time available. The description of the activity defines the outcome expected, the timeframe to be met and the resources that can be used (Earle and Earle, 1999). The inputs and outputs of the activity are shown in Fig. 6.7, the people input and the physical input, which result in the necessary outcomes and decisions. Each activity has certain techniques chosen for it to achieve the desired outcomes.

   Activities are often identified as product, consumer, technical (or processing), marketing and finance, but this does not give the integration that is necessary for product development. For example in Stage 2: Product design and process development:

- First main activity: initial prototype development.
  Specific sub-activities: modelling, product formulation, ball park processing experiments, protective packaging design, consumer panels.

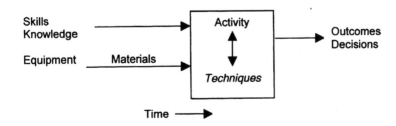

Fig. 6.7   The activity and its inputs and outputs.

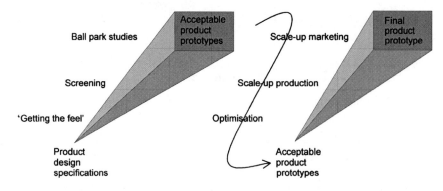

Fig. 6.8 Main activities in Stage 2: Product design and process development.

- Second main activity: developing the final prototype.
  Specific sub-activities: technical development, marketing strategy development, in-home consumer testing, production testing, market predicting,
  costing, finance analysis.

Figure 6.8 shows the main activities for developing the product prototypes and for
developing the final prototype product. The sub-activities can usefully be grouped
into development and testing as can be seen in Fig. 6.9. Design of the product and
development of the process are interwoven with each other and with the technical,
consumer and financial testing and analysis. The sub-activities can vary from
project to project but the ones in Fig. 6.9 are common. Included in Fig. 6.9 are the
possible techniques that can be used in these activities. It is important that the
necessary activities in a project are identified and then techniques chosen to give
the necessary knowledge. These must be within the capability of the company.
There is often a tendency to keep on using the same techniques as it is simpler and
easier for the project team, but they may not be the optimum for the project – they
may be producing unnecessary knowledge or, what is worse, too little knowledge.
The activities may be the same but there should be careful choice of techniques.

Think break

1. In the initial activities of Fig. 6.9 – 'getting the feel', screening, ball park studies –
    the product design and the process development are conducted at the same
    time, and the techniques in the development are experimental designs. Study
    two recent product development projects in your company, and identify the
    activities and techniques used in developing the acceptable product prototypes
    and compare with the outline steps.
2. For future projects, would you change the activities in developing the acceptable
    product prototypes? What new techniques could be introduced in future
    projects?

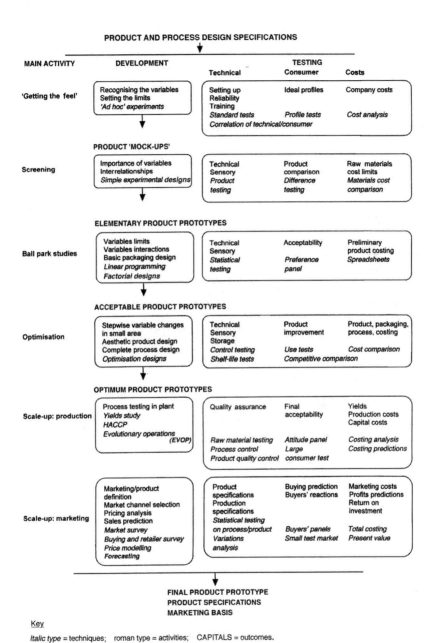

Fig. 6.9 Development and testing in product design and process development
(Source: After Earle and Earle, 1999).

3. In the two company projects you are studying, how were the design and development integrated with the product testing? When was formal testing started?
4. In testing the product prototypes, what techniques were used? How accurate were the techniques in assessing the product qualities? Are there any new techniques that might improve the accuracy of selecting the acceptable product prototypes?

Knowledge and skills can be identified for activities and outcomes. There is a wide variety of skills from specialist skills such as in organising and moderating consumer focus groups and linear programming in product formulation through to generalist skills such as product design and project control. The important point is to identify the knowledge and skills needed for the activities and then identify the source of these skills. There is a basic level of knowledge inside the company, information can be brought from outside the company and used to create new knowledge, the company can create new knowledge through the activity, and also seek new knowledge through strategic research. The basic technical and marketing knowledge within the company may be tacit staff knowledge or explicit knowledge within the company records. This can be split into general knowledge and domain-specific knowledge, i.e. knowledge from general experience and education, and specific knowledge gained through working and education in the area(s) (Boston *et al.*, 1998).

There can be information sources outside the company such as literature, databases, patent information, catalogues, trade journals and of course the Internet. There is obviously a great deal of information available from these different sources, but knowledge is needed in the company to find it, refine it and present it in a usable form. Information has to be transformed into the knowledge needed for the activity; this is the most difficult part, which is often not recognised. Money and time are spent on setting up or contacting databases. But it is left to the stressed person organising the activity to change this into knowledge – is it a wonder that they often fall back on their own or other team members' tacit knowledge and do not use the information bases? The setting up of information bases means medium of presentation, format of presentation, location of delivery, management of delivery; timeliness, accuracy, relevance and cost must be considered – as related to the users of the information in product development (Boston *et al.*, 1998). Information technology can provide a means of collating and distilling meaningful and usable information across disciplines, which is useful as a basis for innovation, but also as a source of on-going relevant information (Ganguly, 1999).

Knowledge and skills can be brought into the company by employing new graduates or experienced people from other companies or from academic organisations. The present staff can be retrained to have the necessary skills. This all takes a great deal of forward planning based on the future product development programme. It is not something that can be done when the project has come to a halt because of lack of specific knowledge or the ability to create new knowledge for the company.

Cooper (1999) identified seven 'blockers' to product development:

1. Ignorance.
2. Lack of skills.
3. Faulty or misapplied new product process.
4. Too confident.
5. Lack of discipline, no leadership.
6. Big rush.
7. Too many projects and not enough resources.

Three of these are lack of knowledge and skills, two are lack of time and resources, and one is lack of project discipline. The lack of knowledge and skills can be in the basic knowledge needed for the project and also in the method of organising the project. The activities needed for the outcomes and decisions cannot be identified, nor suitable techniques to be used in these activities. There is no excuse for this in incremental product development where activities are well recognised, but the important factor in this case is to keep up-to-date with the choice of techniques for the activities. Sometimes people may adopt a new software package for experimental designs but not know the underlying material changes that are occurring in the processing. For example they may be using an experimental design package that plans the experimentation and analyses the results but they may not understand the reactions that are causing the changes in colour, texture and flavour. Certainly in the radical innovation, the pathway may not be clear and there may be a need for significant experimentation before the product design specifications and the pathway for the rest of the project can be set. The radical innovation is entering unknown territory, but skilled knowledge in organising product development projects can keep the project from becoming lost.

Think break

1. Identify the tacit consumer knowledge that is in your company and the people who hold this knowledge. What are the codified sources of consumer knowledge and information available within the company and from sources outside? How does your company create consumer knowledge during the product development project?
2. Is the 'bank' of consumer knowledge within the company adequate for the company's incremental projects? If not, how could it be improved?
3. Identify people in your company who have knowledge in more than two areas of product development. What are their areas of knowledge and how do they integrate these? What are their positions in the company? When and how do they take part in the PD Process in your company?

4.   Identify people either in your company or acting as consultants to the company, who have specialised knowledge in one discipline, activity or technique. When is their knowledge used in the PD Process? How is their knowledge integrated into the PD Process? Is there a need for more specialists – in what areas?

6.6.2   Responsibilities, resources and timing for activities

The kernel of the product development project is making things happen. It is important not only to plan activities but also to see that they occur effectively and efficiently, and are producing the correct new product at the right time and cost of resources. Product development is not a pyramid management activity, the person at the top of the pyramid taking all the responsibility, but for everyone in the project accepting responsibility. But to achieve the outcomes, they need to be given the resources and the time to complete their activities to the level required by the project. Everyone will have some critical activities, which they need to identify and nurture.

Responsibility is an important part of the management of product development. There needs to be clear definition of where responsibilities lie, from the critical decision making of top management to the outcome of the individual sub-activity of the individual project team member. Every part of the product development project can be crucial to the total effectiveness of the product, for example:

- poor optimisation of the flavour may cause an unattractive product to the consumer;
- selection of a variable raw material can cause production problems;
- low resources given by top management may cause lack of knowledge for controlling distribution;
- pressure on timing by marketing can lead to product failure in the marketplace.

So responsibility needs to be identified early in the PD Process as shown in Fig. 6.2 and expanded in Table 6.3, and adhered to throughout the project.

These are management responsibilities but the responsibilities of the individuals in product development also need to be identified. The designer has responsibility not only for creating a product, but also to ensure cooperation with other company staff and also consumers in designing the product. The designer's responsibility does not stop with the product prototype but includes its integration into both marketing and production. The designer has responsibility for communication with the other product development team members to ensure the design fits with the other developments, particularly process development. The supervisor on the production line for the new product not only has the responsibility to produce the product to the specifications but also to see that yields and costs are met. They are responsible for ensuring that

Table 6.3 Responsibilities in product development

Top management
- Clear product strategy and product development strategies
- Prioritise projects objectively
- Acceptance of the product development programme
- Clear definition of decisions and outcomes at the critical points of the project
- Determine and make available the necessary resources
- Determine and ensure the personnel needed in product development
- Make timely, decisive and knowledgeable go/no-go decisions

Product development management
- Product development processes for different innovations
- Coordination of product development programme
- Allocation and timing of resources
- Identification of different activities and their outcomes
- Set standards for quality of activities
- Control times and outcomes for activities
- Education and training of project teams
- Measures and controls effectiveness and efficiency of the product development
- Revamps the PD Process regularly

Project leader
- Identification of techniques for different activities
- Setting the standards for the different techniques
- Encouraging the creativity in design
- Problem solving
- Communication in the team
- Organising time and resources for the team
- Controlling the team's activities to give the desired outcomes

Source: From Earle and Earle, 1999, by permission of Chadwick House Group Ltd.

staff understand the process, process control and product qualities; as well as cooperation between production and quality assurance.

Resources needed for the project are identified, followed by the resources already present in the company and the resources that will have to be brought into the company. The core resources are the complementary assets and organisational capabilities of the company. The complementary assets owned by a company can include

- physical assets such as marketing channels and manufacturing capability;
- knowledge assets such as R&D, patenting and the linking of the buyers and the suppliers;
- psychological assets such as brand image, company image;
- power assets such as high market share and industry dominance (Taylor and Lowe, 1997).

The assets can be physical or functional such as R&D, company/retailer networks, and both technical and commercial knowledge. A company can have assets such as modern manufacturing technology, an understanding of the market, a breadth of market coverage and skill in research and design. It is important that these assets are related not only to the product development programme but also at the level of the activity in the project. There is often non-recognition of these assets by the person at the activity level. Some functional assets such as marketing and R&D are crucially important to product development, and their interaction is vitally important. The assets do not have a value when standing alone; it is the range of assets used together that is important (Taylor and Lowe, 1997). Another important asset is the organisational capability of the company, not just as related to the product development project but the total company organisation – this is particularly important in the initial stages in developing product strategy and product development programmes, and also in product commercialisation and product launching.

It is important not only to identify the assets available as resources in product development, but also the assets that the company lacks:

- Lack of up-to-date technology may limit product development to incremental change.
- Lack of market strength makes it difficult to launch an innovative product.
- Lack of raw material sources can limit the product formulation.

Timing is of course crucial in product development. For over 40 years, planning aids such as critical path networks, Gantt charts, PERT diagrams, job progress bar charts have been used in planning and controlling projects and there is computer software which can be used (Gevirtz, 1994). But the important parts in planning timings are:

- setting the sequence of the activities;
- identifying the critical activities as regards timing.

It is important to identify the activities that can theoretically run side by side, that is in parallel, and the activities that must sequentially follow each other. Sometimes lack of resources may change the parallel activities because there are not the people to carry them out together. So the people resource, and also often the equipment, may change the theoretical sequencing in actual practice. But it is always useful to start with the theoretical, because that will be the fastest track for the project. It is important to realise that by changing the theoretical sequencing of activities, efficiency of the product development will be reduced. Because of its effect on timing, it may be decided to drop an activity and take the risk of lack of knowledge leading to product failure. So there is much balancing of effectiveness and efficiency in planning the timing of activities as shown in Fig. 6.10 (Duffy, 1998). The consumer and the markets pull the balance towards effectiveness, seeking the optimum product that satisfies their needs and wants; the company wants an efficient project that is completed in time and within costs. Product development (project leader and manager) is trying to balance the two wants!

The most important part in timing the activities is to identify the sequence of activities that are critical. If the time taken for these activities overruns, then the timing for the whole project overruns unless there is reduction of activities in

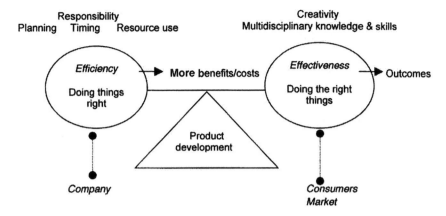

Fig. 6.10 Balancing effectiveness and efficiency (Source: After Duffy, 1998).

later stages. It is very important that care is taken with identifying and controlling the critical activities, early in the project, as losing time in Stage 1 can be very costly in the later stages and also may reduce the chance of success in later stages.

The main activities are split into the sub-activities (tasks) and the completion of each task is called an event (Meltzer, 1996). The length of each of these tasks is arguable, but they are usually measured in weeks. It is important that times for tasks are not too long as they become more difficult to control for timing, but again creativity in design does take time and is difficult to break down into short tasks.

It is important to monitor and maintain the time and resource schedule. Any changes are recorded and their effects on the launching date and also earlier stage completion dates predicted. There can be unanticipated problems such as difficulties in obtaining raw materials or equipment, technical difficulties in the design, patenting problems, competitor actions and even non-availability of top management for decisions. Their effects on the schedule need to be recognised immediately and the flow-on effects predicted. There may be a need to increase the people resources or to put pressure onto some suppliers to get the project back on a satisfactory schedule. Some pre-planning for possible problems needs to be allowed in the timing schedule.

Think break

1. Contrast the responsibilities in your company, for product development, of functional managers including marketing and R&D, and of product development managers. How are these responsibilities integrated? And by whom?

2. In a product development project team in your company, compare the responsibilities of the project leaders and the team members. Do these responsibilities vary among project teams? What causes variations in

responsibilities among teams – differences in type of project, team leader, the composition of the team?

3. What is your company's balance between efficiency and effectiveness in product development projects? Does the balance vary between projects? If so, what is causing the variations?

4. What are the stumbling blocks in your company for increasing the efficiency of product development – lack of people, resources, and knowledge; over-ambitious projects; too many projects; faulty or misapplied PD Process; lack of discipline, no leadership?

6.6.3   Personnel – internal and outsourcing

People are the most important factor in product development. There are the core team members and there is the greater team, including support groups. A great variety of knowledge and skills is required and this needs to be integrated into a complex network supporting the product development. There is no question today that the linear progression from science to innovation has changed into an interdisciplinary relationship (Ganguly, 1999) as shown in Fig. 6.11.

There are four important factors in the network:

1. Integration of the sciences.
2. Integration of the technologies in the total technology.
3. Interaction of the science and technology continuously.
4. Interaction of the science and technology in innovation.

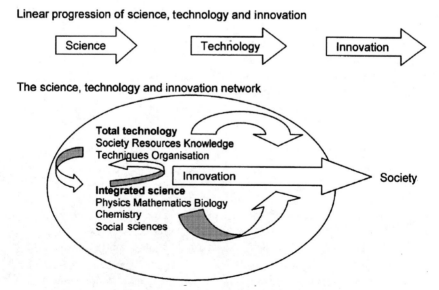

Fig. 6.11   Relationships among science, technology and innovation.

The physical and mathematical sciences are part of all technology, in combination with chemistry and biology in process technology and with social sciences in total technology. The science can be developed separately but often is developed as part of the technology. Technology often cannot wait for the results of the basic science research but has to do the research to solve a basic technology problem. This is true in all areas, for example consumer research cannot wait for research in the social sciences but has to do the research now to solve the consumer problem. So basic research can be in technology as well as in science. Interaction also occurs in the other direction, basic science can come from a theoretical problem but often it comes from a problem identified in the science environment, which includes technology. The latter is often called strategic science, as it has a direction.

Therefore in the innovation process there is a continuous interaction between the development of the product and the science and technology. This interdisciplinary network, which vibrates backwards and forwards during the development, presents a very complex personnel management problem – in selection of people both outside and inside the company with the necessary knowledge, skills and creativity, and then coordinating them into a vibrant, interacting, communicating network. The company needs to have the ability to sustain a leading-edge competence over long periods (Ganguly, 1999), by selecting and educating staff, by careful selection of the outside personnel and building a relationship with them. Sometimes to solve a problem that has arisen, there may be a need to bring in an outside agency or consultant; but to be of value for the people in the network, the agency needs to have knowledge and understanding of the company. All this seems to fly in the face of the company's perceived need for secrecy in innovation, and has led to an increase in intellectual property agreements.

The networks can be inside the company and also connecting the company personnel to academics, technical consultants, research associations, consumer and market research companies, and innovation management consultants. There has been an increasing interaction of academic and government research with industrial research in companies, often encouraged by government organisations and grants. In all Western countries, the increasing need for these networks has been recognised because technology has been progressing so fast that it is difficult for companies to stay ahead. This also happened in the middle to the end of the nineteenth century in Scotland, during the Industrial Revolution, when the communication between academics and industrial technologists was not only close but their work was interrelated. In a time of fast innovation, academic research can get behind industrial research, and industry can be left with basic research problems, which cannot be solved in the time available. As the major problems cannot be solved, product development can make only small incremental changes; this can result in stagnation of the company and perhaps death. This stresses the importance of the interaction between the company research and external research to maintain the rate of innovation. Box 6.2 illustrates some academic and government research with possible applications.

Box 6.2   Application briefs from the *Journal of Food Science*

Better rice formulated with vitamin A

A process for enhancing the content of rice with retinyl palmitate, a particularly effective vitamin A precursor, was studied by researchers at the Department of Food Science and Technology at the University of Georgia. The process was donated in 1997 to the Program for Appropriate Technology in Health by the Coxes of Washington State, who owned the patent. Broken rice is milled into rice flour, combined with a binder and retinyl palmitate and other fortificants, and reformed into rice grains with the same texture as whole rice grains. These are blended with conventional long grained rice at a ratio of 99:1. The present study showed that the retinyl palmitate was quite stable under various cooking procedures. When stored at 23°C for 6 months, 85% of the retinyl palmitate was retained, but at 35°C there were extensive losses, 50% after 24 weeks. Under tropical conditions, this means either the use of controlled temperature storage or rapid turnover or increased levels of fortification to compensate for the loss.

Reconfiguring the fatty acid profiles of dairy foods

In 1970's it was found that by feeding cows a source of high oleic fatty acids, milk with higher levels of oleic acid can be produced. High oleic sunflower oil and canola grain are now available as cattle feed additives and make possible the commercial production of milk with higher levels of oleic acid. Researchers from the Universities of Florida and Virginia Tech have studied cheese making with this milk. By consuming calcium salts of high oleic sunflower oil containing 86% oleic acid, test animals produced milk in which the high oleic fatty acids in the milkfat increased from 26% to over 40%. Latin American white cheese (queso blanco) was made from the milk, and tested for firmness and for sensory differences from conventional cheese made by the same method. No differences were found in firmness, sensory testing showed no significant differences between the cheeses. Latin American white cheeses made with high oleic milk were similar to traditional cheeses.

Source: Reprinted from *Journal of Food Science* 65(5): iv, v. © Institute of Food Technologists, Chicago, Illinois, USA, 2000.

Think break

1. Identify outside agencies and people that your company has involved in product development over long periods of time, and the people in the company who work or liaise with these people. Show how they cooperate in the projects.

2. How could your company employ or educate its own personnel to take the place of the outside agencies? Would this improve the effectiveness and efficiency of product development in the company? How would you show top management the cost effectiveness of doing this?

3. The philosophy and practice of scientific research in universities during the last 100 years have been individualistic with freedom to choose, but the philosophy of science and technology in industry is the creation of clusters of inter- and intra-disciplinary teams with a strategic direction (Ganguly, 1999). In the development of networks between academic institutions and commercial companies, how can these two philosophies be merged to give satisfaction to all the participants and ensure the forward-flow of research and development?

4. In small companies, there are only a few people in product development and they have only certain areas of knowledge. Examples are:
    (a) Small company based on the technical invention of a co-extruder for dough and thick paste, the marketing skills are few. How could this company develop an outside network to overcome this lack of marketing skills?
    (b) Small company formed because a need was recognised in the market for a high-protein drink for endurance athletes; it has little technical product and processing knowledge. How could this company develop an outside network to overcome this lack of technical skills?

6.7 Managing the PD Process

In managing the PD Process for a project, firstly the internal project management is identified and then the external agencies integrated into the internal team. The management for the project as it proceeds through the four stages, audits the outcomes and efficiency, and controls the project so that it is kept on track. The procedure is to lay down correctly the track for the project, and then ensure that the project is not slowed down by track unevenness or indeed goes off the rails and crashes.

6.7.1 Internal project management

Product development is people-driven, and therefore the most important aspect of managing for product development is to activate the product development team and to keep it going forward. This is the underlying energy that drives product development. By better managing and motivating people, the product development performance can improve markedly.

The **product development team** is a critical building block for the effectiveness and the efficiency of the project (Kuczmarski, 1996; Smith and Fowley, 2000). There needs to be the right mix of team members to give the knowledge for the activities and also the organisational capability for integrating the activities to create the knowledge for the required outcomes. These are

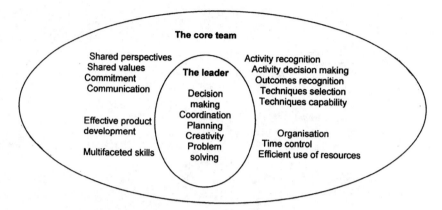

Fig. 6.12   The core product development team.

multifunctional teams and they need full-time leaders. They are often called cross-functional teams, comprising members from various functions and with complementary skills (Cooper, 1999). A truer description is a multifaceted team, having a variety of skills needed in a specific project, but not necessarily related to the functional departments. In the team, there is a need for a variety of expertise and experience, as well as different perspectives on product development. Utilising interdisciplinary project teams whose participants are involved at the onset of a project can greatly reduce development time and improve the probability of success (Gevirtz, 1994).

The actual mix depends on the company and the project. There may be a core team as shown in Fig. 6.12, and then a surrounding team or supporting groups. The core team conducts the activities but they are interacting continuously with the other members. People may move from the surrounding team as the project develops; for example in product design the person with marketing skills may be only an adviser, but in Stage 3: Product commercialisation will become a core team member. It is important that the knowledge and skills needed later in the project are identified, and people with these skills are included in the total team from the beginning.

As shown in Fig. 6.12, each member of the multifaceted team should share a common commitment to the project, with shared perspectives and shared values; being part of the team should be a responsibility that members are given by management. The team members need a mix of creativity and analytical problem-solving skills based on knowledge and experience. They are often individualistic because they have strong ideas on product development, but they need people skills to work in a team. This is not always easy to blend together but it can often be achieved with experience and good leadership.

The project leaders need skills in team-building and conflict resolution. They also need to be creative and good at problem solving; they need to be skilled product developers as well as organisers. In some incremental product development, organisation may be important; but as projects go towards radical innovation, it is the innovative thinker who can bring the team to effective product development. It is the role of the project leader:

- to ensure that the project progresses smoothly, meeting all interim objectives and targets on time and within budget;
- to make sure that the necessary resources are available when and where they are required;
- to act as the primary channel of communication between the project teams, senior management and any external organisations involved (Jones, 1997).

There are several issues to consider in the coordination and management of new product development teams (Holahan and Markham, 1996; Scott, 2000):

- Team structure, team size, membership composition of teams.
- Team leader selection, team management, team control and evaluation.
- Team operations, inter- and intra-team coordination, communication systems, team member access to project databases.
- Team reward structures, team motivation.
- Team training in teamwork and conflict resolution.

The project leader's actions in team management are shown in Table 6.4. The leadership style of the project leader is important as it determines how project members perceive the working climate, learning possibilities and organisational

Table 6.4 Project leader's team management

Setting up the project
- Identify the activities in the project with the team and set up a coordinated project plan.
- Aid the team members in ensuring the appropriateness, accuracy and efficiency of the techniques they choose for the activities.
- Set up a time and resources schedule for the plan.

Managing the team
- Combine the knowledge and skills of the individuals into a combined group knowledge and skills.
- Lead the creativity and problem solving in the group.
- Ensure communication within the group.
- Organise resources for the project.
- Plan and keep the project on the time schedule by cooperating with the team.
- Continuously monitor the quality of the outcomes to ensure the project's effectiveness.
- Keep within budget.

Team outside communicator
- Communicate well with the functional departments.
- Relate the group's aims and the outcomes from the group's activities with senior management.
- Track overall time and achievement lines.
- Be a strong group sponsor.

---

effectiveness of the PD Process. Employee-centred, relation-oriented leadership appears to lead to a positive work climate and a better learning atmosphere in the project, compared with production-centred and change-centred leadership. Supporting innovative learning requires a willingness of the leader to share responsibility and joint experimentation as well as focusing on team objectives. To assemble people in a work group and define work tasks is not sufficient to get synergetic effects – the team must find the climate supportive, trusting and challenging (Norrgren and Schaller, 1999).

The core team has a combined aim and also a plan for the project. Teamwork is essential – members need to have shared values and shared knowledge. Continuous, open, communication leads to an effective and efficient team. Regular meetings are needed to update each other on the individual activities, analyse and compare the results, develop new ideas and keep the project on track. Ideas and results need to be shared and people must feel free to criticise constructively. This core team needs to stay close to the functional departments and to consumers, customers, raw material and equipment suppliers. It needs strong connections with organisations in the distribution chain, and other outside agencies such as consultants and research establishments. Product development is not a closed internal system and team members have to learn how to communicate with the outside and at the same time keep the new knowledge in the project confidential.

There are three important factors to consider in managing the team:

1. Education and training of the team.
2. Over-confidence of the team.
3. Cultural and societal background of the team.

Education and training of the team is important (Cooper, 1999). The team needs knowledge of the PD Process and also the general and specific knowledge needed for the various activities. Team members may be lacking in knowledge of the PD Process, its decisions, outcomes and activities. When people have neither education nor experience in product development, it is not sufficient to have a PD Process on the internal computer network to which they can refer. They need a training course on the PD Process in the company, with examples of past projects as illustrations. The project leader needs an advanced course on ensuring effectiveness and efficiency of the PD Process. The specific and the general knowledge varies according to the project and the person's part in the project, and therefore the educational level of the team members varies a great deal from the young scientist with a PhD to the process worker with many years of experience. If there is not the capability and knowledge for a particular activity such as market research, technical research or consumer research, either team members will need to have further education or the capability will have to bought from outside.

Over-confidence can be a problem in managing product development (Cooper, 1999). There appears to be a tendency in some companies and with some product development personnel to say we do not need to do that activity –

we know it all. Sometimes this may be true, but many times it just shows a lack of knowledge. There appears to be little training, and indeed little research, in judging what knowledge is essential to a project, yet lack of knowledge combined with over-confidence is related to a high risk of failure. Any dropping of critical activities needs to be made in full awareness of the risks and costs involved. For incremental product change, there is often a situation in which:

- extensive tacit and explicit knowledge of the product category exists;
- the marketing strategy and plan need only minor changes from project to project;
- the production knowledge is known and in use within the company;
- the production capacity exists with little need for change to production method.

In this case, the team can consider dropping activities such as scale-up of processing, research for the market strategy and large-scale test marketing. But in radical innovation, where there can be little of this specific knowledge in the company, there can be real dangers in dropping activities such as business analysis before launch and test marketing of the new product with its new production and marketing methods.

The **cultural and societal environment** also affects the organisation and working methods of the product development team. There can be differences in the general society and also among different types of companies so that what is necessary for product development management in one environment may not be applicable in another. Souder's research in various countries in the world identified differences, for example in comparing the USA and Sweden. In the USA, the degree of commercial success was related, in both familiar and unfamiliar markets, to *marketing proficiency, development proficiency* and *customer service efficiency*. But three measures were significant for US unfamiliar products but not for US familiar products – *technical skill adequacy, R&D/marketing integration* and *project manager competency*. In the more innovative projects, there was a greater need for technical skills and for a strong interrelationship between the technical and marketing development. In these, usually large projects with a lot of unknowns, there was a need for innovative and adaptable project management. For Scandinavian product development, *R&D/marketing integration* and *project manager competency* were not related to product success for both familiar and unfamiliar products. This was probably because US-type integration processes and project manager roles may be relatively less important in Scandinavian companies, where collaboration among individuals may be more spontaneous, informal and internally motivated. Their relatively low importance is consistent with the egalitarian Scandinavian cultural emphasis on solidarity and cooperation (Souder and Jenssen, 1999). In comparing Japan and the USA, Souder and Song (1998) identified the greater Japanese belief in technical expertise in product development; which may be related to the culture – the Japanese culture emphasises the position of the technologist, the US culture places more emphasis on the manager. So in deciding on product development project management, it is important to recognise not only the level of innovation – incremental against radical innovation, but also the cultural background of both the company and the society surrounding it. Successful management of the product development project requires careful consideration of the company's internal and external environments. Practices that have proven successful in one company and one society may not be directly transferable to another company in another society.

The product development project needs to be managed in ways that promote the use and the development of each individual's knowledge and skills, and also encourage the coordination among the individuals. The means for managing this may vary from company to company, but they need to be researched so as to obtain the optimum capabilities for the company's product development.

Think break

1. Collaboration in the core project team gives a synergy which produces outcomes that exceed the capabilities of the individuals in the team (Jassawalla and Sashittal, 1998). Discuss the types of collaboration that you have observed in core product development teams and how these have affected both the efficiency of the project and the effectiveness of the outcomes.
2. From your experience, what characteristics of the core team and of the individuals affect the level of collaboration?
3. Collaboration between the core team and the supporting team in the functional departments gives a company-wide thrust to product development. The level of this collaboration depends on company organisational factors such as the priority that senior management gives to product development and the level of autonomy afforded to participants in the PD Process. Discuss the level of collaboration in product development in your company, and how organisational change and also changes in the attitudes of individuals might raise the level of collaboration.

6.7.2   Integrating and managing the work of the outside agencies

The second task is to integrate the work of the outside agencies into the product development project. There are two different groups – those providing knowledge to the company during the project and those providing systems for the commercial development and launching of the product. As shown in Fig. 6.13, outside agencies may provide consumer and market research, design of product and packaging, product testing and consulting in various areas. As the product development project progresses, more and more agencies can be brought in to provide contract processing, physical distribution, market distribution and marketing. The raw materials suppliers and the equipment suppliers fall within both the knowledge and systems acquisitions; they can be supplying knowledge

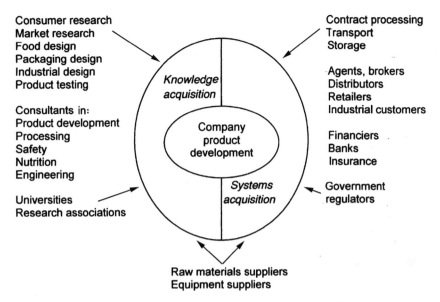

Fig. 6.13   External agencies and company product development.

on product qualities, formulation and processing conditions, but of course in the end hope to supply raw materials and equipment.

Knowledge acquisition methods can vary from casual conversations to binding contracts. The outside agencies can come into the product development project for a one-off piece of work or they can be advising throughout the PD Process. The usual pattern of short-term contracts consists of:

- outline brief from the company;
- proposal from the outside agency;
- discussion and adaptation of proposal;
- contract agreement;
- research reporting;
- acceptance and coordination into product development project.

One of the problems in, say, employing a food designer either to design the basic product or to contribute aesthetic values to the product, is to give accurate descriptions of the consumer and their needs and wants, production/distribution limitations and price/cost limitations. The outline brief needs to include this information as well as the place of the new product in the company and the market. Otherwise there will be endless redesign and discussions to get the design the company will accept, ending in a frustrated designer and unhappy product development personnel. Another problem is with work that takes a long time, for example a research programme with a university that involves postgraduate students. Basically the project is a teaching process for them, and the company must not expect results that can be accepted immediately into the product development project; they will need further development to bring them into the project. Also there are often teaching needs that require the students to do the R&D in a certain way, which is not the accepted procedure in the company. The company takes risks and cuts corners, which the university project cannot do if it is to be accepted academically. Using undergraduate or postgraduate students can be much cheaper than using company staff, but the company needs to realise the limitations, particularly of extended time.

How problems in building systems for the final stages of the PD Process can be solved by outside agencies is shown in Table 6.5. The list is not all-encompassing; many other problems arise; for example importing regulations and clearance of foods at entry will need a qualified person representing the company, the risk of loss of product or deterioration of product in transit will need insurance, and so on. This list is to bring attention to the many problems that have to be solved, and the systems/people that are available which the company can contract to use or actually buy. Little in the product development literature describes research in this area, but it is of course the most costly part of product development and can also have the greatest risk of failure.

The interrelationships between the company and the systems providers are commercial relationships and they are judged as usual by their effectiveness and efficiencies. Very often these relationships are already in place and it is a case of involving them with the new product. During the product commercialisation stage, they have to be brought into the discussions on the production, marketing or financial developments so the procedures cannot only be put in place but have

Table 6.5 Some problems in building systems solved by outside agencies

Problem	Some solutions
Little knowledge of process	Buy a turnkey plant
Need to trial production	Use contract processor
No packaging line	Use contract packer
Process control inaccurate	Contract process control company
Microbiological safety	Contract microbiological laboratories
Need ISO 9001 QA (quality assurance)	Contract QA accredited auditors
No physical distribution	Contract international distribution company
No storage in market	Contract storage company
Poor control over distribution	Contract logistics expert
No marketing system in this area	Sell product to distributor
	Buy marketing company
Poor contact with retailers	Contract food broker, manufacturer's agent
No contact with media	Use advertising agency
Little knowledge of importing/exporting	Sell to exporter
	Use an export agent
Not enough working capital	Borrow from bank
Need investment capital to start	Loan from bank
	Agreement with venture capital company

the support of the outside agencies. There is a balancing of confidentiality on the product with the need to have strong cooperation. Where new agencies have to be sought, the procedures are much more complex, especially working in overseas markets, where there may not be knowledge of the distribution/ marketing systems, culture or even the language. The primary producer can be a very long way from the consumer in the other country, and designing a fresh food for export to an overseas market can provide many headaches as the producer tries to work through a complex network of export agencies/import agencies/trading houses/auction markets/retailers. Reactions to products can take a long time – even years! The larger companies have built up subsidiary companies in the overseas countries, and may even have product development groups working in the country, so they have overcome the hurdles and can proceed with product development in a systematic way. Joint ventures and licensing operations can also overcome the problems in the new country.

Another problem is that foods are biological products, which can deteriorate with time. The distribution system looks suitable on paper but when the distribution development is taking place, there are discoveries of blocks in the system, for example slow unloading of cargo, changes from one container to another in using several airlines and bad vibrations in transport.

Think break

1. What are the major problems that your company has encountered during commercialisation and launching of a radical new product? What outside agencies has the company used to solve these problems?
2. How do you judge outside agencies providing marketing research, engineering consulting, packaging design and advertising design during commercialisation?
3. How are the distribution/wholesaling operations of marketing to the food service industry similar to and different from those of the food retailing industry? If you had been marketing to retailers, and had a line of new products designed for small family restaurants and takeaways, how would you design the distribution for these products?
4. What are some of the problems involved in exporting a new product into an overseas market? What investigations would you make to identify any import controls and also internal food regulations for the food product? What outside agencies could you use?

6.8   Company organisation for product development

Organisation and organisational changes were a significant part of food company management in the 1980s and 1990s; this management structuring and restructuring affected product development. Companies bought or amalgamated with other companies to obtain new brands and new products, and either brought together the product development and R&D groups in the two companies and then reduced the size, or dropped the product development group in one company. Then it was found that the various technologies in the conglomerate companies did not match, so they divested themselves of some areas and went back to their 'knitting' or core group. Other companies decided that they were only in marketing and sold off their processing plants and technologies; some decided that cost-cutting was the name of the game and divested or at least reduced their R&D departments. In all of this reorganising, the processing and marketing technologies were certainly split apart and in many cases were reduced, and product development was absorbed into one or the other. Today there is a need for a more dynamic management system that can grasp the idea of total technology and also be aware of changes occurring both technologically and in society.

There is no right or wrong structure for product development management. The place of product development will be determined for individual companies on the basis of:

- company strategies and objectives;
- industry environment;
- economic climate;
- company's existing product mix;
- level of technical orientation;
- level of market orientation;
- personnel involved;

and dare we say, the prevailing fashion in management!

6.8.1 Formal organisations

On the whole, product development is a misunderstood or perhaps underrated profession. This often results in the product development function becoming an appendage rather than central to company strategic thinking. This in turn leads to product development becoming the domain of marketing or technical, mainly because senior management does not realise what it is and what it can do. Product development can be made to work with almost any organisation providing there is a commitment to product development from top management and a product champion, but the type of product development and its effectiveness and efficiency vary. Marketing tends towards incremental product changes, production to cost reductions and R&D to radical innovations.

Technical
'Technical' is often the home for product development: this may be R&D, production, laboratory or engineering as shown in Fig. 6.14. R&D is often the home for product development in large multifunctional food companies, as it is the base for new scientific and technological knowledge in the company, and also in primary production because of the long time needed to breed new plants

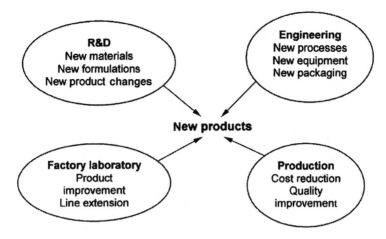

Fig. 6.14   Technical product development.

and animals. The first problem is whether to organise product development within the individual scientific disciplines or to have separate product development groups, in other words split the R from the D (Urban *et al.*, 1987). Within the scientific disciplines, there may be a lack of multidisciplinary research and it may be difficult to impose a tight time frame; the project group may have little real contact with the basic research, and none with consumers and production. A large multinational may try to get over this by having the incremental product development with the individual companies or areas, and the radical innovations in R&D. It is obviously a major problem as one sees the large companies cycling back and forth throughout the years.

Engineering is often the home in companies that are based on process development, especially food processing equipment systems. Process development is often strong in European food companies, because this can be an area for radical innovations. Usually the processes and the equipment are sold to food manufacturing companies, so there is not the same need for consumer input into product development. In the smaller companies, the laboratory or production can be the home for product development. The laboratory can be responsible for quality assurance and product development; product development usually emphasises incremental product improvements. Production controlled product development concentrates on raw material and processing changes, usually with the object of reducing costs, improving yields and improving quality. These are very general categories, and individual companies with different types of enthusiasts for product development can develop radical innovations from the laboratory of a small company.

Marketing
'Marketing' is also often the home for product development, especially in a strongly marketing-oriented company. There may be a product manager who is

responsible for a product area, both established and new products or there may be product managers who are responsible for the established products and a product development manager for the new products. The product development manager is responsible for coordinating all the market and consumer research, and the complete marketing mix for the product, and cooperating with the technical and production people in developing the product and producing it. Marketing also has problems in setting up an organisation for product development as is seen in Fig. 6.15 which outlines the development of the product development organisation in a fictional company 'Rainbow Products'.

Product Development Organisation No. 1
Separate laboratory development activity under production manager. When product ready for market, assigned to a product manager in marketing who took it to the launch.

Success: introduced cake mixes, fruit drinks
Problems: mostly me-too products with reduction in costs, marketed on price cutting

Product Development Organisation No. 2
Formal product planning department, including laboratory development operation and a small-scale production set-up. Product planning manager, reporting to marketing manager, responsible for gathering, screening and evaluating new product ideas. Once the product was successful in technical development, it was handed over to product manager in marketing.

Success: increased sales growth, new products 30% of sales
Problems: rapid increase in numbers of products, product managers could not cope with new and old products, everything getting out of control.

Product Development Organisation No. 3
New product manager appointed, separate from the product managers. Experienced in planning and implementing test marketing and relating the results to the target market, was responsible for market development through to launch and national marketing. The product planning manager reported to the new product manager, who reported to the marketing manager.

Success: reduced number of products in product lines, introduced new types of UHT products.
Problems: withdrawal of a new energy drink after 2 years in the market, new UHT soups took 5 years to develop, copying and launching of a new bread mix by a small competitor during test marketing, decline of rate of growth of sales during last year.

Fig. 6.15   Rainbow Products product development organisations.

The advantages of the product manager system are:

- familiarity with the product area;
- good connections with outside agencies such as advertising agencies, market researchers and retailers;
- real commitment to expanding the product area;
- involvement in all aspects of development;
- direct concern with day-to-day marketing.

Disadvantages of the product manager system for new products are:

- working under great pressure to produce short-term results for all products;
- difficulty in handling complex new products;
- little or no understanding of technological developments;
- difficulty in motivating people in other functional areas.

The product manager often has great difficulty in understanding the radical innovation and its relation to consumer needs and wants. Therefore they tend to produce minor product variations and product line extensions. This is also true when other people in the marketing department are responsible for product development.

Think break

Rainbow Products (Fig. 6.15), 20 years ago was an old-established company selling packaged consumer goods such as special flours, dried peas and beans, baking powder, peppers and spices, when the new general manager decided that the company should increase its range of packaged products. Gradually over the years it has marketed cake mixes, fruit drink powders, dried soups, bread mixes, hot chocolate drinks, health drinks, and recently has expanded into UHT soups and drinks. It has met several product development problems over the past years and reorganised its product development to overcome these problems.

You have been brought into Rainbow Products by the Chief Executive (who started product development when general manager). You have asked him and his staff some questions about their product development:

- Planning of products and go/no-go decisions: new product committee consisting of marketing manager (chairman), new product marketing manager, product planning manager and market researcher. Sometimes the Chief Executive sits in.
- PD Process: generation, screening and evaluation of new products; feasibility and concept testing; product development and consumer tests; marketing plan development; pre-test on product name, packaging, pricing, advertising, promotion; test marketing; national launch. Emphasis on testing at all stages.
- New products: product manager's comment that what is a product modification is regarded as 'new product' by the new product group. They spent two years developing an instant chocolate sponge, which was eventually killed because the market did not want it.

- Control of product development: product planning manager said a problem was that on the new product committee, each member's vote counts as only 1, therefore technical research has little say. One of our new instant puddings never satisfied the committee, and it went back and forth with minor changes for two years.

Now as a consultant answer the following questions:

1. What factors do you think caused the recent problems in the company's new products?
2. Do you think they were caused by:
   (a) lack of knowledge in the company,
   (b) lack of discrimination between incremental and radical new products,
   (c) poor collaboration, decision making, project control?
   Can you identify any other factors?
3. How do you think product development should be planned and controlled in the future?
4. What personnel does the company need in product development in the future?
5. What could be the management structure? Who should be responsible for critical decisions, for the effective and efficient running of the projects?

This example shows how product development is very much influenced by people, their knowledge, attitudes, beliefs and indeed their culture. It is important to recognise this when managing people in product development.

New product department
A new product department is sometimes used to integrate and coordinate the company's capabilities and bear the responsibility for product innovation (Urban *et al.*, 1987). This can work well where there is already good integration between functional departments, but can be left out on a limb if there is competition among functional departments. It certainly focuses the company's product development and also can combine the product, processing and consumer research in the early stages, but it is never large enough to do the marketing and production development in the commercialisation and launching.

All these formal systems can be suitable for incremental product changes, the choice being dependent on the character of the company and its staff, but they are usually not a successful structure for radical innovations.

6.8.2   Dynamic, changing organisation
The radical innovation needs a more dynamic product development organisation, which can change with the, often unpredicted, changes in the project. Some organisational methods are: matrix, subsidiary/divisional, entrepreneurial/ venture, corporate structures.

Matrix organisation is where the product development team member is also a member of a functional department. An individual staff member who is contributing to a product development project will be responsible to the project manager for the daily work on the project but will remain responsible to the departmental head on the standard of work and career direction. This can bring a wide variety of knowledge and skills into the product development project as needed, but can cause uncertainty, hesitation and strife between the two managers.

Subsidiary/divisional is where the product development projects are divided among subsidiary companies, or product divisions, in large companies, and they may be supported by corporate groups in R&D, market research, strategic planning and intellectual property. This means that there is a small product development group embedded in the functions of the subsidiary with specific knowledge, supported by strong groups in the central research. This would appear to be the optimum system for large multinational food companies, developing incremental products in the subsidiary and the radical innovations in the corporate research. But it can build communication problems, which can cause lower technology in the incremental products, and difficulties in technology transfer for radical innovations coming from the corporate research. To keep them as a combined product development structure needs more than the occasional visits between corporate and subsidiary; it needs combined knowledge building by moving staff between groups and joint workshops. Another problem is that the subsidiary may identify a radical innovation, but is not allowed to develop it and so becomes frustrated.

Venture/entrepreneurial seeks to introduce some of the attributes of the small entrepreneurial company into the large, multinational food company. The basic philosophy is to provide maximum responsibility to the venture manager/ project leader who is able to recruit the members of the team, free to use the resources as long as the budget is kept, and organise the activities within the overall aims of efficiency and effectiveness. The team creates the ideas and develops the product through to the launch and, if successful, the team may be allowed to form a company. This gives the opportunity for creative people with management and general business abilities to build a new product area for the company. The company not only has a new product platform but also has someone who has the abilities and the experience to develop new products in the future. If the idea is not developed to full commercialisation, and the project leader returns to their own area, the company has an employee experienced in product development. Venture/entrepreneurial is for introduction of radical innovations and not incremental products.

Corporate structures include new product committee, corporate new product task force or a group of directors on the Board. All of these report to the Board or at least the Chief Executive, and are formed from the senior people in the company. This is bringing innovation into the company at Board level, and is more likely to occur in the new enterprise, rather than the large, long-established company. These groups will set the product strategy for the company, coordinate the projects in the product development programme, monitor the progress of the projects and provide the critical decisions and the resources. A task force may be formed for a major project with high capital costs and risks. With these structures, top management has taken responsibility for product development in the company, and has taken control of it.

In conclusion one cannot say that any one structure is the way to manage product development, but that there are right and wrong ways to manage specific types of product development in specific companies. Incremental product development can have a semi-permanent, slowly changing structure which not only creates knowledge but stores the knowledge either in explicit data sources or in the tacit knowledge in the heads of people who have been in product development over time. Collaboration between people who have the multifaceted knowledge needed for a project is built up over the years. They work closely and develop an extensive shared knowledge. The company becomes a product development team, which splits into small teams for the projects but always feels connected together through the projects. This is very much easier to do in the small company, but the larger companies need to have large teams in different product areas or in different geographical areas, or in different markets. Radical innovations need a much looser, more temporary structure, because they are working in areas of not easily predictable change. Their organisation needs to give the dynamism to drive the project forward to completion.

In all product development organisations, there are some key ingredients:

- A corporate commitment to product development, starting at the top of the company. Product development is a major part of the company culture.
- One person takes responsibility for a project, no matter how large or small.
- The project leader or the product champion, if the project is too small for a team, should have direct access to personnel and their knowledge needed in the project.
- The project leader manages the people working in the project, makes decisions and is accountable for the project.
- Critical decision making is by top management, but all other decisions involve people responsible for the project.

As food enterprises grow from the small company with a few entrepreneurial individuals running or indeed comprising it, the need for more elaborate organisation grows and with that comes the need for explicit frameworks to maintain and expand the activity. Systems will be tried, become accepted and are used often, and then have to be adapted as the company grows. Product development changes a company, and the system for product development needs to change. It must not become a rigid, bureaucratic system, but retain the dynamism needed for successful product development.

6.9   References

BOSTON, O.P., COURT, A.W., CULLEY, S.J. & MCMAHON, C.A. (1998) Design information issues, in *The Design Productivity Debate*, Duffy, A.H.B.

(Ed.) (London: Springer-Verlag).

BURGELMAN, R.A. & MAIDIQUE, M.A. (1988) *Strategic Management of Technology and Innovation* (Homewood, Ill: Irwin).

COOPER, R.G. (1998) *Product Leadership* (Reading, MA: Perseus Books).

COOPER, R.G. (1999) From experience: the invisible success factors in product innovation. *Journal of Product Innovation Management,* 16, 115–133.

COOPER, R.G., EDGETT, S.J. & KLEINSCHMIDT, E.J. (1999) New product portfolio management: practices and performance. *Journal of Product Innovation Management,* 16, 333–351.

DUFFY, A.H.B. (1998) Design productivity, in *The Design Productivity Debate,* Duffy, A.H.B. (Ed.) (London: Springer-Verlag).

EARLE, M. & EARLE, R. (1999) *Creating New Foods* (London: Chadwick House Group).

EARLE, M. & EARLE, R. (2000) *Building the Future on New Products* (Leatherhead: Leatherhead Food R. A. Publishing).

GANGULY, A. (1999) *Business-driven Research and Development* (Basingstoke: Macmillan Press).

GEVIRTZ, C. (1994) *Developing New Products with TQM* (New York: McGraw-Hill).

GRUENWALD, G. (1988) *New Product Development* (Lincolnwood, IL: NTC Business Books).

HARRIS, J.R. & McKAY, J.C. (1996) Optimizing product development through pipeline management, in *PDMA Handbook of New Product Development,* Rosenau, M.D. (Ed.) (New York: John Wiley & Sons).

HOLAHAN, P.J. & MARKHAM, S.K. (1996) Factors affecting multifunctional team effectiveness, in *PDMA Handbook of New Product Development,* Rosenau, M.D. (Ed.) (New York: John Wiley & Sons).

HOLT, K. (1983) *Product Innovation Management,* 2nd Edn (London: Butterworth).

JASSAWALLA, A.R. & SASHITTAL, H.C. (1998) An examination of collaboration in high-technology new product development processes. *Journal of Product Innovation Management,* 15(3), 237–254.

JONES, T. (1997) *New Product Development: An Introduction to a Multifunctional Process* (Oxford: Butterworth-Heinemann).

KUCZMARSKI, T.D. (1996) *Innovation: Leadership Strategies for the Competitive Edge* (Lincolnwood, IL: NTC Business Books).

LORD, J.B. (2000) Product policy and goals, in *New Food Products for a Changing Marketplace,* Brody, A.L. and Lord, J.B. (Eds) (Lancaster, PA: Technomic).

McGRATH, M.E., ANTHONY, M.T. & SHAPIRO, A.R. (1992) *Product Development: Success Through Product and Cycle-time Effectiveness* (Newton, MA: Butterworth-Heinemann).

MELTZER, R.J. (1996) Accelerating product development, in *PDMA Handbook of New Product Development,* Rosenau, M.D. (Ed.) (New York: John Wiley & Sons).

MILLMAN, J. (1996) Mexican tortilla firms stage U.S. bake-off. *The Wall Street Journal*, 10 May, A6.

MULLINS, J.W. & SUTHERLAND, D.J. (1998) New product development in rapidly changing markets: an exploratory study. *Journal of Product Innovation Management*, 15, 224–236.

NORRGREN, F. & SCHALLER, J. (1999) Leadership style: its impact on cross-functional product development. *Journal of Product Innovation Management*, 16, 377–384.

O'CONNOR, P.J. (1996) Implementing a product development process, in *PDMA Handbook of New Product Development*, Rosenau, M.D. (Ed.) (New York: John Wiley & Sons).

SCOTT, G.M. (2000) Critical technology management issues of new product development in high-tech companies. *Journal of Product Innovation Management*, 17, 57–77.

SMITH, R.E. & FOWLEY, J.W. (2000) New product organisations: high-performance team management for a changing environment, in *New Food Products for a Changing Marketplace*, Brody, A.L. and Lord, J.B. (Eds) (Lancaster, PA: Technomic).

SONG, X.M., THIEME, R.J. & XIE, J. (1998) The impact of cross-functional joint involvement across product development stages: an exploratory study. *Journal of Product Innovation Management*, 15, 289–303.

SOUDER, W.E. (1987) *Managing New Product Innovations* (Lexington, MA: Lexington Books).

SOUDER, W.E. & JENSSEN, S.A. (1999) Management practices influencing new product success and failure in the United States and Scandinavia: a cross-cultural comparative study. *Journal of Product Innovation Management*, 16(2), 183–203.

SOUDER, W.E. & SONG, X.M. (1998) Analyses of US and Japanese management processes associated with new product success and failure in high and low familiarity markets. *Journal of Product Innovation Management*, 15(3), 208–223.

TAYLOR, P. & LOWE, J. (1997) Are functional assets or knowledge assets the basis of new product development performance? *Technology Analysis and Strategic Management*, 9(4), 473–488.

TIGHE, G. (1998) From experience: securing sponsors and funding for new product development projects – the human side of enterprise. *Journal of Product Innovation Management*, 15, 75–81.

TOMIYAMA, T. (1998) Concurrent engineering: a successful example for engineering design research, in *The Design Productivity Debate*, Duffy, A.H.B. (Ed.) (London: Springer-Verlag).

URBAN, G.L., HAUSER, J.R. & DHOLAKIA, N. (1987) *Essentials of New Product Management* (Englewood Cliffs, NJ: Prentice-Hall).

VERGANTI, R. (1999) Planned flexibility: linking anticipation and reaction in product development projects. *Journal of Product Innovation Management*, 16(4), 363–376.

7

Case studies: product development in the food system

The four basic stages in the PD Process are the same for all food product development, but there are significant differences in the activities, techniques and timings for new product development in the primary production, industrial food processing, and food manufacturing industries.

Primary production's product development is based on either a breeding process from cultivated varieties or capturing a new species from the wild. The development of new plants, animals and fish takes a great deal of time and depends on times of growing and harvesting. There can be a general product concept based on perceived consumer or industrial wants and needs, and on technical knowledge to identify the possible parents for the new varieties. But it takes generations to develop the suitable variants. This is described in the first Case Study on starting a new apple variety. The industrial ambience is of a farmers' cooperative fruit processing and exporting enterprise working with a national horticultural research institution. The second Case Study looks at another fresh fruit project, on mangoes. This time the emphasis is strongly on the consumer, using statistical and other quantitative techniques to build up the consumers' preference image, and then to use this consumer image and information in assessing current varieties, and moving towards improvements. This is in the framework of government/university research, a national growers' organisation and private exporters and marketers.

Industrial food processing's product development is very strongly processing-based, both in the ingredient supplying and the buying companies. Food manufacture is usually directed towards providing a wide variety of products for consumers, which is continually changing. There are major differences between the activities in the PD Process for industrial and consumer products, as shown in Fig. 7.1.

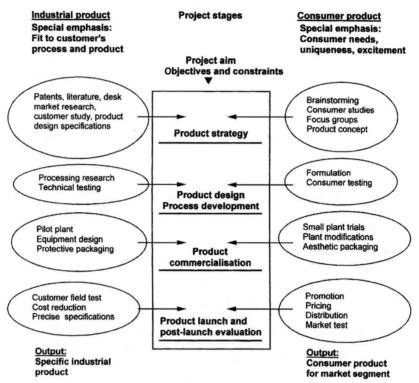

Fig. 7.1   PD activities for consumer and industrial products (From Earle and Earle, *Building the Future on New Products*, © LFRA Ltd, 2000, by permission of Leatherhead Food RA, Leatherhead, UK).

The industrial product, a food ingredient, is usually developed in collaboration with the processors or manufacturers who are going to use the ingredient in the production of their product. These companies may provide the product design specifications or may indicate some of the qualities they need; therefore the PD Process is highly concentrated on process development and the customer does the field-testing. This is illustrated in the third Case Study in which development of an ingredient, a whey protein isolate, sophisticated both in its processing and in its market, is described. Research is carried out by an industry-wide research institute working with a university and a large farmers' cooperative dairy company and dairy marketing organisation.

The fourth Case Study is the development of a consumer product, a new variety of sauces. The development was to establish a new product platform with a number of individual lines, in a large factory unit of a multinational food manufacturing company, and to sell these sauces initially locally, and then for export to major international markets. This product development was totally private enterprise.

In studying these case studies, differences in activities in the four stages of the PD Process can be seen, caused by:

- markets for which the products are designed;
- technology available and used;
- time taken for product development;
- costs of product development;
- priorities set by the various parts of the food system.

The Case Studies cannot be taken as typical of PD Processes because each has its own special features, but they do illustrate generic features.

7.1   Primary production: creating a new apple variety

Why does a consumer choose a particular apple and buy it, or indeed choose an apple at all rather than a pear or a plum? Maybe because it looks bright and attractive, maybe because it looks fresh and crisp, it is available or cheap, or it looks much the same as it always has and earlier experience was favourable. Or maybe because, over and above lots of positive attributes, it is novel and cries out to be tried. And then having bought it, if eating impressions are favourable and it is indeed appealing in taste, flavour and crispness, then the favourable image is retained and it is the variety that goes on being bought especially if the apple is distinctive. So in the striving and seeking to secure and gain market share there is a great potential premium from a desirable new variety.

7.1.1   Product development for a new apple

The area for product development was first identified and then the possible parents chosen. In the next generation, there were many variations, screened on technical analysis of:

- product qualities such as sensory characteristics, nutritional value, composition, use, safety;
- production qualities such as production difficulties/ease, disease resistance, yields;
- handling qualities such as deterioration after harvesting and on storage.

The cultivars in the first generation were screened, and the most suitable for further development chosen for growing in the next season. This further generation grown in the next season was again screened. This time, not only was there technical screening but also the production and marketing experts selected for suitability for production and marketing. This growing and screening took several generations and significant time, in the order of years. On reduction to a few selections, the qualities were related to the market conditions at that time. While senior people in the company made the final selections, they were assisted and guided by consumers and distribution/retail customers who evaluated the new fresh products. Production was started, multiplied with commercial growers/farmers and then the first crop test marketed on a small scale.

Creating a new and successful horticultural variety is a long search through genetic possibility, inheritance, disease resistance, keeping quality, followed by the trade-offs of one attribute with another; and so all of these aspects had to be gone through. This went on over many generations of seedlings which, after the final selection was narrowed right down, has then to grow to trees and bear and proliferate. So the exercise was very much one of seeking to establish just what it is that the target consumers ask from their ideal apple, and then trying to match this against what nature, aided by the skill of the plant breeder, will allow. All this took 15 years or so, making it an especially long-term undertaking. It is also an expensive one in that beyond the visible costs is the inevitability that money risked and spent now can receive no return for all those years. As a project this injects its own special features, while at the same time all the standard routines are still just as necessary as in any other development of a new product.

New Zealand for many years had a 'single desk' exporter and marketer of bulk apples working cooperatively for all of the apple growers. The Horticulture and Food Research Institute of New Zealand Ltd (Box 7.1) carried out its variety development and licensed ENZA as a company to market its varieties.

Box 7.1   Horticulture and Food Research Institute (HortResearch)

HortResearch is extensively involved in developing new plant varieties. We have expert breeding capability in a range of fruiting, ornamental, and agro-forestry crops. Our plant breeders work with industry in order to develop new varieties and rootstocks to provide cultural or market advantages for growers. These activities are backed by extensive in-house scientific capability in sensory science, genetics and plant physiology to assist in making selections to suit the environments, end uses and tastes of different cultural groups both within New Zealand and around the world.

The fruit breeding programmes aim for variety, flavour, texture, storage life, appearance, productivity, pest and disease resistance and climatic adaptation.

HortResearch is a world leader in apple cultivar development. Examples of our success are the Pacific apple series ('Pacific Rose', 'Pacific Beauty', and 'Pacific Queen') marketed by ENZAFRUIT. This new variety development is also well supported by technical back up in orchard production, integrated pest management, postharvest handling and associated capabilities within HortResearch.

Source: Adapted from a publicity letter from Dr Ian Warrington, CEO, HortResearch.

Table 7.1   Timetable for the development of Pacific Rose apples

1st cycle (product strategy)		
	Discussion from 'Pacific' markets of need for blush apples	
6 months	Grown in glasshouses	
		20,000 seedlings
	Expert selection	
18 months		
	Grown in open ground	8000 seedlings
	Selected on resistance to 'blackspot' and 'powdery mildew'	
2nd cycle (product design and process development)		
	Grown in fruit selection orchards	5000 seedlings
4 years		
	Selected on fruit characteristics	
	Grown on two sites	100–200 seedlings
	Selected by plant breeders, pomologists,	
2 years	on fruit and growing characteristics	
	Judged for market suitability	10 seedlings
3rd cycle (product commercialisation and product launch)		
		1 variety selected
2 years	Growing expanded	
1993	Seedlings distributed to growers	1000 cartons
1994	Pomology developed, storage trials	5000 cartons
1995	Multiplied by commercial breeders/growers	22,000 cartons
1996	Commercial production	104,000 cartons

The stages and approximate timing of the development of Pacific Rose are shown in Table 7.1. This indicates the very extended time scale, arising from the intervals necessary for the seedlings of each successive generation to grow so that their fruit can be evaluated.

Plant breeders normally talk about development cycles and these have been arbitrarily related to the PD Process. Because of the nature of developing apples there is not an exact date for launch, but the market is expanded in a rolling launch as the fruit becomes available.

7.1.2   Stage 1: Product strategy

Management decided that the existing varieties had been on the market for long enough, and to provide an edge and a stimulus a new variety was needed. To some extent this is a continuing search. But it gained added stimulus as the older varieties were a bit stale, and market share would surely dwindle as the competition sought to kindle its own novelties. Apart from the very broad concept, a new apple, plant breeders thought back over the whole gamut of experience with apple varieties. They tried to single out characteristics that might be applied usefully to build a new creation. This was reinforced by market insights such as possible gaps in present offerings, fashions as revealed by sales trends, problems exhibited by present varieties, competitors' activities, and so on. They did not know exactly what was wanted but formulated a group of attributes, built on a range of good qualities, and sought to assess these and maximise them by selection from trial seedlings of defined types (cultivars).

For example appearance is a major purchase determinant, and so a target colour and configuration were selected. In this case it was decided that the new apple should be a 'blush' apple, one in which red and yellow colorations shade into one another rather than uniform colour or stripes. Decisions were taken of targets for sweetness, acidity, and acid/sugar ratio, flavour, fruit shape, texture and crispness. Added are those properties that are central concerns to the growers and handlers such as: disease resistance, yield and size consistency, and keeping and storage qualities, and these must be optimised for all apples. A major consideration was the time scale, commitment to perhaps 10–15 years of work overall to build a new variety to commercial market success.

7.1.3   Stage 2: Product design and process development
The first cycle of selection was rapid screening of seedlings, from about 20,000 in glasshouses (six months) reduced to 8000 in the open ground nursery through selection for resistance to blackspot disease (18 months).

In the second cycle, about 5000 per year, after selection for resistance to powdery mildew, were planted out in the fruit selection orchards (4 years). About 2% were selected for fruit characteristics and these were then carried forward to the next cycle. A selection index was set up, and made into a scoring regime. The various desirable attributes were first established and then scored by members of the team as illustrated in Table 7.2, generally on a scale of one to ten, with ten being most desirable. Table 7.2 shows the scores for one apple cultivar. All those cultivars with an 'overall quality rating' less than seven, when aggregated and averaged, were discarded.

In this particular case, the designation moved from a concept, to a tree number, to the final name Pacific Rose (technically the variety name was Sciros, marketed as Pacific Rose). The particular chosen characteristic factors, plus the desirable and more general factors, were pursued through all the generations and the selections. Some of these factors had sometimes, and regrettably, to be traded off to a degree as the selections evolved. To cope with such problems value hierarchies were established, and used to guide selection, and reviewed from time to time.

7.1.4   Stage 3: Product commercialisation
As well as consumer and grower characteristics, consideration had also to be given to vital genetic aspects. These included factors such as the heritability of selected attributes such as mildew resistance, because at some stage large

Table 7.2 BreedBase Report

Family	A040	Seedling R04T119
Crop type	*Apple*	
Fruit shape	*Flat*	

Colour

Background colour	*Yellow*	
Overcolour	*Red*	
% Overcolour	0% |_____3_____|	100%
Colour pattern	*Stripe*	
Colour intensity	Light |___2_____|	Heavy
Lenticel:	Inconspicuous |_____3_____|	Very Conspicuous

Flesh

Flesh colour	*Cream*	
Flesh firmness	Soft |_____7____|	Hard
Flesh crispness	None |_____7____|	Very
Flesh grittiness	Soft |0_____|	Hard

Flesh flavour

Juicy	Dry |_____8__|	Very juicy
Sweetness	Nil |_____5_____|	High
Sourness	Nil |_____5_____|	High
Aroma	Delicate |_____5_____|	Rich
Astringency	Nil |0_____|	High
Bitterness	Nil |0_____|	High

Skin

Skin thickness	Thin |_____5_____|	Thick
Skin greasiness	Dry |___2_____|	Greasy
Skin texture	Non-chewy |_____4_____|	Chewy
Skin flavour	*Not significant*	

Harvest Date	12/3/96	
Storage Days	107	
Weight	176 g	
Maturity	OK	
Eating Quality	Very good	
Attractiveness	Poor |_____7____|	Very good

Overall quality	Very good
Comments	

Note: The numbers and comments inserted represent assessment of a particular seedling.
Source: From HortResearch, Goddard Lane, Havelock North, New Zealand.

numbers of plants will have to be propagated from the successful selection and then established and grown in orchards. Although the key participants were the plant breeders, it was thought to be very important that the scoring be done by a wider-based group. For practical reasons in the initial stages it tended to be a laboratory team but as soon as the earliest stages were completed a wider group was used. The work was monotonous and repetitive so that team numbers are limited but, by their working to a standard scoring system, numbers of selected candidates were reduced to the order of one hundred.

For the Pacific Rose, these selections were then grown in duplicate on two sites and the product apples held 100 days at 0 °C in a cool store and for 7 days at ambient temperatures, to observe storage characteristics. Meetings were held, bringing in other fruit scientists and ENZA staff to widen the vision, and including fruitgrowers and supermarket operators to seek feedback, but still on a largely local basis.

Then in the third cycle the best 10 out of 200 were selected for consumer trials and finally characteristic clusters were assembled where 75% or more of the panel opted for a particular attribute, such as acidity combined with sweetness.

Finally, one variety was chosen by senior management for launching. Trials were then run with selected supermarkets, taking about a thousand cases and trying the market (3–5 years). Pomology work was accelerated, assessing the required optimum growing environment and the hazards. Another important consideration at this stage was naming. The final choice, after a good deal of investigating and agonising, was Pacific Rose. This name seemed to have very many positive overtones and manageable problems. (After its endorsement by the market, it became the forerunner of a 'Pacific' platform of similar apples as the general name and style were clearly found to be very attractive and distinctive.)

During the last two years or so of the commercialisation phase, as well as being checked out locally, small parcels of fruit were dispatched on a trial basis to agents overseas. ENZA has main agents in the UK, Belgium for the rest of Europe, the USA, and in Singapore for the Asian market. These were used for distribution, and also for market intelligence and feedback. The message from these people was very positive. In fact from the marketing viewpoint it was somewhat too positive and was too widely disseminated. This generated an enthusiasm among growers in particular, which stimulated plantings. So in time production threatened to flood the local market and in turn to push the overseas market beyond its powers of initial absorption.

7.1.5   Stage 4: Product launch and evaluation

Finally the trees of the selected variety were multiplied to the extent that commercial growers could enter into production of the quantities needed for initially the launch, and then be ready for the full-scale farming, of a successful new variety. Commercial quantities of the new apples were dispatched to the overseas agents. They in turn fed them to wholesalers and retailers, initially

Table 7.3 Production of Pacific Rose apples (thousands of 10 kg cartons)

1992	1993	1994	1995	1996	1997	1998	1999	2000
0	1	5	22	104	120	173	353	950

selected as those, and in those areas, likely to be most receptive. It was commented that little direct consumer testing was carried out, as results from sample trials correlated highly with agents' opinions which were quicker and cheaper to obtain. The apples were then sent to the trade buyers in retail markets and supermarkets.

The first great hurdle was to get them on the supermarket shelves: once on, customer demonstration and tastings intensified their visibility. It was found that the most receptive area was Asia, the next North America, with the UK a little less enthusiastic. The rest of Europe tailed with comments seeking more consistent fruit quality and more flavour. The customer balance settled to about 40% Asia, 30% North America, and of the remainder most to the UK. The quite dramatic build up of production of Pacific Rose apples is shown in Table 7.3. This also indicates the pressures that arose to move such rapidly increasing quantities through the markets.

It was important to seek to safeguard, as far as possible, the commercial aspects of the development. So plant protection rights were sought for the variety. This essentially was so that the considerable costs of the development could be recouped and also reasonable returns made on the investments of resources and time. It became evident that this protection was significant. Even lawsuits and cloak-and-dagger stuff followed, with some overseas competitors seeking to cash in, unauthorised and without paying, on an obvious success.

It was evident early on that in order to maximise the returns it would be desirable to have overall control of production and marketing. Enthusiasm had brought large early production, with risks of drowning the market and on occasion prices had to be shaded to clear fruit. In hindsight this served to spread and deepen overall consumption, but at the time it looked like expensive advertising if not just losses. Closer matching of production to market would also have allowed more time for the details of growing the variety, with attention to fruit quality and consistency, those prime demands of good supermarket operators. The balance of production and demand is the great intransigent imponderable of all agriculture, and the build-up of supply of Pacific Rose created supply pressures which later variety releases will seek to reduce by closer control of the growing of new varieties.

Another interesting further extension of the development was to seek out, license and harmonise with selected overseas growers, particularly ones who could produce to complement New Zealand production. For example, by spreading some of the growing to the Northern Hemisphere, year-round production was organised so as to even-out supply to satisfy and sustain customer demand.

Later, feedback from the markets was used as a base for the breeding of further members of the variety, and so to build the offering and the acceptance and the sales over an extended platform of similar, but distinguishable, apples. The platform name, 'Pacific' was retained, moving to 'Pacific Queen' and 'Pacific Beauty' to differentiate newcomers as they appeared.

Domestic sales were built up simultaneously, but though important they were only part of the overall business of the industry. They could also be used as a vehicle for sizes, shapes and configurations less attractive to the main line demand which could therefore be selective of premium fruit. So a national and international market was established which became considerable and satisfactory.

This example illustrates how a substantially long-term development of a product with particular problems, those of setting up and evaluating a new horticultural variety, still follows the general principles of product development. One of the problems in developing new plant products for the consumer market is the input of the consumers. At one time the marketing people and the breeders decided that they knew what the market wanted and therefore all testing up to the small test market was done by them or other people in the research station and the company. In recent years, great efforts have been made to bring the consumers in earlier. Obviously they cannot test the many hundreds of samples, but they can determine the concept for the new product. Therefore it is to them that the greatest effort is directed, trying to understand as precisely as possible what it is that they might want from a variety which still has to be produced. Then there is the slow process of selecting and building up fruit, recalling the time consumed between selecting and actually growing the next generation of apple.

Think break

1. Consumers determine market success – reflect on this statement, its accuracy and its implications for fresh fruit product development.
2. Consider carefully and weigh the relative advantages and disadvantages of using available local and expert opinion, contrasted with randomised consumer research, in exploring the required eating characteristics, flavour and appearance in the PD Process for fresh fruit.
3. This study is of a very long-term exercise for a corporate entity, with increasingly limited product flexibility as the development progresses. What are the implications of this for product development management, organisation and operation?
4. How important do you think metrication is as a determinant in decision making in the PD Process? Therefore, because metrication is often difficult, and sometimes very difficult, how much management and technical effort should be devoted to it relative to the exercise of less formal and more qualitative judgement? Short term for a product? Long term for an organisation?

7.2   Development of Thai mango products and their competitive advantage in export markets

Mangoes are an attractive traditional fruit produced in quantity in Thailand. Many Thai varieties have been produced, and from time to time overseas varieties have been grown. In addition to the major local market, quite a substantial export trade has grown up which is of economic importance to the country. The export trade had grown up somewhat arbitrarily. So a clear opportunity was perceived to look more systematically at what was available, and to seek, using quantitative techniques, to describe and determine those varieties that were most attractive to overseas customers, so that overseas sales could be further expanded and marketing improved. It was product development through systematic selection and improvement, with particular reference to consumers and their preferences.

Planning of the development was a collaborative undertaking between the Thai Departments of Agriculture and Agriculture Extension, researchers at Khon Kaen and Kasetsart Universities, and large mango growers and exporters. After discussion, the main brief for the study was determined as:

- select the most suitable ripe and fresh varieties for export to three markets, Japan, China, and the Middle East;
- find the best potential distribution channels; and
- develop suitable brand names for the selected varieties.

This was to be based on consumer preference studies, and also supported by characterisation of the chemical properties such as aroma and volatility, and physical properties such as shape, and stone size and distribution, which would be correlated with the consumer preferences. The outcome would then provide information on the relationships between customers' preferences and measurable attributes of the fruit, which could be used by growers, plant selectors and breeders, and exporters and marketers, to develop the Thai mango export industry.

7.2.1   Study design and development

Both to handle such a substantial project and to distribute tasks to appropriate people, the study was divided into ten activities. Each activity was allocated to a group of researchers, though some members were common to several groups. They worked in appropriate localities, laboratories and departments, and under the overall guidance of the Mango Project Leader. The activities are listed in Table 7.4.

Five varieties of Thai mangoes, major ones being commercially grown, were studied, along with two introduced varieties. Using the same batches of mangoes, tests were carried out on the chemical properties, physical properties, aroma volatiles and consumer preferences, correlating these with the measured properties. These in effect combined into one aspect of the case study. Another

Table 7.4 Mango product development study workgroups

---

1. Mango variety and industry survey
2. Physical properties – size/shape/colour
3. Chemical analysis – constituents/ripening
4. Chemical properties – analysis and flavour
5. Quantitative descriptive analysis – relationships of sensory attributes
6. Consumer preferences – shape/colour/texture/flavour
7. Consumer preferences – correlation with sensory/analytical
8. Consumer preferences – national likes/dislikes
9. Commercial – target markets/channels/distribution
10. Marketing descriptions – brand names/slogans/labelling

---

separate aspect looked at the best brand name and attribute descriptions, for commercial and marketing purposes. Consumers studied were from Thailand, Japan, China, Hong Kong and the Middle East.

For various practical reasons, principally availability of suitable subjects and materials, the numbers of consumers testing varied in different parts of the tests. There was always account taken of the statistical basis and needs of the study, and the results were statistically assessed to justify the conclusions.

7.2.2   Study implementation

Selection of varieties
To start the study, a group gathered information on the varieties of mangoes that were produced in Thailand and also overseas. These were then carefully considered with respect to usage, production technologies, transportation and storage durability, and eating characteristics. The available literature was inspected, and mango growers and the trade were canvassed to gain requisite information so that the group could select the most promising varieties, those most likely to form the basis of a viable and growing industry, for further investigation in detail.

Physical properties of the fruit
One group looked at physical properties which were fruit weight, fruit size, seed size, seed weight and thickness, skin and flesh weight and thickness, skin and flesh colour, and flesh texture.

Chemical analysis
Another group looked at chemical contents and measured moisture, total soluble solids, acidity, sugar, beta-carotene (vitamin A). Also the sugar/acid ratios were noted.

Chemistry of flavours
Flavour constituents were measured including sugars, acids and aroma volatiles. For example, they showed the extent, as the fruit ripened, of the decline in the

sucrose content and the increase in fructose and glucose. Succinic acid was the most prominent acid, while malic and citric acids were also important; individual contents of these acids varied and could be used as indicators of varieties. Aroma volatiles were measured by gas chromatography. Detailed contents were explored, and the total volatile contents were found to vary with varieties.

Consumer preferences

Consumer preferences to determine the degree of liking on a nine-point hedonic (liking) scale, were carried out in a central location test, on varieties of Thai mangoes using Chinese, Hong Kong, Japanese, Middle East and Thai consumers, settled in Thailand. The objective was to select the best Thai mango for export, based on the sensory characteristics most preferred by the target consumers. The sensory characteristics used, for the fruit and the flesh of the mangoes, were skin colour, fruit size, flesh texture and overall liking for the flesh. Overall the results showed that the fruit shape, fruit aroma and skin appearance were more highly correlated with overall liking than the other attributes. A single variety emerged as the preferred one overall, though to some extent rankings of mangoes varied with the attributes.

National preferences

The degree of liking and disliking for sensory characteristics of the ripe fruit and the flesh of two varieties of the mangoes were explored with about 600 Chinese and 400 Japanese tourists in Thailand. Attributes such as skin appearance, fruit colour, fruit size and shape, flesh colour and flavour, and fruit aroma were covered, as well as overall liking; using hedonic scales. From this emerged attribute profiles for the varieties, and one variety preferred by both groups.

Systematic attribute relationship analysis

A systematic comparative technique known as quantitative descriptive analysis (QDA) was employed by another group to look at six of the mango varieties and to build associations between the sensory attributes of the varieties. The results indicated that the difference among fully ripe mangoes was most pronounced on the perception of fruit size, weight, thickness and fruit aroma strength. This showed, for example, that the variety that was emerging as the preferred one was the smoothest, juiciest and most tender, but not the biggest, heaviest or thickest, nor did it have the heaviest fruit odour.

Conclusions about the fruit

Overall from these experimental results a variety emerged which on balance was preferred by the majority of consumers from each of the countries sampled. The preferred variety, both fruit and flesh, was the Num Dok Mai See Thong mango, followed by Rad and Ma Ha Cha Nok. Additionally, and importantly, this choice was for reasons that could be differentiated and substantiated. The sensory results were supported by extensive information on the fruit, the flesh and the association of desired mango attributes. The study results therefore provided clear signals statistically based on consumer responses. The signals were to the growers for plant selection and cultivation, to the trade for technical details in handling, storage and exporting, and to the plant breeders for selection of characteristics on which to concentrate and for further experimentation.

7.2.3   Commercial aspects

Brand image

A strong brand name and image is commercially powerful, so one group was given the task of carrying out consumer research that would enable these to be created most effectively in the target markets. Brand names and brand concepts on the fresh fruits in the market were collected from the literature, from the trade by interviewing experts and exporters, from market observation, and from group brainstorming. The brand concepts so obtained were then used to develop questionnaires for the field survey. The design of artworks, building selected brand names into logos, brand stickers and label materials for packages, was then explored by a group of experts, and referred to the orchardists and exporters. The survey showed that significant attitudes included health, nutritive value, colour, convenience of buying, texture, ease of preparation and of course price. The brand investigation showed that the sensory characteristics concerned with the fruit were the most significant, followed by aspects concerned with the consumer such as nutrition and price and prestige. Box 7.2 indicates the general conclusions that arose from consideration of the brand name image. From this work the preferred brand name that emerged was ThaiMango, and the selected slogans 'Your Fresh Taste' and 'The Fresh Taste'.

Market channels

Another group investigated distribution channels. Their interest took in the target markets:

- potential physical distribution channels;
- patterns of marketing of mangoes;
- volumes and values of these products; and
- potential market channels and the role of fresh mangoes.

Information came from documents, opinions of exporters and mail surveys of importers. Patterns of distribution and delivery investigated included land, sea and air transport. Channels of sales to agents, to trading companies, to institutions, to retailers such as supermarkets, convenience stores and fresh markets, to domestic consumers and to institutions were all investigated. Finally management and financing alternatives, such as joint venture companies, were identified. These patterns of trade often differ from one country to another, and so it was necessary to look in detail at these in each of the countries. Government regulations were very important. Applicable regulations took many forms and included inspection, treatment and certification measures, and also

Box 7.2   Mango brand name research conclusions

The results from a literature review, and observing both local and overseas commercial fruit brands, showed that the important concepts of creating brand name are the source of fruit, good quality, good taste, freshness, nutritious properties and relation to the environment. From interviewing Thai managers and exporters, the important concepts of creating brand name should be merit (goodness), scale, enterprise or company name, and levels of quality.

Experts said that 'no one has created a brand name which is unique and relates to the mango. Most brands have been created for remembering, without adequate concern for the mango characteristics. However brand names may or may not be necessary because they also depend on the selling system.'

Marketing experts suggested that the brand name should be easy to pronounce and remember, the brand mark should be a Thai-identified symbol, and the slogan should relate to buying decision factors.

Attitudes to mango and buying decision factors of Chinese, Japanese and Middle East tourists were surveyed. The results showed that they think of mango in the following ways:

- Mango consumption is good for health
- Mango is nutritious
- Mango is suitable to consume at anytime
- Mango is suitable for everyone
- Mango is available at all times
- Mango is suitable for consumption in every season
- Mango consumption indicates good taste
- Mango is a worthwhile gift

Positive buying decision factors included: without toxic substances, smell of the fruit, taste, colour of the fruit, price, convenience to buy, texture, nutrition, ease of consumption, and availability at anytime.

Source: From Ngarmsak, 2000.

fiscal rules such as entry taxes and tariffs which were generally specific to each country.

7.2.4   Launch

The equivalent of the launch for this case study was a combination: of presentations of results and conclusions, of publicity, and of consequent action by officials and by the various elements of the mango trade.

The first presentation was to the Mango Round Table, on which were experts from the Thai Departments of Agriculture and Agriculture Extension, from Kasetsart University, from large mango orchard owners, and from exporters. After the presentation, and based on the results of the study, the Mango Round Table made the selection of two varieties for the top markets, and allowed one other variety with selected grades only. Subsequently the work was presented to the public, including media and other exporters and orchardists, and it led also to the formation of the Mango Growing Association of Thailand. Other meetings were called, at which planning was started for extension of the areas of production and for continuation of the research. So this product development has injected a new dimension into the mango export programme for Thailand.

Think break

1. Many product development programmes are conducted, sometimes with successful outcomes, using little or no consumer research. They rely instead on historical data and expertise and experts, to provide market predictions. Review the case for and against this approach.
2. One of the problems in consumer research for the food industry is to secure a true 'population' sample. In this case some use was made of expatriates and tourists rather than home residents for the sampling. To what extent might this make the results skewed in some way and less valid?
3. Producers of raw materials, and in particular agricultural raw materials, have some special difficulties with new product development. To what extent do you feel that their needs can be fitted by a standardised PD Process, and to what extent might it be better for them either to have a standard process of their own, or to set up *ad hoc* processes with each particular situation?
4. This and many other consumer surveys reveal cultural differences, which should be taken into account in the PD Process as they may influence the success of the outcomes. Reflect on whether cultural differences should, or need to, influence the management of product development, and if so in what aspects?

7.3   Industrial products: PD Process and management for whey proteins

The New Zealand dairy industry is basically a farmers' cooperative. Within it there are dairy companies which process the milk, and they are represented on the NZ Dairy Board which markets the dairy products worldwide. There are several subsidiaries but two important marketing companies are New Zealand Milk Ltd and Whey Products New Zealand Ltd. The latter is responsible for marketing milk proteins as industrial products. There are several industrial milk protein categories, for example, caseins and whey proteins; these are divided into further categories according to their properties and also their uses.

Traditionally whey is the liquid remaining when curd for cheese making has been strained off, taking a proportion of the milk proteins. It contains a number of useful constituents including lactose or milk sugar, a little fat and the whey proteins, together with nearly 20 times their weight of water. In manufacture of whey powder, both the lactose and the water can be separated from the proteins. Because the proteins are not a single entity they themselves can be fractionated. So, specific protein fractions can result, each with special characteristics. These specific fractions can be characterised in terms of physical, chemical, functional and nutritional properties that offer the potential for new food products (Huffman, 1996).

Product development, in this case, lies in the scope for both designing the processing, which can include separation of protein fractions and other manipulation, and in finding worthwhile markets. Thus the product concept is essentially technologically defined as technical product characteristics and processing capability. The first of these uses the knowledge that whey proteins have been shown by research to be nutritionally important and desirable in the human diet, and the second the capacity to produce in quantity a range of these proteins to a close functionality specification. These protein products have then to be tailored in the PD Process to meet the needs of a market, which is identified and explored.

The main development group was the New Zealand Dairy Research Institute at Palmerston North, which also has overseas laboratories including one in California, and which is a substantial research and development unit of the NZ Dairy Board. To supplement their technical resources and skills, they worked with Massey University through the Food Technology Research Centre on technology, including product evaluation and model testing, and through the Chemistry Department on ion exchange resins. They cooperated with regional companies of the NZ Dairy Board in the USA, Europe and Japan particularly for market assessment, and with a dairy company, Kiwi Dairies Ltd, on aspects of the processing. Coordination was strategically vital, because of the long development period and so many groups involved in the development, and was effected by Whey Products New Zealand Ltd, another NZ Dairy Board subsidiary. The sequence of the development and the activities of the different groups are shown in Fig. 7.2.

7.3.1 Stage 1: Product strategy development

In the pre-design and development phase it was necessary to assemble knowledge, tacit and written, from a variety of sources within the various parties. This included:

- On the technical side –
  - (a) heat behaviour of whey proteins,
  - (b) properties and potentialities of ion exchange resins, which could separate even more tightly defined protein constituents,

ALACEN 895 PROJECT TIMELINE

MARKET

NZMP USA — customer contact

NZMP Japan / NZMP Europe / NZMP Australia

- Sample Evaluation / Target Application
- Sample Evaluation
- Sample Evaluation / Detailed Market Evaluation

WPNZ

- Market study ##
- Prelim Economics ##
- Site Selection ##
- Proposal Preparation ##

R & D — NZDRI/MASSEY Joint Venture

Food Tech Res.

- Fundamental Technology Research
- Sample Manufacture / Technology Optimisation
- Sample Evaluation
- Process Optimisation and Process Design / Product Optimisation
- Sample Evaluation
- cont. ......

ENGINEERING

- Process Design and Costing
- Process Design
- Plant Installation cont. ......

MANUFACTURE — Kiwi Dairies

- Process Design
- Process Design
- Plant Installation cont. ......

NOTES

- ## WPNZ decision to proceed with detailed project evaluation
- ## Decision to proceed with Ion exchange
- ## WPNZ decision to proceed with Kiwi Dairies
- ## WPNZ Bd approval for stage I

TIME

June 94	J	A	S	O	N	D	January 95	F	M	A	M	J	J	A

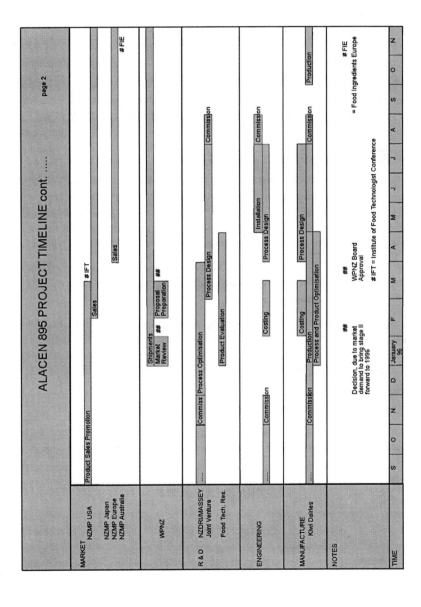

Fig. 7.2 Whey protein isolate time line.

    (c) expected functional properties of various fractions that might be prepared, and

    (d) existing or achievable manufacturing processes which might be challenged or improved by new ingredients.

- On the marketing/commercial side –
  - (a) selected market, its demands, characteristics and expected growth,
  - (b) available raw material, whey, its increasing supply and competing uses,
  - (c) expected financial returns and their stability,
  - (d) security of the process against competitive inroads if the project proved successful,
  - (e) anticipated profitability of the venture.

So on the market side, the originality and the prospective impact of the product, and its critical formulation were established, together with its relation to technical advantages and to market imperatives. The NZ Dairy Board's international marketing network was used together with the vertically integrated structure which gave access to a very wide range of expertise to work with existing and potential customers, and to find what they wanted and would buy. On the processing side the use of ion-exchange resins for whey protein manufacture had been limited and small, worldwide, so extension of this was an essential part of the scheme and its costs had to be investigated and firmed up.

From these preliminary investigations the product concept emerged as:

- unique, competitively robust, acid beverage component powder;
- high-protein, low-lactose/cholesterol/fat;
- made from a wholly natural product, whey.

This would be designed for the expanding sports market in the USA, Europe and Japan. It would be produced by a process for which the basic data were predictable or accessible, the necessary technical developments should be achievable, and the raw materials were available. The venture should be profitable.

7.3.2 Stage 2: Product design and process development

A number of critical product qualities emerged from the product concept that required process investigation and development. Although some whey protein powders had been made previously, they were a substantial distance from the demanding specification that was deemed necessary for the proposed product concept. This was for a higher (90%) protein content, and for much reduced fat and lactose. To achieve these, increased use could be made of membrane separations. These would be used to remove the larger fat globules, but ran the risk of carrying some protein with the fat that would reduce the yield. So it was necessary to investigate preliminary modification of the flocculation of the fat globules before the microfiltration steps. Lactose, being a small molecule, would pass through the membranes with the protein, but it could be broken down into simpler sugars using enzymes. Then came exploration of the ion exchange by which, after altering the acidity, the protein could be fastened and removed from the liquid whey on the ion-exchange resin and then subsequently unfastened and detached from the resin by altering the acidity again. All of these steps required careful exploration so that they would not only work in the laboratory, but also could be designed for full-scale plant operation, controlled sufficiently tightly to meet the specifications, and then transferred to a working factory for manufacture.

Along with the chemical composition, perhaps the most vital element of the product specification was the functional properties of the product, those properties that would make it uniquely suitable for its intended use and clearly superior to the competition. Important properties were:

- very high solubility in acid solution, so that a clear, high-protein liquid with an acidic taste would result on dissolving the powder;
- sufficiently bland flavour of its own so that added flavouring can determine the taste of the drink; and
- demonstrated nutritive value because it is this that convinces athletes of performance-enhancing potential.

All of these had to be built into a process that could adjust to a natural raw material, fresh milk, that varies compositionally throughout the dairying season, and still continue to produce from it large quantities of whey protein isolate to the very tight specification under which it would have to be sold.

7.3.3   Stage 3: Product commercialisation

An early survey of the market had indicated that there was a real opportunity in the acid beverage market for drinks, which at the same time as satisfying thirst, offered on a credible base increased endurance or performance to sports people. A prime key target was the USA, a large wealthy market where sports and exercise were widespread and glamorous. Investigation showed a potential business that could grow to over $10 million of sales annually. Samples were prepared and sent for assessment to markets in the USA, Europe and Japan. These confirmed the concept of initial concentration on the USA and that the financial outcomes should be favourable. A strategy was prepared to pursue a staged plant development programme so that production and demand could increase in parallel.

The Food Technology Research Centre set up a model beverage system, developed by the New Zealand Dairy Board's North American subsidiary, and this was used to screen further samples as they emerged from the technical programme. They were also screened in the prospective markets in North America and elsewhere. Feedback from the screening was used to guide the continuing product and process developments. The feedback also confirmed good market acceptability for the product. With that, the directors of Whey Products New Zealand Ltd approved sufficient capital expenditure for a manufacturing plant at the site of one of the major operating dairy companies.

There was a great deal of design and planning to be undertaken, some of the plant being more or less standard equipment but some of it requiring novel features including high precision in operating and control detail. Engineers were engaged to design and to build and commission the required plant, which incorporated new and proprietary technology to meet demanding specifications. Final development of the ion exchange process was continued to meet the necessary deadlines. Also work was continued on aspects of the product protein functionality that was so critical to success on the market. Activities had to be undertaken in parallel and in sequence, and the necessary information, and a satisfactory trial product, had to be available as and when needed to move smoothly to the ensuing steps of the programme.

Coordination of the entire project was from Whey Products New Zealand Ltd. It was able to call on the vertically integrated New Zealand dairy industry as might be needed from time to time. The team included expertise in customer requirements, marketing, protein chemistry, ion-exchange technology and technical aspects of proteins for acid beverages, industrial whey powder manufacture, and product evaluation and model food system testing. A full manufacturing scale plant was built, installed and commissioned. Production duly started on time and on budget.

7.3.4   Stage 4: Product launch and evaluation

From late 1995 sales promotion information was fed to the chosen market. This was essentially beverage manufacturers. It was expedited through the close relations the NZ Dairy Board subsidiaries overseas had built up with leading beverage manufacturers and with major food processors. Shipments of product were started in about October 1995 so that commercial quantities would be available to US customers in 1996. A major promotion was as a featured product at the (US) Institute of Food Technologists' annual meeting at New Orleans in March 1996 where a group from New Zealand, including technical and marketing staff, was available to explain and demonstrate the product. There was extensive advertising in the food trade literature, and in handouts (see Box 7.3). An article on technical features of the product appeared in the American journal, *The Food Technologist*, in February 1996, setting out the background to the manufacture, and explaining the functional properties of the product, particularly those that were seen to offer major advantages.

The results of this were seen in the sales of 100 tonnes, all of the available product, in 1996. Confidence from this success led to the stepping up of production facilities and capacity towards 700 tonnes per annum at the first site, and the planning of expansions to 1000 tonnes annually. Financially, the returns were significant in lifting the value of standard whey protein concentrate from around NZ$4000 per tonne to around NZ$15,000 per tonne. Although the production costs were of course higher, and the development costs to this point were over NZ$1 million, the overall returns for the industry were very satisfactory from what, not too many years earlier, had been seen as almost a waste stream.

Box 7.3 New Zealand Milk Products unveils whey protein isolates

To meet the growing interest in and demand for whey protein isolates, New Zealand Milk Products will be introducing the whey of the future at IFT (Institute of Food Technologists' Annual Meeting): ALACEN Whey Protein Isolates.

ALACEN Whey Protein Isolates are more than 90% high-purity protein, with less than 1% fat. In addition to superb nutrition, they provide excellent functionality – complete solubility plus acid and heat stability with a bland flavour. Clear RTD beverages benefit from ALACEN Whey Protein Isolates' unique transparency in solution, and foods requiring stringent nutritional labelling benefit from the WPI's low fat and low lactose levels. The instant versions of ALACEN Whey Protein Isolates are ideal for applications such as dry-mix beverages.

Source: Adapted from New Zealand Milk Products *NEWZ*.

Development was continued into both product quality and manufacturing improvements. It came from the technical product and process developments, which continued, and was augmented by comment and experience from customers. A reliable product was built up, with a good market. Whey protein isolate has continued to be a successful major product. As well as being the first of other specialised whey products for the New Zealand Dairy Board, it has, as an ingredient, itself made possible new and innovative products for many beverage and food manufacturers.

Think break

1. Discuss whether this whey protein isolate product was market- or technology-driven. Was this important? Do you consider it made any difference to the development pattern? To the launch?
2. Why do you think the promotion was largely directed to food technologists? Could other promotional targets have been usefully added? Substituted?
3. List the major criteria you think are essential to success in the health food market. Taking your local environment, do you think that a product such as this one would be successful, and why? What developments in the local culture might make it more successful?
4. What special features can you instance that distinguish food ingredient development from that of other food products?

7.4   Consumer products: new products and a new platform in variety sauces

For a large food manufacturer with an established market and reputation, a continuing line of new products is a vital dynamic element in strategy for growth and the future. Wattie Industries had been built up over about 30 years as the largest food processing company in New Zealand with also a substantial export business. It had a varied line of products including canned and frozen lines, and a major market share with a solid, quality, customer base. But its success and size then attracted various manipulations and reorganisations, over quite some years, and ultimately the international US company, H.J. Heinz, bought it. Today trading under the name Heinz Wattie's Limited it has become an important part of their international production resource with particular emphasis, outside of its local market, on Australia and Japan.

The activity in Hastings, New Zealand, located over three sites, employs about 1800 people at peak and for example annually produces about 40,000 tonnes of canned soups, baked beans and spaghetti for Heinz, Australia, and about 200,000 tonnes totally. It operates the largest hydrostatic cooker in the world, and the current canned food production rate is about half a billion units per year. A current major growth driver is the Japanese market; about NZ$100 million has been spent on the plant in the last five years, much of it on sorting and handling equipment but also on up-to-date processing facilities. They have a product development team on site of over 40. The scene is of a large production unit of a large multinational company looking for new consumer food products on selected international markets.

The new product chosen was a line of speciality, variety sauces, and an outline of the PD Process that was used is shown in Table 7.5.

7.4.1   Stage 1: Product development strategy

In mid-1997 it was decided to look for a new sauce product to modernise the brand and open a new platform. At the time, the company had two basic tomato-based sauces with wide sales and a commanding market share, and some hot meal sauces packed in cans. There was nothing on offer in a more up-market, adventurous, range. Brainstorming produced the creative idea from which was born a concept and an advertising campaign; the product was then developed to fit the concept. So a brief emerged for a 'modern, quirky, fun sauce, of premium quality flavours to enhance experience and add some spice to life.' It could be benchmarked to potential competitors, and targeted to enter a smaller, highly fragmented market which at the time in New Zealand displayed 300 separate products from 31 brands. A different product was needed; more up-beat and up-market with several variants and designed to fit the concept. Six flavours were started, and the original six, with some slightly altered benchmarks, were what finished as market products. For a company that had built its reputation on dependable quality, everyday, best-value, products, it was a major marketing

Table 7.5 Activities in sauce PD Process

Product brief: 7 July 1997
Product strategy – inception of initial concept – preliminary product development work and planning – formulation of product development brief and project plan.
Decision: Acceptance to proceed as a Project by Product Manager

Product design: 9 July 1997 to 1 October 1997
Product design and process development – preliminary surveys and ball park costings – recipe formation, assessment and refinement – laboratory and ingredient and engineering assessment and experimentation – preliminary packaging – label information – cooking procedures – quality assessment and control procedures – to a full product and process specification.
Decision: Assessment and approval of plant related expenditures and project continuation.

Factory trials: 25 September 1997   **Finished product assessment**: 29 September 1997
Product commercialisation – factory trials – feedback and attention to shortcomings and problems – trial samples prepared and checked – factory operational planning – marketing planning – sales forecasts – final costings.
Decision: Acceptance of formal specifications and Approval to proceed to launch, by Senior Management.

Production: Started 22, 23, 24 October 1997   **Launch approval**: 22 October 1997
Launch: November 1997
Product launch and evaluation – factory production – presentation to sales and trade – marketing of products.
Review and continuation – feedback from sales, marketing, retailers – review of lines – withdrawal of less successful items and planning of additional items on the platform – further development and launches.

---

excursion. It was also a substantial challenge to their traditional formulation and packaging patterns though it seemed it would not present too many new problems in production. The brief therefore demanded an unusual product, justified unusual packaging, and cried out for an unusual brand name. It was a new platform in an extended environment.

The brief was assigned to a product manager from the marketing group, and presented to a product development team on 7 July 1997. The required time scale was very short, four months.

7.4.2   Stage 2: Product design and process development

Major innovative issues arose in formulation, including product characterisation and scaling up from batches of a few litres at the laboratory stage, through about 500 litres in the pilot plant to the thousands of litres in the batches in the production plant. Maintenance of the chosen desired flavour balance from the initial concept recipes to plant formulation involved much careful experimentation. The maintenance of final acidity after processing was critical to keeping quality and was difficult, especially with some of the sauces. Problems arose in aspects such as sauce viscosities and behaviour of starches and thickeners, in separation of constituents such as oils in emulsions on standing, and liaison with and checking of suppliers to secure ingredients with low mould counts so that product shelf lives would be adequate. But the solutions were not so obvious and needed a good deal of laboratory and pilot plant work to find them.

Intensive action commenced on preparing the commercial products. Six separate and attractive sauces finally emerged in the initial platform. These had all to be formulated and set up to give a full product specification for production. This work started on 5 August and continued through that and the following month, reaching agreed products and plant procedures on 12 September.

Packaging and package design presented special problems. In the available time, it was not possible to design and make new bottles, so after exploring all possibilities, long fat-necked bottles, from a line of soft drink bottles that were available to the manufacturer, had to be used. Bottle capping with a hot fill containing recognisably large ingredient pieces had to be explored and accommodated, labels needed designing, and a deep anti-tamper sleeve organised. This deep sleeve turned out to have an additional advantage: the capability of the filler was somewhat limited but the deep label concealed any variation it produced. The different sauce varieties had slightly different specific gravities and all bottles were filled to the same nominal volume. This meant that the customers for the heavier varieties received a systematic advantage, or, put another way, the company was consistently giving away product with the heavier sauces, providing a strong inducement for the further development in due course of a more precise filler. Also, following on a product demonstration to the trade, it was decided to move the bottle tray configuration from 4 × 3 to 5 × 2, so as to improve display, and this required a last minute reorganisation and redesign of the corrugated board trays and cartons and their assembling lines. So packaging was a busy scene.

There were effectively three teams in the group working on the project; the product manager's team (in Auckland), and the product technology and packaging technology teams in Hastings. Their work had all to be coordinated and combined, drawing on the full knowledge of all members of the staff with appropriate expertise. Cooperation over a wide range of people and skills was excellent and contributed very substantially both to the successful outcome and to the speed with which it was reached.

7.4.3   Stage 3: Product commercialisation

Product factory trials were conducted, starting on 25 August, using members of the development team along with other local staff, as they were available. Innovation was needed to move from a substantially manual process line to a much more automated one. Some ingredients presented problems, for example plum pulp to maintain the desired consistency for a high-class product. There was extensive testing of the factory product, with the necessary adjustment of detailed procedures and formulation and ingredients to reach the texture, appearance and flavours desired for the product. Quality assessment and statistical process control procedures, that were substantially available, could be adapted and changed in detail to accommodate the special features of the new products.

Major attention continued to be devoted to the packaging. The hot-filling of a sauce with particulates into a difficult necked glass bottle was a new experience for the team, as was providing the deep plastic-wrapping round the screw caps. This involved checking and upgrading of skills and equipment, and careful attention was required to the glass capper and the in-line labeller. Finished product assessment could finally be undertaken by 29 August.

During this time there had been major activity on the marketing side. One very significant issue was the generic name of the new product platform sauces. That finally chosen was suggested by design consultants, and was 'A Bit on the Side'. The choice was the subject of some controversy. It was a departure from the tradition of largely straightforward descriptive titles. As a new adventurous product, displaying zip to a younger adventurous age group the title needed pep; but so much? In the event it was the platform name chosen, along with appropriately spicy individual sauce names incorporating rather minimal description, on mildly funky but clear labels, to maintain both interest and distinction for each of the six sauces.

7.4.4 Stage 4: Product launch and post-launch evaluation
The platform name was also strongly incorporated, and somewhat suggestively, in the publicity for the product. This was mainly by prominent billboards in the largest New Zealand (Auckland) market, just prior to and during the actual launch. It used a clever stratagem suggesting all manner of eager candidates for a 'Bit on the Side'; initially without revealing that a sauce was involved at all, and then completing the billboard by adding a picture of the labelled bottle, as illustrated in Fig. 7.3. The promotion certainly aroused curiosity and drew

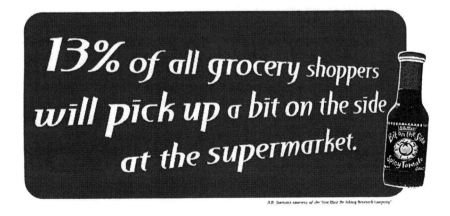

Fig. 7.3 'Bit on the Side' sauce.

> ## Box 7.4   Wattie's gets saucy
>
> Business category manager, Rose France, says the market was ready for the new range with cheeky perceptions.
>
> > The time was right for a product and presentation with an identity of its own ... We wanted to produce something that was fresh, innovative and a little bit of fun to appeal to a new generation of Wattie's consumers – the 18 to 35 year olds and families with older children ... Product development required enormous patience – it's a trial and error process that involves endless cooking and tasting before testing and Bit on the Side tested very well within its target group ... Handling the glass bottles and neck seal was a first for the company and the results really come down to a great team effort ... We wanted something that was truly unique and differentiated our product – but while there might be slight innuendo about the name, it's really about reinforcing the way the sauces should be used and adding extra zest to the tastebuds.
>
> Source: From a Heinz Wattie's house magazine.

attention. The implication was that the competing sauces were everyday. These new Heinz Wattie's products were for 'experiential' eating as illustrated in Box 7.4, which it was hoped would carry buying forward into repeats, and towards habits. There was also some limited TV advertising, featuring on brief cookery clips but at a very prime time spot. It was all well targeted. Sales of the new products rose rapidly, within three months, to brand leader, and the products have held a prime position on the New Zealand market since that time. Obviously a gap in the market was correctly identified, and filled, by satisfying products.

A subsequent development was the withdrawal of two of the sauces. These two were closest to the volume-market, and to Heinz Wattie's previously existing sauces which so many customers had found to be adequately satisfying. Perhaps there was not enough differentiation from these still very popular, and cheaper, products. To add to the offering and coverage however, more new flavours were added to the platform, giving the range indicated in Table 7.6.

After two years' success in New Zealand, 'Bit on the Side' sauce, with four products in the range, was introduced to the Australian market. After trials, the recipes had been modified and the flavours adapted to meet different consumer expectations. There was some both qualitative and quantitative consumer research. But the situation was rather different from that in New Zealand. There was a more advanced variety sauce market, better developed. There was TV promotion at the launch. But the market impact was substantially less than that in New Zealand. Analysis attributed this to the proliferation of sauces available in Australia and the segmentation of the market, to the campaign not building adequate initial awareness, and to the range offered being not large enough on the shelves there to

Table 7.6   Heinz Wattie's 'Bit on the Side' sauce range

New Zealand: Launched 1997 – Sweet Chilli, Java Satay, Oriental Plum, Spicy Tomato, Gourmet BBQ (later deleted), Ketchup later deleted
Added, 1999 – Sweet Mustard, Spiced Apricot, Cracker Cranberry
Added 2000 – Cool Mint, Absolutely Apple, Salsa (four varieties)
Australia: Launched 2000 – Sweet Chilli, Oriental Plum, Java Satay, Gourmet BBQ
Added 2000 – Del Gourmet BBQ

---

impact sufficiently. Also it did not have the local momentum of the Wattie's brand that had helped carry it forward in New Zealand. The launch and subsequent history showed less impact and yielded smaller market share.

Overall the development has had success, both for itself and for indications of new avenues for further product lines. The impressively tight timetable, which was achieved by the product developers, is shown clearly in Table 7.5. Market share in New Zealand has been well retained, the line is established on the supermarket shelves, and occupies a new slot for Heinz Wattie's. There are intentions to carry the concepts and lines forward to the Japanese market.

Think break

1. Heinz Wattie's put the four new salsa products on to the existing product platform. Would it have been more effective to have started another product platform?
2. How do you see that further innovations could be built from this on to a new product platform?
3. This was an extremely fast, major product development, from brief to launch. What do you consider the essential elements allowing this to be achieved? What, if any, additional activities might have been able to improve the outcome?
4. If you were asked to launch a similar product on your home market, how would you go about it?

7.5   Some brief comments on the case studies

These case studies were selected to illustrate the PD Process in different but common food industry situations. They are not typical in that no one case is ever typical, but they show and demonstrate much that has been considered in this book, reinforce the claims that the concepts are practical, and briefly set out the way in which real product development problems have been handled.

The first case study looked at a fruit, fresh apples, that is quite a major commodity, moving from New Zealand to world markets with relatively little processing. For fresh fruits, product development into new varieties can be a powerful tool in gaining and retaining market share, and the aim of the project was to develop a new type of apple which could lead to a number of varieties. A great deal of expertise had been built up and this substantially guided the project, though it was appreciated that this has vulnerabilities and increasingly inputs from the consumers are being sought. Modern technology has opened up possibilities for more organised and sophisticated technical developments for the growing processes, but this type of product development has special features of its own and in particular a long time scale which many food companies would find very hard to contemplate.

The second case study looked also at a fruit in which the primary concerns were to bring better returns from a significant export trade. Being very much a consumer product, the work was largely based around modern statistical techniques of consumer research. The study met two major objectives. One was to guide the shorter-term decision making in seeking a product that made the best of available fruit and its organisation on to the markets. The other was to generate information that can be used in the longer term to guide possible future breeding and improvement of the fruit lines.

The third case study demonstrated a step in a continuing programme for the generation of new and more valuable specialised food ingredients from a major food raw material. In this, highly sophisticated processing was employed, which had to be developed so that it was successful not only in production but also in the market. The basic information came from the literature, and this was further generated and extended, and industrially implemented, by the development technologists. There was much technical work to be done, both technical development in the product and in the processing, and in the technical sales. There were also quite major design and commissioning to be undertaken and with them capital expenditures, and marketing development. The resulting high-grade, highly specified ingredient had to be produced and exported to match into expensively promoted manufactured foods with elaborate and demanding acceptance criteria.

The fourth case study was a more typical one of a food manufacturer, a large well-established one, wishing to diversify into a new product and product platform. In this case the information employed was largely in-trade and in-house. A substantial product development and marketing organisation was in place, but there were still plenty of challenges. They included the designing of a rather different product and product image, the setting up and handling of packaging with problems new to this factory, the industrial line reorganisation needed, the possibility of adventurous marketing which was cleverly exploited, and, not the least, a very tight time scale.

7.6   Acknowledgements

Grateful acknowledgement is made to the sources of information for these case studies. In particular, sincere thanks for their time and trouble to: Dr Ian

Warrington, CEO, and Mr Allan White, Portfolio Manager for Pipfruit New Varieties, of HortResearch; Mr Tim Allen, New Product Development Manager, of ENZA; Dr Tipvanna Ngarmsak, Mango Project Director, of Khon Kaen University; Drs Allan Anderson, CEO, and Mark Pritchard, of The New Zealand Dairy Research Institute; Mr Gerry Townsend, Product Development Manager, and Ms Suzanne Weston, Product Development Technologist, of Heinz Wattie's.

7.7 References

EARLE, M. & EARLE, R. (2000) *Building the Future on New Products* (Leatherhead: Leatherhead Food R.A. Publishing).

HUFFMAN, L.M. (1996) Processing whey protein for use as a food ingredient. *Food Technology*, 50(2), 49–53.

NGARMSAK, T. (2000) *Development of Mango Products and Their Competitive Advantage in Export Markets* (Bangkok: Thailand Research Fund – translated from Thai).

8

Improving the product development process

Best practice in product development is a dynamic target. Not only are new practices being developed and refined but the differences in organisations demand the tailored application of these practices. There are eight basic principles and four basic stages in product development which are true for all companies, all projects and at all times. But the company philosophy, knowledge, skills and assets change; and these changes cause changes in the types of product innovations and the activities in product development. Successful companies recognise that product development is an important strategic issue that demands constant attention. There is a need to evaluate the product development performance and the product development success rate (product development efficiency and effectiveness), and then combine this evaluation with the company's strategic direction to determine and organise improvements in both the effectiveness and efficiency in the future. This is not simple because creativity and criticism are two opposing thought processes. Creativity, vital to product innovation, goes into the unknown and makes mistakes; the product development evaluation looks for mistakes and criticises them. Emphasis on mistakes leads to conservative product development; emphasis on creativity leads to wild product development; the successful companies intertwine the creativity and the evaluation in the project.

Product development is unique to the company and is related to the company's history, philosophy and knowledge, but the company's position relative to the best practice in the related industry and market is an indicator of the company's past and present product development effectiveness and efficiency. From this evaluation can be built up strategic plans for improving product development. As shown in Fig. 8.1, product development effectiveness and efficiency are improved together to give the strategic product success

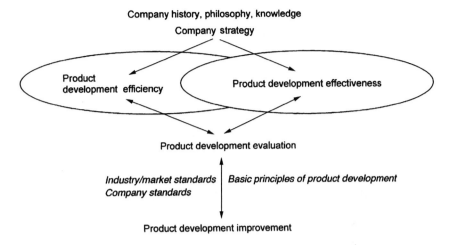

Fig. 8.1 Evaluating and improving product development.

indicated by the company top management. Product development effectiveness and efficiency are inextricably linked.

8.1 Key messages

There is no single recipe for successful product development. During the previous chapters, a number of basic principles were identified which are common to all successful product development as shown in Table 8.1.

8.1.1 Link to overall business strategy

'Doing the right things' must always be central to any product development effort. The product development strategy must be in harmony with the overall business strategy. It should both receive and provide direction to the business strategy. The balance of reactivity to proactivity will depend on the nature of the organisation and its overall goals.

Table 8.1 Basic principles of product development

Link to overall business strategy
Balanced product development portfolio
Clearly defined product development strategy

Appropriate systems and processes for project management
Appropriate human and physical resources
Committed and supportive organisational climate

Understanding the consumer, customer, market and society

8.1.2   Balanced product development portfolio

The business goals and strategy should define the key criteria to be used in preparing the product development portfolio. These include:

- degree of novelty or newness;
- level of technology;
- target market;
- level of risk;
- desired return on investment;
- time frame.

Preparing a balanced portfolio of new product development, consistent with business aims, is a critical part of product development management.

8.1.3   Clearly defined product development strategy

The product development strategy should provide:

- total clarity about the relationship between the portfolio of product development projects and the overall business strategy;
- clear definition of the portfolio of new product development projects relative to business selection criteria;
- indication of the costs and timeframes involved to achieve the desired outcomes of the portfolio;
- indication of the resources required to achieve the desired outcomes – what resources are required internally and what should be out-sourced.

The product development strategy is linked to, and indeed is the basis for, the tactical strategy that organises the product development programme and the individual product development projects. In a number of companies there is often a communication blockage between the product development strategy and the tactical strategy that determines the work of the designers, engineers, marketers, production and other personnel involved in the project. This can reduce both the effectiveness and efficiency.

8.1.4   Appropriate systems and processes for project management

Having decided on what things to do (the portfolio), it is important to have the appropriate systems and processes to support individual projects – 'doing things right'. There are four clearly identified basic stages in the PD Process – product strategy development, product design and process development, product commercialisation, product launch and evaluation. But there are differences in the activities, decisions and outcomes in the different projects, although there are significant ones that occur in many projects.

Although the PD Process is important to the successful completion of projects on time, in budget and in line with the initial target, it would be wrong to force all projects into a standard process. The PD Process is unique to the company, level of innovation, and the level of technological knowledge. The company can design different PD Processes for product improvements and major innovations, for consumer products and industrial products, and also make some changes between product development projects (de Brentani, 2001). The chance and costs of product failure can also cause the company to make changes in activities; for example, the low cost of project failure may lead to significant short cuts in market analysis and business analysis. The choice of activities also depends on the company's level of risk. If the company is not afraid to live with product failures, it may omit many activities; if the company does not want to risk product failure, it will include activities that increase its knowledge of the technology and the market. In creating both new products and new services, a platform-based approach can be used, which relates directly to the design of systems and PD Processes (Meyer and DeTore, 2001).

8.1.5   Appropriate human and physical resources

All the best systems and processes can be worthless without the right resources. People, above everything else, make product development successful. *Knowledge* of technology, market, consumer, product development activities and decision making, and the *skills* to use this knowledge in practice are the basis of successful product development. Capable and committed people, who are able to work in teams, across functional boundaries, will make systems and processes work for them. Systems and processes will rarely change people. There is a need to recognise the tacit knowledge of individuals and teams, as well as the knowledge bases both within and outside the company. Most important is the ability of the individual and the team to create new knowledge during the project.

8.1.6   Committed and supportive organisational climate

Perhaps the most important aspect of all in determining successful product development is the organisational climate. Historically, this has received relatively little attention in the product development literature and yet it has the potential to have the greatest impact on product development outcomes. Climate includes:

- clarity of direction;
- management commitment;
- team commitment;
- flexibility;
- standards;
- rewards.

The decision making by top management at the beginning and throughout the project must be timely and based on knowledge; from this the project management and the team need to see clear directions which are not changed without further knowledge and discussion.

8.1.7   Understanding the consumer, customer, market and society

If the needs, wants, attitudes and behaviour in the target market and in the society in general are not identified and understood, and then interwoven into product development practice, then product failure can occur either in the short or long term. The food industry has a history of introducing innovations over the years that cause suspicion by the general public and the consumers, so that food regulations are used to control the product. The immediate customer, whether industrial user or retailer, needs to be integrated into the PD Process from the initial stages of developing the product concept to the final evaluation after launch. In developing new consumer products and indeed in all food product development, the final consumer who buys and eats the food is an integral part of product development.

Think break

The authors have summarised what they identify as the basic principles of product development from the preceding seven chapters.

1. Do you agree with their list? Have you identified any other basic principles? Would you drop some of their basic principles?
2. Compare with other principles in the literature, e.g. Cooper and Kleinschmidt's (1995) factors found to drive new product success.
3. For your own company, list the basic principles for product development at the present time.
4. How have these principles changed in the past and how do you predict they will change in the future?
5. List the basic principles for product development for your company for the next decade.

8.2   Evaluating product development

Conducting a post-development review of a specific product development project and a regular review of the product development programme, is a very good way of learning what is excellent, all right and bad in the company's product development. For the product development project, the initial product strategy needs to be compared with the final total product in the market; the final product characteristics with the consumer needs and wants; the efficiency of the product development project with the overall implementation of the launch. For the product development programme, some important measures are:

• ratio of major innovations to incremental products;
• key differentiating factors in products and services;
• number of new products in a time period;

- programme complexity – the size of the programme and the interrelationships between projects;
- commercial constraints on the programme;
- company pressures on the programme.

In recent years there has been an increasing interest in developing methods for evaluating product development. For example, the assessment tool and methodology (ATM) of Barclay *et al.* (2001) measures the complexities and newness of a product and relates them to the PD integrating activities and process. Clark and Wheelwright (1993) developed a method for auditing the individual project. Cooper and Kleinschmidt (1995) developed a tool aimed at identifying the firm's critical success factors in product development. It had two sets of measures for the product development programme: programme profitability and programme impact on the company. They separated companies using these measures into:

- *high-impact technical winners* with highest product success rate and % sales from new products, but not so high profitability
- *dogs* with poorest performance on all measures
- *solid performers* with highest profitability and second highest product success rate, lower % sales from new products than high-impact technical winners
- *low-impact performers* with mediocre product success rates and low impact of new products on company sales.

There have been the general industry comparisons described in Chapter 1, for example Griffin (1997), which have useful measures and results to compare with your company's results.

This comparison of the company's product development effectiveness and efficiency with those of other companies or of the industry in general is known as 'benchmarking'. Benchmarking the company's current practices against the latest findings in the literature and through comparison with other companies is an essential part of overall product development management. The application of best practices to our specific situations and the on-going measurement of performance ensure a basis for continuous improvement.

8.2.1   What is benchmarking?

Benchmarking is a process of continuous evaluation to achieve a competitive advantage. It measures a company's products, services and practices against those of its best competitors or other acknowledged leaders in their fields. It can be a specific area such as the benchmarking of the new product concept against the competing products (Rudolph, 2000), the company's technology against the most technically advanced company, the company's innovation strategy against technology predictions. But mostly there are multiple measures in benchmarking.

Benchmarking can be at different stages of the product development project, for the overall product development project and the product development programme. There can be short-term and long-term benchmarking; for the short term, Hultink and Robben (1995) identified product-level measures such as speed-to-market, launched on time, development cost; in the long term, customer acceptance (met revenue goals, market share goals and unit sales goals, percentage of sales by new products) and financial performance (attaining goals for profitability, margins, return on investment). Four factors were equally important for short-term and long-term success: customer satisfaction, customer acceptance, meeting quality guidelines and product performance level. Finally benchmarking must be related to possible improvements; there is no point in extensive benchmarking in areas where the company or personnel cannot make improvements because of lack of people, knowledge and assets. Benchmarking and continuous improvement need to be linked. Zairi's (1998) comment is worth remembering when benchmarking

> the impact of its application is more for changing attitudes and behaviours and raising commitment through better education, awareness and inspiration from model companies. Benchmarking is perhaps the best means for servicing the human asset by continuously supplying new ideas to sustain superior performance levels.

Over recent years benchmarking has become a fashionable tool for many organisations. Like many such tools, one has to question the rigour and objectivity with which many benchmarking exercises are carried out and, in turn, the value that is captured from these exercises. Benchmarking is not a tool (the many methods suggested for benchmarking are tools), but it is a method of increasing knowledge and skills of all people involved in product development from the top management to the junior team member, so that product development is more effective and efficient.

8.2.2   Basic steps for benchmarking product development
There are some basic steps in benchmarking, shown in Fig. 8.2, which need to be followed to maximise the return on any investment in benchmarking (Zairi, 1998; Czarnecki, 1999; Barclay *et al.*, 2001).

Clearly define the benchmarking objectives
Before beginning a benchmarking study, the organisation should be clear on what the subject is to be; what are the desired outcomes; who will use the results; and how will the results be used to benefit the organisation in the future. It is all too easy to embark on wide-ranging data collection, which, in the end, provides very little useful information for the organisation and its specific requirements.

Determine the sources of benchmarking data
The benchmarking can be internal and using internal data sources, but usually the comparison is with companies within the specific industry or in industry in general. Sources include the following:

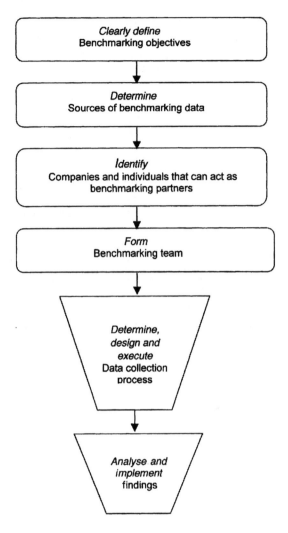

Fig. 8.2   Basic steps in benchmarking.

- Published materials. Case studies, industry surveys and research into R&D management methodology and practices provide an excellent source of primary data. These materials can also be used to prepare a list of potential benchmarking partners.
- Databases. For example those prepared by the Industrial Research Institute (IRI) in the USA, provide an excellent source of information on level of spending on R&D, number of R&D employees, number of patents granted, etc. Regular summaries of these data are presented in the *Journal of Research – Technology Management*.

Identify companies and individuals that can act as benchmarking partners
This partnership can range from an infrequent exchange of general information about company operations and practices to regular meetings where details relating to company practices are shared with a common aim of improving the overall practices of both partners. It is important not to be limited to similar companies in the selection of partners. Look to the best practices, wherever they occur. The object is to learn from the leaders, those with world-class operations and techniques.

Form a benchmarking team
Although an individual can conduct a benchmarking study, a team effort will usually get a better return. Involving a cross-section of people with different skills and organisational responsibilities will provide benefits not only in the capture of the information but it will also enable greater acceptance and more rapid assimilation of best practices into the organisation.

Determine, design and execute the data collection process
There are various ways of collecting benchmarking information, including telephone interviews, postal surveys, face-to-face meetings and desk-top research. Clearly, the type of information required, the level of detail and the available budget will determine the methodology. The best approach is probably to start with the relatively simple desk research to provide the background material and further clarify the critical information required from surveying or from face-to-face meetings.

Analyse and implement the findings
Very often there will be a number of valuable findings from the study. It is important to prioritise these and to focus on the implementation of a manageable few. Commitment and support from senior management is an important starting point. It is also essential that besides showing that these are 'best practice initiatives' there must also be clear evidence to all who are involved in the implementation that there is real benefit to their business.

It is important in setting up a benchmarking or assessment system that it should be (Barclay *et al.*, 2001):

- relevant to the users' needs;
- fairly comprehensive;
- capable of a variety of approaches;
- both educational and action-oriented;
- capable of being used in total or selectively;
- able to 'force' the development and implementation of action plans.

In other words, set up a system that does not overwhelm with information, but gives the knowledge needed to lead to product development improvement. In Box 8.1 is shown the development of benchmarking in the New Zealand Dairy Board, which shows some methods that can be used in benchmarking the total

Box 8.1   The New Zealand Dairy Board (NZDB) experience

1993   An audit of the NZDB by the Boston Consulting Group (BCG) required for statutory reasons and covering all parts of the business. The major recommendations for R&D focused on the improvement of the product development process in achieving greater speed to market and greater success rates.

1994   Development of a phase-gate process for product development and implementation across all parts of the organisation.

1995   Consolidation of PD Processes with increased emphasis on 'doing things right'.

1996   Recognition that future gains in R&D effectiveness would most likely come from 'doing the right things' in addition to 'doing things right'.

1997   A small cross-functional benchmarking team was formed initiating a three-pronged approach:

- **Decision practices**. An internal survey of the performance and areas for improvement against key decision practices required for 'best practice'. The decision practice framework was defined by the Strategic Decision Group (SDG) in California (see Matheson & Matheson, 1998). Analysis of the survey data by SDG pointed to a number of specific areas for improving the decision practices that lead to 'doing the right things'.
- **ProBE survey**. Developed by Robert Cooper and Scott Edgett at the Product Development Institute Inc. (Ontario, Canada). An internal survey designed to evaluate product development performance against 11 critical success factors was used to identify areas of strength and weakness relative to industry average results and those of the top 20% of firms in the Product Development Institute database.
- **Secondary data.** A range of published materials including annual reports, management journals, the Industrial Research Institute (IRI) R&D database were used to provide background information on industry and individual company performance.
- **A set of prioritised initiatives**, centred around 'doing the right things' was recommended. These focused on linking R&D to business strategy including technology planning and portfolio management.
- **Implementation** of these initiatives was begun.

1998   A second BCG audit of the NZDB confirmed most of the recommendations of the internal benchmarking team and endorsed their implementation.

Box 8.1 (continued)

1999   A further benchmarking study was started with Arthur D. Little, focusing on the use of metrics to track R&D performance. The scope of the project was widened to include all parts of the innovation process and not only R&D. A suite of metrics was developed based on lagging, real-time, leading and learning indicators. These metrics were implemented in the business units of the NZDB.

product development in a company. This benchmarking development shows in sequence the aims of firstly product development efficiency (doing things right) and then product development effectiveness (doing the right things); and also the use of different evaluation methods and different consultants. The important part in benchmarking is to choose the correct measures or metrics.

Think break

The NZ Dairy Board is a large company and is able to employ a range of consultants. If you were a small or medium-sized company:

1. Discuss the ways you could measure product development effectiveness and efficiency.
2. How would you select and use suitable methods of benchmarking product development for your company?
3. How could you identify the essential product development activities, outcomes and decisions for the successful business performance of new products?
4. How could you design suitable product development processes for your company?

8.3   Innovation metrics

Increasingly we are being required to justify the expenditure on innovation. How effective is it? Does it meet the organisation's objectives? What is the return on the investment? There is very little doubt that justification of expenditure on innovation is necessary, just like any other element of organisational expenditure. But all too often the measures that are used only provide information about past performance. They contribute very little to our understanding of why that level of performance was achieved; to our improvement of innovation practices; or to our prediction of the future value of our current innovation efforts.

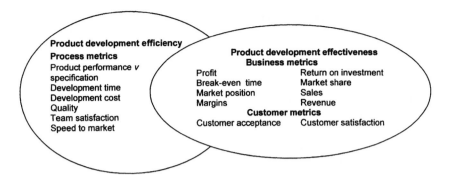

Fig. 8.3   Product development project and programme metrics.

Firstly, what do we mean by measures and metrics? Measurement applies to anything that has a quantifiable characteristic; a metric is a quantifiable characteristic, which one can measure against (Dimancescu and Dwenger, 1996). Metrics such as 'new sales ratio' (contribution derived from new products/services), 'R&D intensity' (spend on R&D as a percentage of sales) and 'number of patents granted' provide some insight into an organisation's innovation performance, but it is historical. Greater benefit can be achieved from having a range of metrics that provides both forward and backward looking information. *Static metrics* are data gathered only after the event has occurred; *dynamic metrics* are real-time data feedback usually related to a goal.

Barclay *et al.* (2001) suggested process metrics for internal efficiency; business metrics and customer metrics for external effectiveness as shown in Fig. 8.3. Arthur D. Little (personal communication) proposed a metric suite based around the timing of the information and on specific areas of focus. They suggested that metrics be designed with a framework of time and the holistic dimension.

8.3.1   Timing of information

Metrics can be measured at different times in the product development project and programme. They can be on information from past projects, or from the present project or they can be predicted for the future product development performance, as shown in Table 8.2.

Table 8.2   The time of using metrics in product development

Lagging metrics provide information on past performance.
Real time metrics provide information on the current performance.
Leading metrics provide information on the likely future performance.
Learning metrics provide information on the rate at which an organisation is improving its performance.

Learning metrics or motivational metrics translate business objectives into meaningful and motivating measures that teams can work against. A specific performance gap may have been identified, and then the goal is to gradually reduce it over time. For example, in initial production runs with a new product, rejections usually lie between 10 and 20%, then the aim is to gradually reduce this until it lies near the level of standard production which is 1%; or it could be the time to market for a new product which may be 24 months and the aim is to reduce this gradually to 15 months. A reasonable time frame for these reductions needs to be set (Dimancescu and Dwenger, 1996). Metrics are meant for continuous improvement of product development performance; historic data may set a basis but it is the continuing measurement during the development of the project and the programme that give the more useful metrics.

8.3.2   The holistic dimension

Metrics can be applied to all areas of the product development programme and the project:

- **Strategy**. Is innovation aligned with business objectives, strategy and vision?
- **Process**. Do the innovation processes support successful execution and outcomes?
- **Resources**. Are the desired level of resources being applied to innovation?
- **Culture and organisation**. To what extent does the culture, climate and organisational structure support innovation?

Examples of metrics within the Arthur D. Little framework for analysing the product development programme are shown in Table 8.3.

Table 8.3   Examples of innovation metrics in the product development programme

	Lagging	Real time	Leading	Learning
Strategy	Contribution from new products/services	% of growth targets met through innovation	Value of portfolio	Increase in revenues from new products
Process	Number of patents per year	% milestones on time	Forecasted project completion time	Reduction in breakeven time
Resources	R&D spending as % of sales	% outsourcing	Forecast resource allocation	Changing demand for specific capabilities
Company culture	Client survey feedback	Number of ideas logged	Innovation climate surveys	Change in critical climate dimensions

8.3.3   Metric selection criteria

The set of innovation metrics should be selected according to the specific needs of the organisation. In selecting measures, they must be economical to collect, understandable to the people who are going to use the results, learning focused, externally focused, actionable, broad in scope and accomplish the stated objectives (Czarnecki, 1999). Innovation metrics should be regularly reviewed and changed as the direction and priorities of the organisation change. Some basic criteria for metric selection are:

- use a matrix approach, selecting a few metrics from throughout (as shown in Table 8.3);
- support the weakest link in the current innovation systems;
- emphasise real time or leading measures where possible;
- select metrics for which results point directly to actions;
- focus on simple and obvious measures that clearly support business imperatives;
- select those that are easily measured consistently over an extended period.

Some pitfalls for choosing metrics are predominance of short-term, financial, efficiency, economy and functional measures. It is important to select metrics not only because data are easy to find and they are within the capability of the benchmarking team and the understanding of top management. Metrics must also be relevant to the improvements to be made.

8.3.4   Integrating innovation metrics into the business

The application of innovation metrics will be successful only if they are 'bought into and truly owned' by the business or business unit. All members of the business management team must see the benefits from the metrics, both to themselves and to their business unit. There are four steps in this integration:

identifying the growth gap, defining the innovation programme to meet the growth gap, defining an appropriate set of metrics for each project, measuring and tracking performance over time.

Step 1: Identify the growth gap
The first, and most important, step in the application of innovation metrics is at the strategic level where the required contribution from innovation is defined against future business targets:

• What is the total business growth aspiration?
• How much of this growth will come from organic growth?
• How much can be expected from mergers and acquisitions?
• What is the value of the innovations currently in the pipeline?
• What is the growth gap that must be filled by new innovation?

This is illustrated in Fig. 8.4.

Step 2: Define the innovation programme to meet the growth gap
Determine the value and timing of the current innovation portfolio to ensure that it provides the required contribution to meet the growth gap. The total innovation portfolio value is made up of the sum of contributions from all innovation projects.

• What innovation projects are planned?
• What is the time of delivery of these projects?
• What is their predicted revenue and earnings before interest and tax (EBIT)?
• What is the total predicted value of the current innovation portfolio?
• Does this value meet the growth gap aspirations? If not, what further innovations are required?

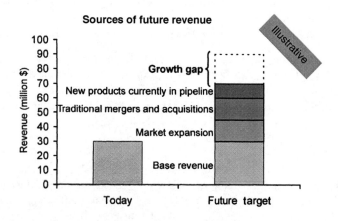

Fig. 8.4 Identifying the growth gap.

If satisfied with the current innovation portfolio value, use this value and the schedule of realisation of the value as a leading metric.

Step 3: Define an appropriate set of metrics for each project
A set of leading and real time metrics should be defined and used to measure the progress of individual projects.

- Define the individual contribution of each project to the total. This becomes a key target for a leading metric for the project.
- Define a set of real time metrics and targets for each project. These might include milestones on time, expenditure against budget, etc.

Step 4: Measure and track performance over time
Lagging metrics should be used to measure past performance. This should be compared with the predicted performance as indicated in the leading metric targets. If the overall innovation portfolio is on target then the innovation contribution to the growth gap will be achieved.

- Use lagging metrics such as current return from products developed over the last five years to measure past performance.
- Compare this performance to the targeted performance required to meet the growth gap.
- Study the underlying reasons for differences between what is achieved and the target. Learn from past mistakes and successes and apply this learning to improvement in the overall innovation practices.

Over recent years a great deal of time and effort has been focused on the improvement of new product development (NPD) management. Not only are we seeing an abundance of research literature on the subject but we are also seeing significant emphasis on the management of research and development activities as a senior management function in many companies.

Think break

Consider your company:

1. Step 1. Identify the growth gap.
2. Step 2. Define the innovation programme to meet the growth gap.
3. Step 3. Define an appropriate set of metrics for a project for an incremental product, and for an innovation.
4. Step 4. How would you measure and track performance over time?

8.4   Striving for continuous improvement

It is no longer enough to have a creative group of product developers. Success comes from having a fully integrated NPD function, supported by first rate practices and processes, and focused on the business goals of the company. This is illustrated in Fig. 8.5. For the total company product development function, the business strategy is connected to the product development programme which is interrelated to the individual product development projects. Benchmarking can signify changes to the business strategy and this is then transferred to the product development programme and to the individual product development projects. Or the benchmarking study may have been on individual projects and the results are recognised in the business strategy, or in the product development programme which is transferred to the product development projects. Continuous improvement based on benchmarking is an interactive process. It is directly connected with the basic parts of the PD Process with the specific standards for decisions, outcomes, activities and techniques being set by different levels of management, but interconnected. There must be focus on the effects on the market and also on teamworking and general company cooperation in product development.

8.4.1   Steps in continuous improvement

The information and knowledge gathered during the benchmarking exercise have to be converted into efforts that will result in improved product

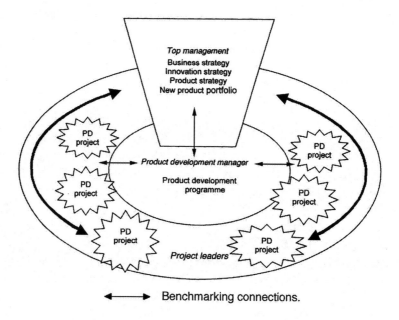

Fig. 8.5   A fully integrated NPD function.

development performance. Four important steps are (Codling, 1996):

1. Communicate benchmark findings.
2. Adjust goals and develop corrective improvement plan.
3. Implement the corrective improvement plan.
4. Review progress and calibrate.

Communication is important if cooperation and indeed commitment to the proposed changes by individuals, teams and management are to be built. The relation of the benchmark results to the proposed future changes in the product strategy, the product development programme and the individual projects need to create a vision for the future of product development in the company. There needs to be the opportunity for self-assessment and discussions by all involved in product development to build up the improvement plan. The plan is not an edict from top management on the verdict of outside consultants, it is the working together of management and key individuals in product development to create the change in product development. The time frame for the changes and how people will be involved in the changes need to be communicated.

Goals for the product development strategy, programme and individual projects are usually changed or at least adjusted by benchmarking. In particular, the new decisions and outcomes at the various stages in the product development project and for the overall project need to be identified by both top management and the project managers. These outcomes are going to be the goals for the activities in the product development project, and will affect the choice of activities and techniques. The activities and techniques are also going to be changed by the product development efficiency benchmark results, and by the resources made available by top management for the changes. There are two ways of improving performance (Barclay et al., 2001):

- **Refinement change**: product development is treated as a business process and is continuously reviewed and refined.
- **Radical change**: a major shift in PD activities and process usually prompted by poor results or a shift in strategy and/or market.

If the refinement change is followed continuously and wisely, the radical change which is costly in resources, people and time, can be avoided.

Implementation is not easy. The action plans need to include descriptions of the proposed action, time scale for introduction, resources required and available, knock-on effects in other areas of product development and in the functional departments, measures of performance of the change, expected outcomes (Coughlan and Brady, 1995). But most important, is to identify the key people and to have them cooperate in developing the action plan and putting it into practice. The management has to recognise the complete plan, identify the staffing and how the changes are to be guided.

Reviewing the progress is important. This means reviews undertaken at recognised intervals not just at the end. Is the plan stalled because of resistance by team members, lack of resources, lack of knowledge or lack of cooperation between the product development team and the functional departments? Are the changes being introduced too fast so that people do not understand what is to be achieved and how to achieve? The benchmarks set to measure the progress need to be achievable with the time and resources available. There will be fundamental differences between projects, and they cannot be set common benchmarks such as reducing the time by six months – easy to achieve in an incremental product development project, often impossible in a major innovation. There needs to be opportunity to make changes to the action plan, if the short-term benchmarks are not being achieved and there is no hope of attaining the long-term benchmarks. Product development is new and creative, so it is not always possible to predict accurately in the action plan.

8.4.2   Product development project

The product development process and its decisions, outcomes, activities and techniques are going to be improved from the results of the benchmark study. The milestones in the project need to be set, and then followed in the project to see if they have been accomplished partially or completely. The targets for the later stages and the whole project may need to be reviewed as the project proceeds through the various stages, because of the new knowledge and achievements in the early stages. The benchmark metrics are accepted into the project and used during the project, and adjusted if necessary. For example, in past projects, the product quality may have been identified as low because of poor packaging and storage properties; this means more creative and controlled package design together with more extensive storage tests, and metrics of packaging quality such as improving reject level on the production line or in distribution, and lengthened storage life of the new product. In putting the benchmark improvements into practice, the most important factor is to have the cooperation and commitment of the multifunctional team. This means their ownership of the project goals, cooperation across the team and good team leadership (McDonough, 2000), as well as the resources and knowledge to make the changes, and top management support. Self-assessment of team members is the most important basis for product development improvement; there can be comparison between projects in the company, and also with other companies as shown in Box 8.2. In the large company, it can be between different projects, and in small companies working in 'clusters', it can be between the member companies of the cluster group. It is important to have self-assessment in company projects before cooperating with other companies. A difficulty is in identifying suitable companies for the comparison, gaining access to these companies' information at a useful level of detail, and deriving useful guidance for the company's product development from this comparison (Coughlan and Brady, 1995).

Box 8.2   Self-assessment and benchmarking product development in five Irish firms

The main objectives of the study were:

- establish benchmarks of current practice in the management of the product development process in five manufacturing firms drawn from differing industries in Ireland;
- increase awareness of areas of choice in the management of product development among manufacturing firms in Ireland with a view to improving their management of the product development process.

Each company selected two recent product development projects for assessment. Each project illustrated development in different situations or different approaches to development. The projects represented different degrees of product change and manufacturing process change. Six of the ten projects fell into the category of incremental or derivative projects, four of the projects were platform or next-generation projects.

The self-assessment and benchmarking approach consisted of three generic phases: data gathering and initial self-assessment; communication of insights both within and between the firms; development and discussion of action plans. All three phases required the active participation of up to ten staff members in each firm, drawn from the product development projects under review.

Arising out of the research each firm identified a range of performance limiting practices in its development process, which had caused schedule delay or cycle time extension through:

- insufficient up-front technology planning and development,
- reacting to short-term resource shortages,
- accepting productivity limiting practices,
- inadequate product and product line planning,
- allowing requirements to float,
- reliance on major versus incremental changes.

The issues were concentrated in the areas of market focus, teamworking, transfer of manufacturing, leadership, resourcing and performance evaluation.

Source: After Coughlan and Brady, 1995.

8.4.3   Product development programme

In improving the whole product development programme, a new innovation/new product strategy needs to be formulated and a plan to achieve it developed. This is a much more fundamental change to the company's product development. It is important not to set the achievement levels so high that the company and the individuals cannot achieve them in a reasonable time with the present or agreed expanded assets. Goals need to be reasonably flexible, to allow for adjustment as the new programme proceeds and for any environmental or internal company changes. Programmes are often organised for 3–5 years, but future predictions up to 10 years should be made. There is a need for constant monitoring of the programme benchmarks, and making changes when necessary. The product development programme is a dynamic organisation and must allow controlled (but not wild) changes. The corrective improvement plan needs to monitor/ check/review the impact of the product development programme changes on the outcomes and the critical product development success factors.

Learning from the projects is important and their benchmark data need to be incorporated into the product development programme so that continuous improvement of the efficiency and effectiveness of the programme can occur. The natural tendency in a company is to go forward into the next project without reviewing the product development programme to see what can be improved. The knowledge must be absorbed into the company through the product development programme. Organisational and individual learning are the outcome of benchmarking projects and the knowledge learnt must not be lost. Learning from product development projects is one of the most difficult things that a company can do.

There are two objectives in programme improvement: to be better at designing new products and processes, and continually to build and improve the company's procedures, processes, leadership skills, techniques and methods in order to do things faster, more efficiently and with higher quality (Clark and Wheelwright, 1993). Building the development capability is also another important objective.

8.4.4   Product development and business strategies

The company's top management needs to have an increased awareness of the areas of choice in product development management and the performance limiting practices in the company. It has to know not only how to modify the business and product strategies in response to changes in market and competitive actions; but even more important to act proactively through its own diagnosis of the need for change in technology or/and consumers. Management also knows from its own examination of the company as to how the company is performing in product development, who are the key individuals on which the product development is based, but it needs to also identify the knowledge and lack of knowledge in the company, and the financial and other resources needed. When top management has to introduce consultants and make drastic changes, then it knows that its management of product development has been poor. By continuous improvement integrated throughout the company, management can prevent this happening.

A company, and indeed an industry, can choose its own improving standard of development through the four levels of PD practice (Coughlan and Brady, 1995):

Lowest level: Product development is not managed and encouraged.

             Basic procedures, management and motivation are in place.

             Product development is managed and encouraged as a key objective for the firm.

Highest level: 'World-class' development performance is the norm.

Think break

You have now read eight chapters on product development.

1. What are the most important factors that you have identified to improve product development effectiveness in your company?
2. What are the factors that you have identified to improve product development efficiency in your company?
3. If your company does not conduct benchmarking of individual development projects, do you know why it does not? How might your company overcome hindrances and stumbling blocks to make project benchmarking a standard tool?
4. How does your company create and store technical knowledge from previous projects, to make it available for present and future projects?
5. How does your company create and store customer/consumer knowledge from previous projects to make it available for present and future projects?
6. How does your company improve product development? Can the method of doing this be changed to bring product development to a higher level?
7. What is the overall standard of product development in your company?
8. How can the standard be raised?

8.5   References

BARCLAY, I., DANN, Z. & HOLROYD, P. (2001) *New Product Development: A Practical Workbook for Improving Performance* (London: Butterworth-Heinemann).

CLARK, K.B. & WHEELWRIGHT, S.C. (1993) *Managing New Product and Process Development* (New York: The Free Press).

CODLING, S. (1996) *Best Practices in Benchmarking* (Houston: Gulf Publishing Co.).

COOPER, R.G. & KLEINSCHMIDT, E.J. (1995) Benchmarking the firm's critical success factors in new product development. *Journal of Product Innovation Management*, 12, 374–391.

COUGHLAN, P. & BRADY, E. (1995) Self-assessment and benchmarking product development in five Irish firms. *Journal of Managerial Psychology*, 10(6), 41–47.

CZARNECKI, M.T. (1999) *Managing by Measuring: How to Improve Your Organization's Performance through Effective Benchmarking* (New York: Amacom).

DE BRENTANI, U. (2001) Innovative versus incremental new business services: different keys for achieving success. *Journal of Product Innovation Management*, 18, 169–187.

DIMANCESCU, D. & DWENGER, K. (1996) *World-class New Product Development: Benchmarking Best Practices of Agile Manufacturers* (New York: Amacom).

GRIFFIN, A. (1997) *Drivers of NPD Success: The 1997 PDMA Report* (Chicago: Product Development & Management Association).

HULTINK, E.J. & ROBBEN, H.S. (1995) Measuring product success: the difference that time perspective makes. *Journal of Product Innovation Management*, 12, 392–405.

McDONOUGH, E.F. (2000) Investigation of factors contributing to the success of cross-functional teams. *Journal of Product Innovation Management*, 17, 221–235.

MATHESON, D. & MATHESON, J. (1998) *The Smart Organisation: Creating Value Through Strategic R&D* (Boston, MA: Harvard Business School Press).

MEYER, M.H. & DETORE, A. (2001) Perspective: creating a platform-based approach for developing new services. *Journal of Product Innovation Management*, 18, 188–204.

RUDOLPH, M.J. (2000) The food product development process, in *New Products for a Changing Marketplace*, Brody, A.L and Lord, J.B. (Eds) (Lancaster, PA: Technomic).

ZAIRI, M. (1998) *Effective Management of Benchmarking Projects* (Oxford: Butterworth-Heinemann).

Index